Machine Vision for the Inspection of Natural Products

Springer-Verlag London Ltd.

Mark Graves and Bruce Batchelor (Eds.)

Machine Vision for the Inspection of Natural Products

With 245 Figures

Springer

Mark Graves, MEng, PhD
Spectral Fusion Technologies, 45 Roman Way, Coleshill, Birmngham B46 1JT

Bruce Batchelor, BSc, PhD, DSc, CEng
Department of Computer Science, University of Cardiff, Cardiff CF24 3XF

British Library Cataloguing in Publication Data
Machine vision for the inspection of natural products
1.Natural products - Analysis 2.Manufactures - Analysis
3.Computer vision
I.Graves, Mark II.Batchelor, Bruce G. (Bruce Godfrey), 1943-
670.2'8563
ISBN 978-1-4471-3918-8 ISBN 978-1-85233-853-4 (eBook)
DOI 10.1007/978-1-85233-853-4
Library of Congress Cataloging-in-Publication Data
A catalog record for this book is available from the Library of Congress.

ISBN 978-1-4471-3918-8
http://www.springer.co.uk

Originally published by Springer-Verlag London Berlin Heidelberg 2003
Softcover reprint of the hardcover 1st edition 2003

Typesetting: Electronic text files prepared by editors

69/3830-543210 Printed on acid-free paper SPIN 10835229

Rejoice in the Lord always.
I will say it again: Rejoice!
Let your gentleness be evident to all.
The Lord is near.
Do not be anxious about anything,
but in everything, by prayer and petition,
with thanksgiving, present your requests to God.
And the peace of God, which transcends all understanding,
will guard your hearts and your minds in Christ Jesus.

I dedicate this book to my wife Esther, my son Daniel and to my parents Jack and Marian.

Mark

I humbly offer my contribution to this book to the Glory of the Almighty God and dedicate it to my dear wife, Eleanor; my Mother, Ingrid; my late Father, Ernest; my children, Helen and David; my grand-children, Andrew, Susanna, Louisa and Catherine.

Bruce

Preface

Imagine the consternation caused by the newspaper headline *"Glass found in Baby Food"*. Food safety is always an emotive issue, although most of us probably don't give it much thought when we are actually eating. When you ate breakfast this morning, were you anxiously checking the food as you ate it? For example, did you consider the possibility that the cereal might contain broken glass, or metal swarf? Were you concerned that the bread might include body parts of a dead mouse? The fruit might have been contaminated by bird pecks. Did you check? A recent analysis of nominally "boneless" chicken meat, purchased at UK supermarkets, showed that bone fragments are roughly 30 times more common than the retailers claim. Although bone is a "natural contaminant", it is nevertheless unwelcome and potentially dangerous; hard, sharp foreign bodies in food can damage teeth and soft tissue in the mouth and gut. They can cause choking, even death! Contaminants such as these are all too common, as even a casual glance at a local newspaper will show. Reports of legal action over contaminated food products appear regularly in local (not national) newspapers. The threat of litigation over real or imagined injury caused by negligence is a problem faced by all food manufacturing companies, and retailers. English law requires that companies take *all reasonable steps* to ensure that their products are both safe and wholesome. Of course, responsible companies exceed the legal minimum requirements in order to build/maintain a reputation for supplying high-quality produce and to minimise the risk of injury to their customers. However stringent the quality checks it imposes, a company is sometimes obliged to recall a large batch of a food product because some serious contamination has been discovered in samples from it. This is both expensive and damaging to the company's public image. It is also detrimental to the reputation of other brands of the same type of product. There is and there always will be a need for improved instrumentation to detect foreign bodies in food. Non-critical defects in food products have a lesser impact, affecting financial, aesthetic or social parameters.

Detection of foreign bodies in food is just one example of many potential applications where the use of Machine Vision can help improve the quality, safety usefulness and aesthetic appearance of natural materials. This technology, called *Machine Vision*, has previously been refined to a high level of sophistication and has been applied extensively in engineering manufacture. It combines optical, infra-red, ultra-violet, or x-ray sensing, with digital video technology and image processing. A system combining these component technologies (and others) has to be designed very carefully, as there are many pitfalls that can all too easily spoil its performance. The essence of a good design is harmonious integration, so that all parts of the system are able to perform at or near their optimum level. No part of a Machine Vision system should ever be forced to work close to the limits of its range of reliable operation because the designer has neglected one part of a system, by concentrating too much on another. The cabinet is as important as the computer; the lens is as important as the software and the lights are as important as the mathematical procedures (algorithms) that it uses!

The following is a working definition of Machine Vision that will be used throughout this book:

> *Machine Vision (MV) is concerned with the engineering of integrated mechanical-optical-electronic-software systems for examining natural objects and materials, human artefacts and manufacturing processes, in order to detect defects and improve quality, operating efficiency and the safety of both products and processes.*

In addition to inspection, Machine Vision can be used to control the machines used in manufacturing and materials processing. These may perform such operations as cutting, trimming, grasping, manipulating, packing, assembling, painting, decorating, coating, etc. In the following pages, we shall encounter non-food applications too. Plants, cut flowers, timber, decorative stone (e.g. marble), textiles, knit-wear and leather-ware are typical examples of highly variable objects that concern us here. Natural products and materials are processed and incorporated into a wide variety of industrial products, such as cigarettes, brushes, carpets, floor and wall tiles (both mineral-based and cork), bricks, abrasives (sheets and wheels), china and fine porcelain.

When browsing through the technical literature, the reader will soon encounter the term *Computer Vision* (CV) and will realise that it and *Machine Vision* (MV) are used synonymously by many authors. This is a point on which we strongly disagree; we are firmly convinced that these two subjects are fundamentally different. Some university researchers, working in what we regard as Machine Vision, oppose our view, since the computational techniques employed are similar. On the other hand, many industrial designers of vision systems simply ignore much of the academic research in CV, claiming that it is irrelevant to their immediate needs. In the ensuing pages, we shall see that *Machine Vision* is a practical and pragmatic subject that applies techniques borrowed from *Artificial Intelligence* (AI), *Pattern Recognition* (PR) *and Digital Image Processing* (DIP), as well as Computer Vision. While Machine Vision makes use of numerous algorithmic and heuristic techniques that were first devised through research in these other fields, it concentrates on making them operate in a useful and practical way. This means that we have to consider all aspects of a vision system, not just techniques for representing, storing and processing images inside a computer. This is the essential difference between MV and CV, which naturally enough, is concerned almost exclusively with the information processing that takes place inside a computer. The problem of nomenclature arises because MV, CV, DIP (and sometimes AI and PR) are all concerned with the processing and analysis of pictures within electronic equipment. (We might refer to these collectively as *Artificial Vision.*) We did not mention computers explicitly in the definition of Machine Vision, because it does not necessarily involve a device that is recognisable as a computer. MV allows the image processing to take place in a conventional computer, specialised digital networks, arrays of field-programmable gate arrays (FPGAs), multi-processor systems, optical/opto-electronic computers, analogue electronics, and various hybrid systems. Although MV, CV and DIP share a great many terms, concepts and algorithmic techniques, they require a completely different set of priorities, attitudes and mental skills. The dichotomy between CV and MV may be summarised thus: Computer Vision is science, while Machine Vision is engineering.

Until very recently, Machine Vision was applied almost exclusively to the inspection of engineering components, manufactured by processes such as casting, stamping, pressing, moulding, rolling, turning, milling, extrusion, etc. These produce close tolerance artefacts, usually made in metal, plastic, ceramic, glass, rubber, wood or paper. On the other hand, some products, such as food, textiles, leather-ware and natural products (seeds, nuts, fruit, vegetables, etc.) exhibit wide variations in overall size and shape, internal structure, colour and surface texture. It is possible to define a number of metrics that reflect our intuitive concept of variability, or conformability. When we study the list of successful Machine Vision applications, it soon becomes apparent that almost all systems that have been installed to date are dedicated to inspecting products with a low variability score. On the other hand, there is an outstanding need for fast non-contact inspection systems that can ensure the quality and safety of a wide range of raw materials, semi-processed organic and mineral products and highly variable manufactured goods.

Even in industrial manufacturing, there are numerous inspection tasks where the scene/object to be examined is uncontrolled and, as a consequence, is highly variable. A notable example of this type of application is to be found in solder joints on printed circuit boards; solder flow and adhesion to a surface depends on microscopic features (surface texture and fine-detail shape) and contamination, neither of which can be controlled easily during machining. Manufacturing processes that rely on the flow of semi-fluid materials are nearly always prone to produce highly variable products. On the very day that I (BGB) wrote this Preface, I visited a factory where adhesive is applied by a needle injector, depositing an irregular "worm" of sticky black glue onto a smooth metal surface. The latter is produced to a close tolerance but the adhesive bead is not. The complete assembly is a high-precision home entertainment product. Many "high tech" products, like this, contain parts that have low-tolerance features embedded within them. The skill of an industrial design engineer is to hide them, to produce a close-tolerance assembly.

When an artefact is made in a mould, or cut to precise dimensional tolerances, there is one far-reaching principle that we can employ to design an inspection system:

If the product is not within its specified tolerance band, it is faulty and should be rejected.

Natural products do not have "design tolerances"; we do not, for example, have a specification for any dimension, or other feature, of a banana! (In some cases, rule-based criteria have been formulated to recognise natural products. This has been done, for example, so that the General Agreement on Tariffs and Trade (GATT) can be applied fairly. It is also necessary for the grading of fruit, vegetables, etc. In recent years, certain UK newspapers have poured scorn on EU regulations regarding the identification of fruit, such as bananas. This was mis-placed, because no account was taken of the need to control the price and protect poor farmers.) For this and other reasons, we are inevitably faced with some fundamental and severe difficulties in defining the tasks that a Machine Vision system is expected to perform. Here are some examples of ill-posed questions:

- How do we define a colour class, such as *yellow*?
- How do we formulate the rules for judging the aesthetic appearance of a slab of marble, or a wooden panel?
- How do we define an objective criterion for identifying the shapes of well-formed loaves, or cakes?
- How do we specify the texture of a piece of high-grade leather, suitable for making shoe uppers?
- What does a good "worm" of adhesive look like?
- How do we identify an unhealthy plant, such as an azelea, impatiens, etc.?

As so often happens, we must formulate questions appropriately before we can expect to obtain sensible answers. The questions just cited are not in this category, whereas the following are better:

- Does the shape of the loaf that you have given me resemble the shapes of the "good" samples you presented earlier?
- Does this apple satisfy the rules to be classed as Grade I?
- Does this object satisfy the GATT criteria to be classified as a banana?
- Does the texture of this piece of leather resemble one already seen on a piece of "good" leather?

Asking the right question is vital for Machine Vision, whether it is to be applied to engineering products or natural products.

Highly variable objects, such as natural products, are our chief concern in this book. However, they are notoriously difficult to characterise properly. A major problem is caused by the lack of consistency of opinion about what descriptive features are important. We can postulate various suggestions about meaningful measurements but, in the end, each one is supported by nothing more substantial than an opinion. The precise interpretation of qualitative terms varies from person to person. For example, the word "tall" has different meanings for women over 1.85 m and men that are under 1.5 m. Even if we set this problem aside, there is an even more difficult one lurking: how do we combine all of the various pieces of evidence available to us, so that we can to reach an appropriate decision about the acceptability of an item? How do we know what is acceptable and what is not? What authority do we consult, to know what is a "good" apple and what is not? When we are unable to formulate explicit rules for calculating an *accept/reject* decision, we must derive discriminatory criteria in some other way. Self-adaptive learning is one of the more effective techniques available to us and might be applied, for example, to design a classifier based on texture or colour. In more general terms, the computational procedures that we employ for inspecting highly variable products are likely to rely on Artificial Intelligence techniques and Pattern Recognition techniques. We shall see that heuristics, rather than algorithms, are our prime tools. The criteria by which we can judge them are ill-defined, so the concept of an algorithm is, in any case, a spurious one.

The growth of interest in Machine Vision is due, in large part, to the falling cost of computing power. The domestic video market is also exerting a major impact on the subject. LED light sources, plastic aspheric lenses, good-quality, low-cost cameras, with a variety of interface standards (CCIR, RS170, Ethernet, IEEE 1394 ("firewire") and USB), are all available at low cost. They are all exerting a strong positive influence on our subject. (Notice the use of the present tense - the show is not over yet!) This has led to a proliferation of vision products and industrial installations in recent years. It has also enabled the development of cheaper and faster machines, with increased processing power. In many areas of manufacturing, serious consideration is being given now to applying Machine Vision to such tasks as inspecting, grading, sorting, counting, monitoring, controlling and guiding, etc. Automated Visual Inspection systems allow manufacturers to monitor and control product quality, thus maintaining/enhancing their competitive position. Machine Vision is also being used to ensure greater safety and reliability of manufacturing processes. The confidence being gained by applying Machine Vision to engineering manufacture is now spilling over into industries such as food processing, agriculture, horticulture, textile manufacturing, etc., where product variability is intrinsically higher. Thus, we are at the threshold of what we predict will be a period of rapid growth of interest in Machine Vision for applications involving natural products.

No Machine Vision system existing today, or among those planned for the foreseeable future, approaches the interpretative power of a human being. However, current Machine Vision systems are better than people at some quantitative tasks, such as making measurements under tightly controlled conditions. These properties enable Machine Vision systems to out-perform people, in certain limited circumstances. Vision systems can routinely inspect certain products at very high speeds, whereas people have considerable difficulty making consistent judgements in these circumstances. Machine Vision systems exist that can inspect peas individually at a rate of 16 tonnes per hour, which is well beyond human capability. On many tasks, a Machine Vision system can improve efficiency substantially, compared to a human inspector. A machine can, theoretically, do this for 24 hours/day, 365 days/year. Machine Vision can be particularly useful in detecting gradual changes in continuous processes (e.g. tracking gradual colour or texture variations). Gradual changes in shade, texture or colour are unlikely to be detected by a person. On the other hand, people are better at making difficult decisions, based on novel and incomplete data. Machine Vision and human inspectors will both have a place in the food processing factory of the future but the balance will, we predict, shift slowly towards the machines in the next 10 years. The purpose of this book is to accelerate this movement, so that human beings can spend their lives doing more interesting things than watching fish, rice, peas or potatoes moving past on a conveyor belt, at high speed.

The stated aim of Machine Vision is to improve product quality and safety, enhance process efficiency, reduce waste and avoid human inspectors being placed in danger. Few will doubt the desirability of these aims but there is the ever-present spectre of unemployment resulting from work of this kind. We hope that by liberating the workforce from tedious inspection tasks, they can be re-employed in way that enhances human dignity. There can be no excuse for blaming Machine Vision for making people redundant; people make other people redundant.

We would like to thank our fellow authors who have provided us with so much excellent material to fill these pages. They have waited patiently while we have completed the editorial work on the manuscript. Our colleagues have all provided us with exciting material to read, new ideas to discover and have reminded us convincingly that this subject is growing rapidly. The cumulative experience of the contributors to this book is formidable, over 500 years of experience is to be found between these covers! We count ourselves privileged to be in such august company. We are proud and grateful for the links of friendship that we have established with our co-authors. The wide geographic spread of interest in this subject is also apparent from the list of authors' addresses This is a technology that will eventually affect us all in the developed world and will one day have a significant impact on developing nations too. We hope that this book will make it happen a little sooner, with fewer false starts and greater confidence.

We are pleased to acknowledge the assistance of Mrs Terrie Hately and Mrs Esther Graves, both of whom have worked hard in preparing the camera-ready document you are now reading. Our colleagues at Springer Verlag have helped us enormously by answering our many questions and helping us to turn a vague idea into a finished product.

Finally, we would like to thank our respective wives for their unending love, patient encouragement and support. During the final stages of the preparation of the manuscript, Mark's wife, Esther, gave birth to a baby boy. He has no excuse now for not changing nappies! On the other hand, Bruce is now free to enjoy his grand-children and... write another book!

Contents

Chapter 1

Editorial Introduction

As far as we are concerned in this book, Machine Vision is that branch of Systems Engineering relating to the design and construction of integrated opto-electronic devices for examining raw materials, natural and man-made objects and dynamic processes, such as manufacturing operations. Machine Vision is used to detect defects and improve quality, operating efficiency and the safety of both products and processes. During the last two decades, Machine Vision has been applied quite extensively in manufacturing industry, particularly for automated inspection. It has been used principally in areas where closer tolerance products and well-controlled scenes are being examined.

The subject is now mature enough to consider broadening its range of application to natural products, food products, and other situations in which there is a high degree of variability. (Where it can be measured quantitatively, variability is commonly found to be several orders of magnitude greater in natural products than in industrial artefacts.) These new application areas present certain difficulties for a vision system that have not been significant until recently. New types of algorithms for image processing are needed. Normally, these are required to be much more intelligent than industrial systems need to be. Greater use is made of the computational techniques encompassed by the terms Artificial Intelligence, Pattern Recognition and Fuzzy Logic. Whether it is applied to industrial manufacturing, food processing, agriculture, fishing or mining, Machine Vision is necessarily concerned with much more than computational methods and their implementation; vision engineers must also be prepared to design the lighting, image formation (optics), image sensing (cameras), mechanical handling (transport mechanisms), human–machine and machine–machine interfacing. Making all of this work together in a harmonious manner is essential to the success of Machine Vision. Indeed, systems integration is the very core of our subject; undue emphasis on any one aspect to the detriment of any other is a sure recipe for failure.

Chapter 1

Like Two Peas in a Pod

B.G. Batchelor

"Even the vision of natural objects presents to us insurmountable difficulties."

Anonymous, *Oxford English Dictionary*

1.1 Advantages of Being Able to See

Sight is unquestionably the most valuable of the senses available to people and many of the higher animals. The ability to see allows a mobile organism to hunt for food, search for a mate, look for a place of refuge and avoid predators and other potential dangers - all with the minimum expenditure of energy. At this very moment, you are using your marvellous eye–brain complex to learn what I was thinking about on a certain evening in January 2001. Did you watch television last night? If so, you were entertained/informed/educated visually. Didn't you choose the clothes you are wearing in part because they convey a visual message about your character and feelings to the people you think you might meet today? When you put cream in your coffee, you use your eyes, to judge how much more to pour into the cup. When you go to a supermarket, you select fruit, vegetables, meat, poultry, etc. visually, relying on their appearance to assess their quality. We inspect our clothes, homes and cars visually, for dirt, grime and stains. We decorate our homes, using wall-paper, paint, carpets, curtains, pictures and ornaments, to create a visual environment that suggests tranquility, elegance, or excitement, as we think most appropriate. Many of the games we play rely heavily, sometimes entirely, on our ability to see. We conduct commerce almost exclusively by exchanging messages written on pieces of paper, or computer screens. We teach children using the same media. We do so much with our eyes, throughout the whole period of wakefulness, that the threat of the loss of eye-sight is dreaded almost as much as any other human affliction. We use our eyes to move around safely, at work, in our leisure time, to enhance our personal relationships and to learn.

This book is concerned with the design of artificial vision systems that attempt to emulate one particular human activity: inspecting and manipulating highly variable natural objects and human artefacts. Animals and human beings clearly demonstrate that vision is fast, safe, hygienic and versatile. A machine that can "see" would possess these same advantages. Building a machine that can sense its environment visually, to perform some useful function, such as industrial inspection or controlling a robot, has been the subject of day-dreams for many years. However, it is only very recently in human history that we have been able to contemplate this with any realistic chance of fulfilling our ambitions. During the last quarter of the 20th century, our subject, which is called *Machine Vision*, evolved from an exotic technology, created out of the dreams of academics, into one that is of considerable practical and commercial value, and which now provides assistance over a wide area of manufacturing industry [1,2]. Hitherto, Machine Vision has been applied in industry to tasks such as inspecting close-tolerance engineering artefacts, during or shortly, after, manufacture.

Recent advances have shown how Machine Vision technology can be extended into those areas of manufacturing where natural materials and other highly variable objects are processed. Our discussion in these pages is aimed at those areas of manufacturing industry where wide tolerances are encountered, as well as activities, such as agriculture, horticulture, fishing, mining, etc., where similar conditions apply. However, we should not lose sight of the fact that many so-called high-precision industries are plagued by important areas of uncertainty. Of course, the electronics industry is renowned for working with high-precision components (integrated circuits, printed circuit boards, etc.) However, it is also concerned with highly variable entities such as solder joints, flexible cables and components with flying-lead wire connectors (resistors, capacitors, diodes, coils, etc). Much of the discussion in this book is relevant to applications such as these.

1.2 Machine Vision

As we have stated, this book is concerned with Machine Vision applied to highly variable objects. We shall concentrate primarily on the automation of visual inspection of natural materials and products with an ill-defined shape, size, colour or texture. Food manufacturing forms an important area of application but there are applications of this kind in a wide range of industries, including: agriculture, fishing, horticulture, mining, catering, pharmaceuticals, clothing, foot-wear, etc. In addition, many tasks in engineering manufacture require the handling and processing of highly variable objects. Of course, we create certain difficulties for ourselves by working with highly variable objects, since both Machine Vision and Robotics have hitherto achieved their greatest success with close-tolerance products. By broadening the scope of the subject now, we are implicitly acknowledging that Machine Vision has reached a certain level of maturity and that it is able to reach into areas of application that have hitherto been considered to be too difficult. There are significant commercial and practical benefits to be obtained. For example, there is a strong and continuing demand for new instruments that can improve the quality and safety of all products, especially food and pharmaceuticals items [3,4]. If we are successful, Machine Vision will bring significant economic, health, safety and social benefits to several areas of industry that have hitherto been inaccessible to it.

When we apply Machine Vision techniques to the inspection/handling of manufactured goods, we know roughly what to expect. This is certainly true for high-precision objects, such as plastic mouldings for automobile dashboards, electrical switches, videocassette cases, etc. It is also true for many types of mass-produced food items. For example, one loaf, confectionery bar, meat pie or cake, of a given kind, should be "similar to" all other items of the same nominal type. Even livestock exhibits some similarities but these may be harder to define objectively. Most pigs of a certain breed will have "similar" colouring patterns. More generally, natural products of a given class usually possess some degree of "similarity" to each other. Indeed, this helps us to define the concept of a "class". In recent years, Machine Vision, and the allied subjects of Pattern Recognition and Artificial Intelligence, have all developed to the point where we can broaden the concept of "similarity" from the limited idea necessarily imposed in earlier years. We shall touch on this in Chapter 3.

It is difficult to provide a completely satisfactory definition of Machine Vision, because it is concerned with a very diverse range of technologies and applications. However, we shall define the subject in the following way:

> *Machine Vision (MV) is concerned with the engineering of integrated mechanical-optical-electronic-software systems for examining natural objects and materials, human artefacts and manufacturing processes, in order to detect defects and improve quality, operating efficiency and the safety of both products and processes. It is also used to control machines used in manufacturing.*

Such a terse all-embracing definition requires a good deal of explanation and there are several important points that we should emphasise:

Our subject is *Machine Vision*, not *Computer Vision.* We maintain that these two subjects are fundamentally different in the mental attitudes they embrace, even though they are clearly related [1,5].

- We shall concentrate upon applications in industry, taken in the broadest sense.
- Machine Vision is a branch of Systems Engineering; it is not a science. On the other hand, Computer Vision is the science of storing and manipulating pictorial data inside a computer.
- Machine Vision is more likely to be used to verify than recognise. Hence, a Machine Vision system will probably address questions such as "Is this widget well formed?" rather than more general questions like "What is this?".
- Integration and harmonisation of several different areas of engineering technology are essential to the success of every Machine Vision project. We shall see that a successful system almost always requires us to integrate ideas from mechanical engineering, lighting, optics, sensors (cameras), analogue and digital electronics, software, human factors, etc. (Figure 1.1).

Figure 1.1. Visual inspection. Human and Machine Vision systems require the same essential components: mechanical handling of the objects being examined, lighting optics, image sensor (eyes, or camera), image processing system (human brain, or electronics), knowledge database, "accept" and "reject" channels for the sorted product. (Original caption: *Checking the spelling of sea-side rock.)* This is a confectionery bar that often has the name of a holiday resort running internally, along its whole length. (Cartoon reproduced by kind permission of Mr C. Besley.)

1.2.1 Model for Machine Vision Systems

The formal definition of Machine Vision given above does not really help the newcomer to understand what this subject is really about. To do this, we present the outline of an archetypal vision system in the form of a block diagram (Figure 1.2).

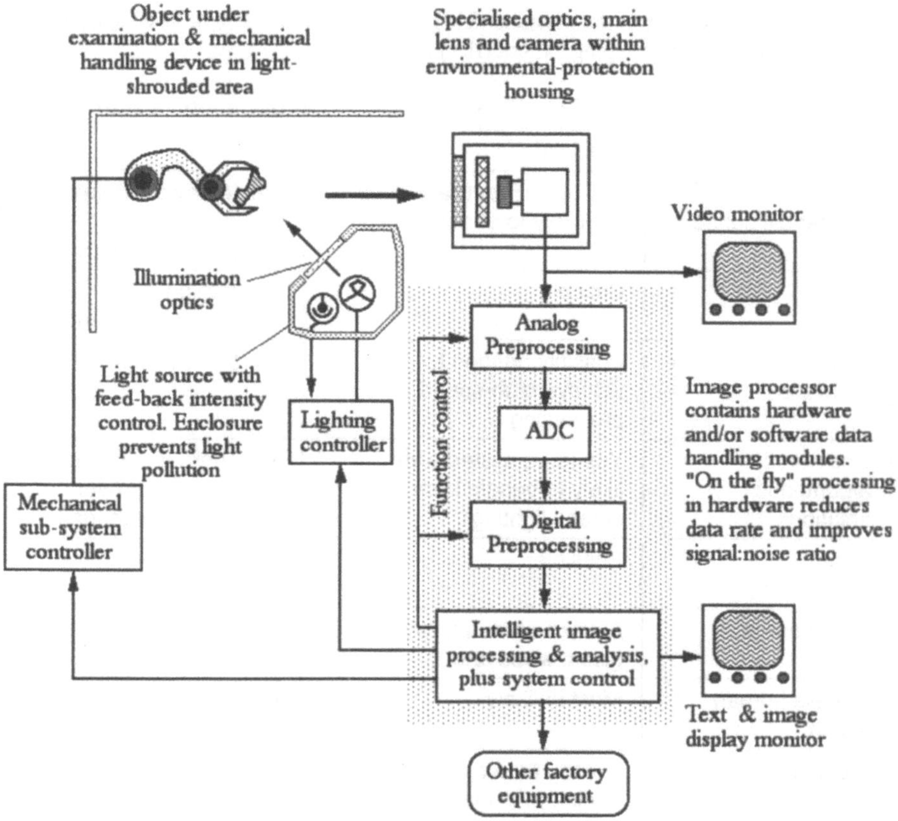

Figure 1.2. Archetypal Machine Vision systems applicable to a wide range of industrial applications.

Implicit in this diagram is the fact that Machine Vision is a multi-disciplinary subject and necessarily involves designers in mechanical, optical, electrical, electronic (analogue and digital), and software engineering and mathematical analysis (of image processing procedures). Less obviously, several aspects of "soft engineering" are also required, including human–computer interfacing, work management, and QA procedures. Hence, a team of engineers with a range of skills is needed to design a successful Machine Vision system. Integrating such varied technologies to create a harmonious, unified system is of paramount importance; failure to do so properly will inevitably result in an unreliable and inefficient machine. This point cannot be over-emphasised. This is why we insisted in the very first pages of this book that Machine Vision is quite distinct from Computer Vision. The point is made more forcibly in Table 1.1. The distinction between Machine Vision and Computer Vision reflects the schism that exists between engineering and science.

Table 1.1 Comparing Machine Vision and Computer Vision. Entries in the central column relate to the factory-floor target machine (TM), unless otherwise stated, in which case they refer to an interactive prototyping tool-kit (IPT) [2].

Feature	**Machine Vision**	**Computer Vision**
Motivation	Practical	Academic
Advanced in theoretical sense	Unlikely. (Practical issues are likely to dominate)	Yes. Many academic papers contain a lot of "deep" mathematics
Cost	Critical	Likely to be of secondary importance
Dedicated electronic hardware	Possibly needed to achieve high-speed processing	No (by definition)
Use non-algorithmic solutions	Yes (e.g., systems are likely to benefit from careful lighting)	No. There is a strong emphasis on proven algorithmic methods
in situ programming	Possible	Unlikely
Data source	A piece of metal, plastic, glass, wood, etc.	Computer file
Models human vision	Most unlikely	Very likely
Most important criteria by which a vision system is judged	a. easy to use; b. cost-effective; c. consistent and reliable; d. fast	Performance
Multi-disciplinary	Yes	No
Criterion for good solution	Satisfactory performance	Optimal performance
Nature of subject	Systems Engineering (practical)	Computer Science, academic (theoretical)
Human interaction	IPT: vision engineer TM: Low skill level during set-up. Autonomous in inspection mode	Often relies on user having specialist skills (e.g., medical)
Operator skill level required	a. IPT medium/high b. TM must be able to cope with low skill level	May rely on user having specialist skills (e.g., medical)
Output data	Simple signal, to control external equipment	Complex signal, for human being
Prime factor determining processing speed	IPT: human interaction TM: speed of production	Human interaction. Often of secondary importance

While Figure 1.2 does not indicate all of the areas of technical specialisation necessary to build a successful Machine Vision system, it is a lot better than the simplified diagram shown in Figure 1.3, which is often used as a model to introduce Computer Vision. As an absolute minimum, a Machine Vision system must contain:

a. some means of presenting the object to be inspected to the camera;
b. lights;
c. camera;
d. an electronic circuit card to digitise the signal from the camera;
e. computer, or dedicated electronic image-processing hardware;
f. software, if a computer is used for image processing;
g. actuator: this may be anything from a simple *accept/reject* gate, to a multi-axis robot.

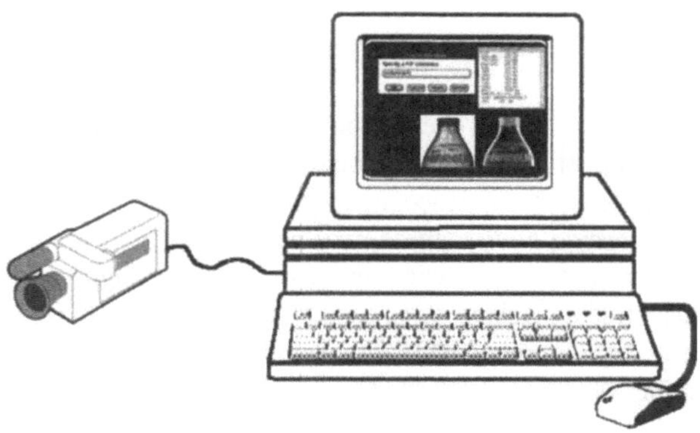

Figure 1.3. An over-simplified view of a Machine Vision system. There are no lights. The camera and computer are not protected and the computer provides output only to its own screen. However, this is a valid model for many Computer Vision systems.

The naïve view exemplified in Figure 1.3 is dangerous because it ignores the multi-disciplinary nature of Machine Vision. As we progress through the following pages, the true nature of the subject, as an engineering discipline, will become obvious.

1.2.2 Applications Classified by Task

Vision systems are currently being used extensively in manufacturing industry, where they perform a very wide variety of inspection, monitoring and control functions. Those areas of manufacturing that have benefited most in the past include electronics, automobiles, aircraft and domestic products (from furniture-polish and tooth-paste to refrigerators and washing machines) [1,2]. Vision systems have also been used in the food industry, agriculture and horticulture, although to a smaller extent [6]. The potential for applying Machine Vision technology in these areas is huge, although the penetration of these markets to date has been somewhat less than in "hard engineering". Present and projected applications of Machine

Vision to natural products may be classified conveniently according to the function they perform. No particular commercial importance is implied by the order of the following list:

a. *analysing the content* of a mixture (e.g., count how many sweets of each kind are put in a box; estimate the fat to lean ratio of a batch of butchered meat; check the mixing of granulated materials).
b. *analysing the shape* of whole products as a prelude to processing them using robotic manipulators (e.g., fruit, vegetables, animal carcasses, etc. might be trimmed/packed by a visually guided robot).
c. *analysing texture* (e.g., bread, cork floor tiles, wood panels, etc.).
d. *assembling food products* (e.g., pizza, quiche, meat pies, etc.).
e. *checking aesthetic appearance* (e.g., loaves, cakes, quiches, trifles, wood veneer, marble tiles, etc.).
f. *cleaning* (e.g., selective washing of cauliflower, broccoli, leeks, lettuce, cabbage, etc.).
g. *coating* (e.g., chocolate coating of confectionery bars, icing (frosting) on cakes).
h. *counting* (e.g., counting cherries on the surface of a cake, counting bubbles on top of a trifle, estimating the number of healthy leaves on a growing plant.).
i. *decorating* (e.g., cakes, chocolates, trifles, pies, etc.).
j. *detecting foreign bodies* (these may be "natural" and intrinsic to the product (e.g., seeds, nut shells, twigs, etc.), or truly "foreign" (e.g., stones, rodent remains, broken machine parts, contact lenses, false teeth, etc.)).
k. *detecting surface contamination* (e.g., mildew, mud, bird excrement, etc.).
l. *estimating the size or volume* (e.g., fruit, fish/animal/poultry carcasses, meat fillets, etc.).
m. *grading* (e.g., identifying premier-quality fruit and vegetables).
n. *harvesting* (there is a huge variety of tasks of this type).
o. *identifying loss of integrity* (e.g., rotting of fruit/vegetables, withering due to disease or lack of moisture, bird pecks, insect holes).
p. *measuring linear dimension* (e.g., fruit, vegetables, animal carcasses).
q. *packaging* (e.g., fragile/variable products, cream cakes, meringues, etc.).
r. *sorting* (e.g., fruit from leaves; fish by size/species on a trawler).
s. *spraying* (e.g., selective spraying of insecticides and fertiliser).
t. *stacking/packing* (e.g., carrots, oranges, frozen chickens, boxes, jars, bottles, cans or tubs, flexible bags).
u. *trimming and cutting* (fish and vegetables all require robotic cutting). Butchery and automated shearing of sheep are also included in this category.

This does not include the multitude of tasks that can arise in "hard engineering" industry, where low-tolerance features occur, giving rise to highly variable images to be processed. The following list gives a hint of the wide variety of inspection tasks of this kind that exists:

- adhesive strings/drops
- complex assemblies of components
- electrical coils that have flying wires attached

- feed stock for automatic machining
- flexible pipes
- gobs of semi-molten glass (prior to forming jars, bottles, etc.)
- moulds loaded with powders, granulated or semi-fluid material
- material-feed hoppers
- powdered material prior to melting, sintering, etc.
- rubber/plastic gaskets, water and oil seals, etc.
- solder joints
- spray cones, flames and smoke plumes
- sprue, swarf, remnant material (for example, after punching)
- unfired ceramics, including tiles, table-ware, etc.
- welds
- wiring looms

The fact is that even so-called high-precision manufacturing is not as tidy and well controlled as we would like. The main reason is that raw materials are unpredictable in shape, size, and colour. Another important factor is that it is sometimes expedient, or necessary, to apply a low-tolerance manufacturing technique (e.g., applying adhesive) to high-precision parts. The engineer's skill is manifest in doing this in such a way that the "untidy" parts are not apparent to the customer. While the main emphasis in this book is on inspecting and manipulating natural products, very similar Machine Vision methods are needed in applications like those just listed.

1.2.3 Other Applications of Machine Vision

Apart from manufacturing industry, artificial vision systems find application in a number of other situations, including but not limited to:

- document processing (optical character recognition/verification and document authentification);
- security and surveillance (identifying intruders in secure spaces);
- medicine and health screening (e.g., of cell samples for genetic screening, identifying cancer cells);
- military (e.g., target identification and fire control);
- forensic science and finger-print analysis;
- research (e.g., astronomy, bio-medical, particle physics, materials engineering);
- traffic control/monitoring (both pedestrian and motor vehicles).

Our discussion in these pages cannot possibly cover such diverse areas of application, since they present such very different requirements. One of the characteristic features of Machine Vision is that the equipment design is strongly constrained by applications requirements. Indeed, we may take this as the central criterion separating Machine and Computer Vision. In order to do justice to these and other potential applications of Machine Vision, we would need to repeat the analysis described in these pages, as each one presents a unique set of characteristics and challenges. It is only when we acknowledge that generic solutions are probably not appropriate for difficult practical applications that we are likely to see real progress. For example, vehicular traffic control requires a system that is able to tolerate glinting, image degradation due to rain and fog, brilliant ambient light (sun-light) and almost complete darkness. It is unreasonable to expect the image processing sub-system to cope with all of these difficulties unaided. By designing the image acquisition sub-system carefully, the problems can be minimised. The point is that the solution lies outside the image processing.

1.2.4 Machine Vision Is Not Natural

As indicated in Table 1.1, Machine Vision does not set out to emulate human vision. At some time in the future, it may be both possible and convenient to build a machine that can "see" like a human being. At the moment, it is not. Today, an industrial Machine Vision engineer is likely to regard any new understanding that biologists or psychologists obtain about human or animal vision as interesting but largely useless. The reason is simple: the "computing machinery" used in the natural world (networks of neurons) is quite different from that used by electronic computers. Certainly, no person can properly look introspectively at his/her own thought processes in order to understand how he/she analyses visual data. Moreover, there is no need to build machines that see the world as we do. Even, if we could decide exactly how a person sees the world, it is not necessary to build a machine to perform that task in the same way. (In the natural world, there are clearly several different and successful paradigms for visual processing including: insects, molluscs, fish, and mammals.) In any case, such a machine would have the same weaknesses and be susceptible to the same optical illusions as we do.

None of the authors of this book have set out to copy the way that people see things. Indeed, the author of this chapter feels that we should resist any temptation to do so until our knowledge of human/animal vision is much more complete. It would very useful if our understanding of biological systems were sufficient to enable us to build better Machine Vision systems. Unfortunately, with our present limited knowledge, this is just not the case.

1.3 Product Variability

As we have already noted, most of the successful applications of Machine Vision have concentrated on the inspection of high-tolerance components. However, the theme of this book is the application of Machine Vision to highly variable objects, such as natural products, processed food items and other artefacts of ill-defined size, shape, colour or texture. In this section, we discuss the nature and significance of the differences that exist between precisely made objects, such as engineering piece parts and these more variable objects.

The title of this chapter is based on a common English proverb; we often say that two similar objects are *"as alike as two peas in a pod"*. Yet, peas in a pod can be very different from one another (Figure 1.4). They can differ very considerably in size, shape, colour and surface texture. Some peas are likely to be bright green, while others have a brown or yellow tinge, or are nearly white. Their skins can be split and some peas have insect holes through them. They can also differ in shape; some peas are almost square, having been squashed by their neighbours in the pod. Most peas are nearly spherical, while others may be ovoid. Natural objects, like peas, frequently vary from one another in many different ways. In the light of these remarks, it may seem that by stating that *"objects A and B are as alike as two peas in a pod"*, we are not making a very strong statement about their similarity. In fact, we need to revise our concept of similarity, since all of the peas growing in a given pod are genetically identical; they all contain the same macromolecules and they have very nearly the same nutritional value. Natural variations in physical size, colour, shape and texture all have to be *tolerated* by an inspection system. Peas are never more than 15 mm in diameter. They are never blue, striped or hairy. They do not have a sponge-like surface, nor do they move of their own volition. This is the nature of their similarity, which is fundamentally different from that relating two cars of the same model, or two bottles made in the same mould.

Figure 1.4. Peas are not alike in form, colour or surface texture.

In this section, we demonstrate and, as far as possible, quantify the differences in tolerance values that exist in engineering artefacts on the one hand and natural products on the other.

1.3.1 Linear Dimensions

Before we begin our discussion in earnest, we need to define a measure of variability. Let X be some conveniently measured linear dimension, such as the diameter of an apple, the length of a banana, or the circumference of an orange. Let X_0 be the average value of X, for a certain class of objects, C. Furthermore, let δX be the maximum observed deviation from X_0. (Hence, all estimates of X for objects in class C lie in the range $X_0 \pm \delta X$.) The degree of conformity of objects in class C is defined as V_C where

$$V_C = \log_{10}(X_0/\delta X) \tag{1.1}$$

Statisticians would, no doubt, argue that we should instead relate V_C to the standard deviation of X, rather than the total range of this variable. However, the logarithm in the definition of V_C makes this change relatively unimportant. Furthermore, as far as both manufacturers and customers are concerned, the range of variation ($X_0 \pm \delta X$) is often more important than statistical measures. If an object, such as a loaf, is too large, it will not fit into its packaging. On the other hand, if it is too small, the customer will not buy it! The present measure of conformity (V_C) is conceptually simpler than one based on statistical parameters and is perfectly satisfactory for our present purposes.

We shall see that V_C is large for high-tolerance (engineering) products and small for highly variable objects (e.g., food and natural products). In the following discussion, remember that almost all of the successful applications of Machine Vision to date have been based on products for which $V_C \geq 2$ (Figure 1.5).

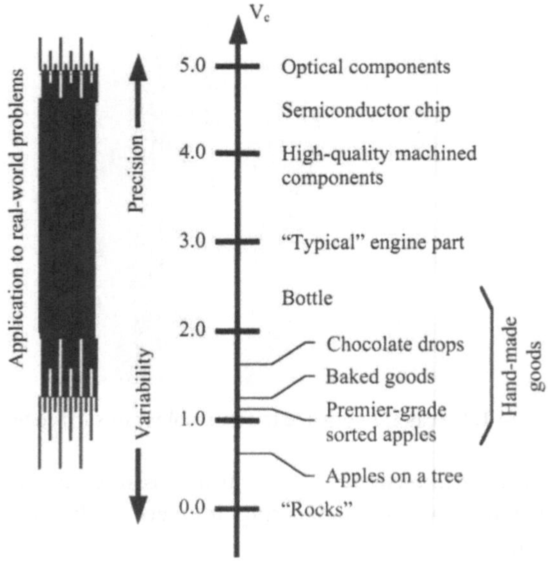

Figure 1.5. The measurement of conformity in human artefacts and natural products. All values of Vc are approximate. Hitherto, almost all applications have been based on products for which $Vc > 2.0$.

Among the very highest precision objects made in industry are mirrors and other optical components. A good-quality plane mirror has a surface ground to about $\pm\lambda/10$, where λ is the working wavelength. Suppose a mirror has a diameter of 50 mm and is expected to reflect green light ($\lambda = 0.5\mu m$), then, $V_C = 5$. (It is more meaningful to compute V_C in this way than to compare λ to the thickness of the mirror substrate.)

Electronic circuit boards combine high-precision components (semiconductor chips) and low-precision features (e.g., resistor placement and solder joints). Integrated circuit chips with a conductor width of 2 μm (i.e., $\delta X = 0.2$ μm) and a diameter of 8 mm have a conformity score $V_C = 4.6$.

A "typical" engine part for an automobile engine might be 10 mm long and have a tolerance of 10 μm. In this case $V_C = 3$.

A bottle or jar has a slightly greater degree of variability. A wine bottle is about 300 mm tall and its variation in height is roughly 1 mm. (Although it is made in a mould, the bottle is liable to sag under its own weight, before the glass cools to become "solid".) Hence, $V_C = 2.4$.

Now, consider more variable items. A jam doughnut is about 150 mm in diameter and its variation in diameter can be as much as 10 mm. (It is not truly round either.) Hence, $V_C = 1.18$. Baked goods, formed by cooking semi-fluid dollops of dough on a plain tray, can vary by even more than this. The baking process, combined with the ill-defined size of a "dollop" can produce values of V_C less than 1.0. The author measured two packets of chocolate drops, of different sizes, yielding measurements of V_C equal to 1.54 and 1.56. Bread rolls, croissants and many other baked goods and items of confectionery are highly variable, giving values of V_C equal to 1.0 or even less.

Natural products, such as apples, of a given variety, typically vary by a large amount, giving values of V_C as small as 0.4–0.5. Modern supermarkets tend to avoid selling apples of highly variable size, by the simple expedient of grading them before they are put on sale. (Grading apples represents a potential application for Machine Vision.) In a series of measurements of premier-quality graded apples, a value of $V_C = 1.10$ was recorded.

Agriculture, as distinct from food processing, is concerned with very high degrees of variability, since growing animals and plants can differ greatly in size. A farmer cannot properly control the size of fruit on a tree; he/she must pick whatever happens to be there. Hence, V_C is likely to be very small, possibly under 0.5. Of course, farm animals also vary in size. For adult pigs of a given breed, $V_C = 0.5$.

There is another important class of objects that we have not mentioned yet but which deserves attention: hand-made goods.[1] Consider a hand-thrown pot, such as a jug, or vase. No two vessels, even of the same nominal type, are exactly alike in size and shape and we might expect V_C to be about 1.3–1.9. Hand-carved and painted artefacts produce a roughly similar range of variation, as does hand-cut/engraved glassware. Exceptionally, human beings can make objects for which V_C is as large as 2.3.

---

[1] It is not immediately obvious why a hand-made object, such as a carved wooden bowl, painted ceramic figurine, or cut-glass decanter should be examined by machine, since it is inspected by its creator, both during and after manufacture. There is an unfulfilled requirement for machines that can objectively examine fragile and highly variable objects of this kind before and after transportation to a warehouse or retailer.

There is another even worse situation in mining and quarrying, where it is almost impossible to define the size of a "typical lump" of rock. Here, it is meaningless to quote typical values for our measure of variability, V_C. Let it suffice to say that gravel, lumps of rocks and coal, etc. are very variable ($V_C \leq 0.3$) and therefore require non-specific handling techniques.

1.3.2 Shape

Although shape cannot be quantified sensibly using a single parameter, in the same way that size can, we are able to define a range of measurements, any one of which is sensitive to the variability that occurs in nature. For example, the so-called *shape factor*[2] is one such shape descriptor that is popular among vision engineers. Five ivy leaves (Figure 3.17) produced values of shape factor varying from 0.425 to 0.546. Similarly, five chicken "butterflies" (Figure 3.21) gave values between 0.421 and 0.597. On the other hand, a croissant produced a very similar value (0.633) to that obtained from a slice of bread (non-lidded tin loaf, 0.631). The fact that intra-class variability is comparable to and may even exceed inter-class variability merely indicates that shape factor is not a particularly good measurement of shape. Other shape measurements may be more sensitive to the particular types of shape variation that are important in a given application.

Despite the differences that exist among objects like ivy leaves, human beings are readily able to classify them as belonging to the same general class. Of course this merely confirms that human beings are well suited to recognising natural objects, despite their highly variable shapes. This is just one example of the general principle that tasks that a person finds easy may be quite difficult for a machine (and *vice versa*). In Chapter 3, a more sophisticated way of describing shape is described. This represents a blob-like object by a tree. However, we shall show this is unstable in many situations, which may inhibit its use in practice.

When designing Machine Vision systems, we have to assume that the shapes of both natural products (e.g., fruit, vegetables, nuts, individual leaves, decorative domestic plants, etc.) and certain manufactured items (baked goods, confectionery, meat pies, joints of meat, etc.) will vary considerably (Figure 3.19). We have only to pick up a handful of fallen leaves, or buy a kilogram of carrots or potatoes, to observe the high level of variation of shape that a vision system must tolerate. Of course, while shape variation is usually manifest in three dimensions, it is only observed in two dimensions, if we employ a single video camera. A stereoscopic vision system or 3-D scanner may see the full effects of shape changes but we do not need either of these to understand the nature of the problem that shape variation causes. It must be understood that conventional (e.g., geometric) concepts of shape are totally meaningless in this context. For example, "round" fruit is not truly circular, in the mathematical sense. Furthermore, straight lines, right-angle corners,

---

[2] The *shape factor* of a blob-like object is defined as 0.079577 ($1/[4\pi]$) times the ratio of its area to the square of its perimeter. This has a maximum value of 1.00 for a circle and has low values for spider-like objects. For example, a hair comb gave a value of 0.01532.

rectangles, parallelograms, ellipses and other geometric figures do not exist in nature, except in the special case of crystalline structures.

The unpredictable shape of certain natural objects creates severe problems for an automated mechanical-handling system. Fixed tooling is out of the question for tasks such as trimming lettuce, shearing sheep, or automated butchery. On the other hand, multi-axis robots require complicated multi-sensor feedback control techniques to cope with highly variable object shapes. Even seemingly simple vision-based tasks relating to natural objects can require the use of complex equipment. For example, removing spot-like surface defects on nominally round fruit or vegetables (e.g., "eyes" on potatoes) requires the use of a 4-axis (X,Y,Z,θ) visually-guided robot. Trimming rhubarb also requires a 3-axis robot (X,Y,θ), while placing decorative features on top of a cake requires 4 degrees of freedom, (X,Y,Z,θ). Everyday tasks, such as removing the leaves from carrots, or trimming leaks, are often quite difficult when we try to automate them. We should bear in mind too that in many cases we must treat natural products with great care, to avoid damaging them. By clever user of guide rails, shaped conveyors, etc., it is sometimes possible to reduce the complexity of a robot handling highly variable products. For example, a single-axis robot could be used to trim the ends of carrots that have already been aligned, perhaps using a slotted conveyor. Nevertheless, the complexity of the handling equipment is usually higher than it is for high-precision objects. The general rule is that, if we know what an object is and where it is, simple handling techniques are possible, even desirable. Variations in shape make this prerequisite difficult to achieve.

Many shape-analysis techniques that work satisfactorily on one example of a given type of object may not work with other samples of the same product kind. In other applications, ill-defined shapes are present that are amenable to only the simplest and most robust methods of analysis. In some situations we may even have to change the method of analysis as the shape changes. A child's toy illustrates this point (Figure 3.14). Organic materials and objects frequently change shape as they grow. Hence, we must be prepared for a similar situation to occur when we are working with natural products

1.3.3 Why Physical Tolerances Matter

There are several reasons why physical dimensions matter:

a. Objects to be inspected must be fed through the inspection machine, whatever their size and shape. The mechanical handling system must accommodate all variations of these parameters without jamming. Clearly, this is much more difficult if these objects are of uncertain size and shape.
b. In view of the preceding point, the system for mechanical handling of highly variable objects must either be very intelligent, or very simple, such as a plain-top conveyor belt. Of course, a conveyor belt is far simpler and cheaper than a multi-axis robot but has one serious drawback: it is unable to hold the object in any particular pose before the camera. The design of robust mechanical

handling systems for highly variable objects, especially plant products with flexible foliage, requires considerable skill and ingenuity.

c. The camera may not be able to see certain features. Certain objects (e.g., bread rolls and some cakes) are actually made on a conveyor. It is therefore quite difficult to examine their bases. Certain objects will always fall into the same pose, if dropped haphazardly onto a flat surface (e.g., a plain conveyor belt). Other objects may have two or three stable states, while objects such as apples have one or more quasi-stable states. As a result, measurements of certain parameters, such as the diameters of apples, may show statistical bias, unless care is taken to orientate the objects first. The conveyor itself may obscure some features. It is common practice to use shaped (e.g., cupped) conveyors to stop delicate objects such as eggs, fruit, etc. from banging into one another. Inevitably, this causes some restriction on viewing angles.

d. Varying shape presents problems by obstructing illumination, or viewing. At its simplest, this means that certain features are obscured, either by being out of the camera's line of sight, or by being in a shadow. The latter can *always* be solved by coaxial illumination and viewing but this may not yield sufficient contrast to be of practical use. Obtaining images from the unpredictable form of untrained foliage presents a particular challenge and a high level of intelligence is needed to perform operations such as pruning, picking fruit, selective spraying, etc. (Figure 3.20). In general, the lighting arrangement needed for highly variable products is problematical, since the optimal illumination configuration for one sample may be far from ideal for another. The problem is made worse by the fact that samples of the same type of object may lie in different poses. The range of options for maximising image contrast by good lighting is rather smaller than that available for well-defined, high-tolerance products. The use of intelligent lighting is another possibility for coping with high levels of product variability but has, as yet, received scant attention.

e. When object size and/or shape is highly variable, choosing the most appropriate lens can present problems. The field of view must be large enough to accommodate the biggest objects of a given class. Yet, this may mean that the smallest examples may not be seen easily. The problem is worse than we might imagine at first, because natural products tend to be delivered haphazardly for inspection. It is impossible to predict how they will fall to rest., since they roll, fall into a quasi-stable pose, or even change shape as they come to rest (e.g., soft-heart lettuces, certain types of cut flowers). All of these produce an uncertainty of position that must be accommodated by increasing the size of the camera's field of view (Figure 1.6).

f. Changes in object size alter the contrast in x-ray images. The thicker the sample, the denser an object is to x-rays and the lower the contrast on small foreign bodies (Figure 3.21).

g. Finally, the image processing operations needed to inspect natural and other highly variable objects ($Vc \leq 2$) often differ fundamentally from those used for high-tolerance parts. ($Vc \geq 2$). The former have to be more intelligent, since they have to cope with greater levels of uncertainty about all aspects of the object's appearance: size, shape, pose, position, texture and colour. This will be discussed in detail in Chapter 3.

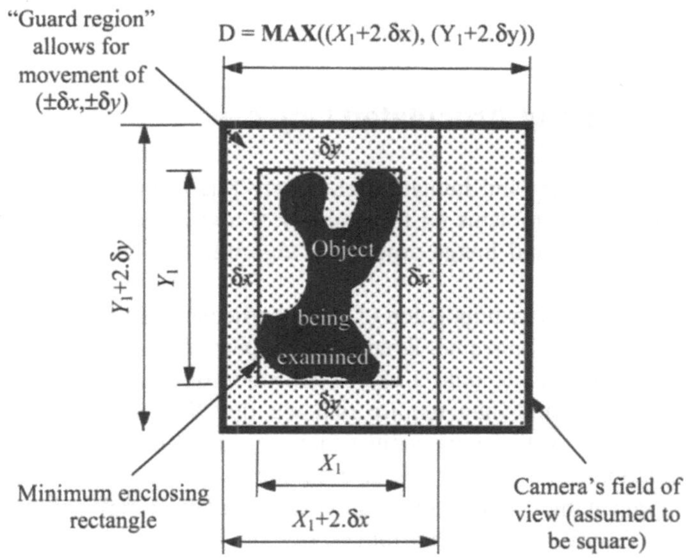

Figure 1.6. Uncertainties in the position of the object being examined require that the camera's field of view be larger than the object itself. If the object orientation is not controlled, the field of view must be even larger.

Perhaps the most significant problem of all is the lack of a clear objective criterion for what is an acceptable product. Bakers have devised rules of thumb for identifying malformed loaves or cakes. These are used principally to train inspection personnel and were derived by bakers, trying to predict what their customers want. The best practice is, of course, to ask a large number of customers what they will accept. However, this is expensive and inaccurate. Precision in such an exercise is notoriously difficult to achieve, because there is no such thing as a "typical" customer. Judgements are made sub-consciously, based upon a large variety of factors, such as packaging, cost, and even location on a supermarket shelf. In some companies, opinion about what is acceptable may even come from casual remarks made by the CEO's wife! However, the final decision about what constitutes an acceptable loaf or cake is based on purely subjective criteria by individual customers in a supermarket. Strict rules of thumb exist for grading and classifying fruit and vegetables. Indeed, these are fundamental to the worldwide GATT trade agreements. These include, but are not restricted to, specification of shape, size and colour. Criteria for grading living plants and animals are almost inevitably expressed by rules of thumb, derived by nursery-men, farmers, whole-salers, retailers, etc. who try to anticipate what the end-customer wants. Aesthetic judgement about decorative items, such as wood panelling and marble tiles, is difficult to automate but criteria expressed as rules of thumb probably represent the best explicit statements that we can achieve. To a software engineer, the words "rule of thumb" suggest the use of heuristics, rule-based decision-making techniques and expert systems. However, as we shall see later, self-adaptive Pattern Recognition techniques, including Artificial Neural Networks, are appropriate when a human inspector can only say that an object is either *"good"* or *"faulty"*.

1.3.4 Flexible and Articulated Objects

Many natural products and human artefacts are totally or partially flexible and present considerable problems for the designer of a vision system. The mechanical handling sub-system and image processing algorithms have to be designed with particular care.

Tasks involving the manipulation of simple articulated objects may be far more difficult than we might imagine. Even a simple pair of scissors alters its topology as it is shut and opened (Figure 3.15). A collection of linked levers is even more difficult to inspect and may require that we flex the assembly to examine it properly. This may well require sophisticated sensing (vision and force feedback) and manipulation (robotics), as well as a high level of intelligence and knowledge about the application. The posture of an articulated object may becomes unstable, causing it to fall over, as it is (un)folded. Certain features may become invisible, for example when an articulated arm is folded. Flexible objects and articulated mechanical systems are often fragile and therefore require special care in handling. Non-contact opto-electronic sensing is, of course, ideal, provided that the inherent difficulties of working with flexible/articulated systems can be overcome. Unfortunately, we are not yet able to build an effective "multi-purpose" handling system for flexible and articulated objects; each application has to be considered separately.

It is fairly easy to design a mechanism to transport carrots, once their leaves have been removed but they do not oblige us by growing according to our whims! Vegetables and fruit may well be delivered to an inspection system with their leaves and stalks still attached. Of course, it is likely to make the design task easier for the vision engineer if they have been removed beforehand. Many types of fruit and vegetable can be handled successfully by cleverly designed mechanical systems but these are expensive, and prone to jamming and damaging the product. In addition, food-trimming machines require frequent cleaning and may be quite wasteful by removing too much "good" material. It would be better, therefore, if we could use vision to minimise the physical contact between the machine and the product that it is processing.

Gardening and horticulture present some interesting but very challenging applications, including pruning, grafting, taking cuttings, replanting seedlings, etc. These are difficult tasks for a machine and consequently require intelligent image analysis techniques and careful design of the robot and its end-effectors. Micropropagation (Figure 3.20) is a well-constrained task of this type and offers immediate financial and social benefits. (Human beings have to wear facemasks and overalls to avoid infecting the tiny plants as they are working.)

Broadly similar problems occur in engineering industry, for example, when handling objects that have trailing tubes, wires or cables (Figure 3.16). Industrial parts-handling applications create a wide spectrum of difficulties.

Some objects such as certain cuts of meat and fish, foliage, leather, cloth, etc. are, of course, flexible. Sometimes a vision system is required specifically to guide a robot to handle and process such materials. This type of task still presents a formidable challenge to both vision engineers and robot designers. Hence, on many occasions, we still have to rely on human beings to deliver flexible materials to an inspection station. The chicken-breast "butterflies" illustrated in Figure 3.21 are fed to the x-ray system in this way [3,4].

Once again, it may be impossible to achieve optimal lighting everywhere. At first sight, structured lighting (Figure 1.7) seems to provide a possible way to acquire detailed information about the three-dimensional structure for certain applications. However, there are considerable problems, due principally to glinting, occlusion, and the limited working speed of light-stripe triangulation techniques.

1.3.5 Soft and Semi-fluid Objects

Machine Vision has an important role to play in inspecting soft and semi-fluid materials, such as uncooked dough, pastry and cake mixture, minced meat, whipped cream, jam, various gels and pastes used in the food and pharmaceutical industries, etc. Many industrial artefacts contain materials that are semi-fluid at some time during the manufacturing process.

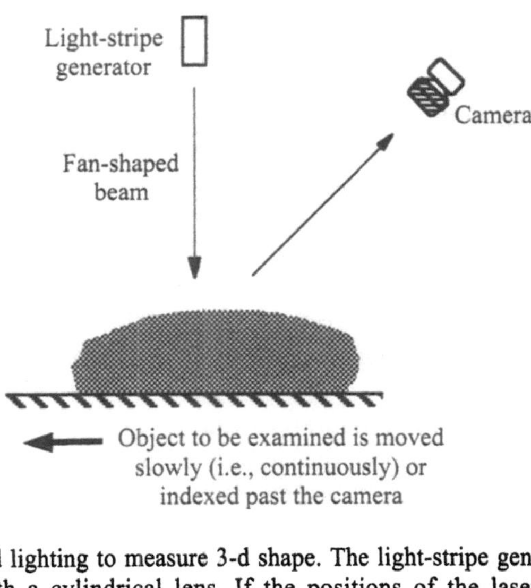

Figure 1.7. Structured lighting to measure 3-d shape. The light-stripe generator is usually a diode laser fitted with a cylindrical lens. If the positions of the laser and camera are reversed, the data is more difficult to analyse.

Examples are uncured foams, molten solder, globs of adhesive, gobs of molten glass, etc. These all require specialist handling techniques, as we must not disturb a semi-fluid dollop unnecessarily, simply to suit the whim of a vision engineer.

A characteristic that is peculiar to this class of materials is that their composition can be changed very easily and quickly. To avoid drying and curing semi-fluid material, infra-red and ultra-violet should be removed from the illuminating radiation, using standard optical filters. Sometimes in particularly dirty factory environments, an air purge is used to maintain clean optical surfaces. In this situation, we must ensure that it does not blow dust/moisture onto the damp surface. An ill-directed air purge can cool the surface too quickly and blow dirt onto it.

However, if care is taken to avoid the problems just mentioned, Machine Vision is ideal for inspecting semi-fluid materials, as no physical contact is needed. Thus, the shape of a lump of unbaked dough, or a dollop of whipped cream, is not altered by the inspection system. Furthermore, there is no possibility of introducing micro-organism contamination to food or pharmaceutical products. These facts provide powerful motivation for using Machine Vision in these industries. However, there is one fundamental problem: *What does a "good" lump of dough look like?* How do we define an acceptable range of shapes for dollops of cream on a cake or trifle? This the classical "ground truth" problem encountered when designing Pattern Recognition systems. We mentioned this topic in Section 1.3.3, and shall return to it again in Chapter 3.

1.3.6 Colour Variations

Colour Variations in Natural Products

One of the greatest challenges encountered when applying Machine Vision to natural objects lies in the recognition of colour. Natural objects tend to display very subtle variations of colour, which usually changes continuously across an object's surface. While sharp colour gradients sometimes occur in nature, they are usually less common and less pronounced than those found in polychrome industrial artefacts. Natural objects of the same nominal type may differ from one another, with significant changes in colour. On the other hand, damage or disease in plant or animal material may be manifest by very subtle colour changes. (Sometimes infrared and ultra-violet illumination helps to improve image contrast.)

Colour Variations in Manufactured Products

Subtle colour changes occur on the surfaces of many industrial artefacts as a result of processes such as drying, oxidation, rust formation, chemical corrosion, micro-organism attack, etc. Detecting these conditions can be just as difficult for a Machine Vision system as any task involving natural products. Highly sensitive sensors have been developed for industries such as textiles and automobiles, where high colour fidelity is essential. However, these usually measure the colour either at a single point, or averaged over a region. The use of a video camera for monitoring subtle colour variations across a large area, such as a complete car body, has not been very successful, to date.

Pseudo-natural Colour Recognition Task

Many engineering artefacts are either monochrome, or polychrome with colours that are distinct in both space and hue. The most notable exception to this is to be found in high-fidelity colour printing, which attempts to emulate natural colouration. For this reason, we should regard colour printing as a "natural" scene, although two important differences exist:

- We know very precisely what colours to expect at each point in a printed colour image. Of course, we do not in a natural scene. In principle, this allows direct, point-by-point comparison to be made between two samples of colour printing. However, this approach is not possible as a basis for inspecting true natural products.
- Colour printing uses very few pigments, perhaps only three or four, to approximate continuous colour variations. It is necessary therefore to view the printing at an appropriate resolution; if we wish to see individual dots of coloured ink, we need a very high resolution. To view the same printing in approximately the same way that we do, we must use a much lower resolution to obtain spatial integration.

Traditional Colour Theory

A common practice in books and papers on Colour Theory is to annotate the standard CIE Chromaticity Diagram by placing a spot to indicate the colour of, say, a ripe tomato, lemon or lettuce. Unfortunately, this excessively naïve view is unhelpful for our purposes and can be misleading. In reality, a ripe tomato should be represented by a large diffuse cluster, not a single point, as there is likely to be a considerable variation in colour across its surface. In theory, the ripening process of a tomato could be mapped onto the Chromaticity Diagram by drawing a thick arc, starting at green, travelling through yellow, to orange and then red (Figure 1.8). However, the standard treatment of colour is inadequate to represent the subtlety of natural phenomena, such as tomatoes ripening. The difficulty lies not, so much in the representation of colour, as the analysis of the data. No general theoretical model yet exists. As a result, colour recognition is probably best approached using Pattern Recognition learning methods. We shall discuss this in more detail in Chapter 3.

Problems in the Naming of Colours

The core of the colour recognition problem lies in the fact that we do not know what a label such as "yellow" really means. Of course, you "know" what "yellow" is, because you were taught from the cradle to associate that word with a certain class of visual stimuli. Since then, you have had your mental concept of "yellow" fine-tuned by years of exposure to other people, each of whom has a slightly different idea. The person who taught you what "yellow" means did not teach me, so my concept of "yellow" is slightly different from yours. Proving this is easy: simply ask a few people to draw limits of what they perceive to be "yellow" on the spectrum, or any other rainbow-like pattern. Figure 1.9 shows the results of one simple study of this kind. Why are there such wide differences in the definitions of colours? We all agree on certain "core" colours being given names such as

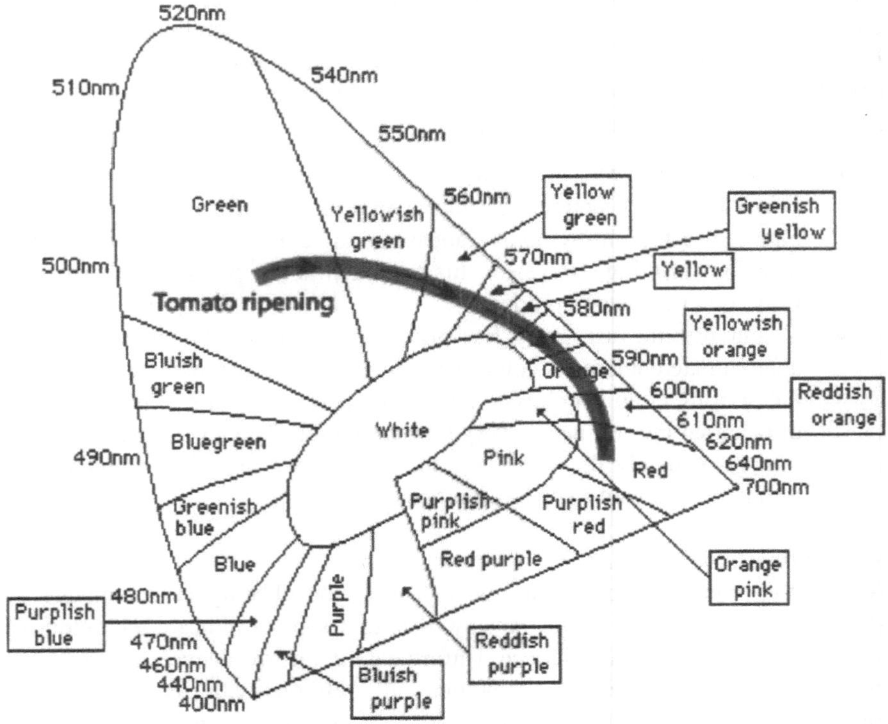

Figure 1.8. In this monochrome form, the Chromaticity Diagram gives the false impression of being able to discriminate accurately between colours. (It is often printed like this, to avoid using expensive colour-reproduction techniques.) The segments are labelled with names such as "bluish green", "blue green" "greenish blue", etc, even though these are not universally agreed colour categories. The diagram does not indicate the location of commercially important colours, such as "mould green", "butter yellow", "leaf green", "well cooked cake", or "ripe tomato". The process of ripening in a tomato is represented diagrammatically by the broad grey curve starting at "green" and progressing to "red".

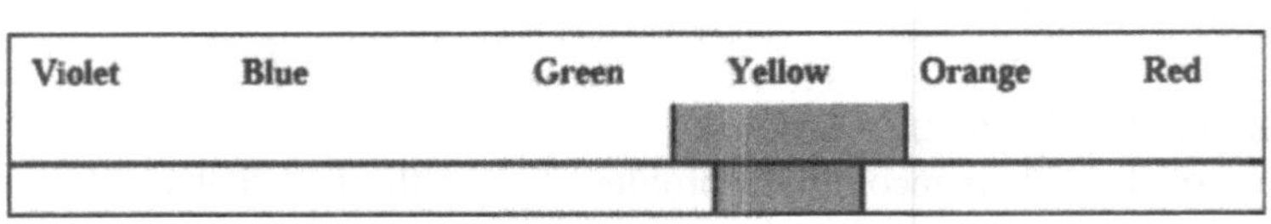

Figure 1.9. Problems in describing colours by name. Two native English speakers were asked to draw the limits of what they perceived to be "yellow" on an image of the spectrum, displayed on an LCD computer screen. The author added the colour labels "violet" - "red".

"yellow", "red", "turquoise", etc. However, we differ in our naming of those colours at the boundaries between these classes. There are certain reasons for this:

a. Different languages name colours in different ways. Figure 1.10 explains the differences in the naming of certain colours in Welsh and English. There are even significant differences in the use of colour names between

Welsh speakers from North and South Wales. Major differences exist, for example, between English and Japanese, as well as between English and Chinese.

b. Certain diseases, brain injury, ageing and drugs can all cause changes in colour perception [1].
c. Specialist occupations, such as printing, weaving, etc., lead to refinement of previously learned colour concepts.
d. Ambient lighting. The grey skies of northern Europe and the subsequent heavy reliance upon electric lighting provides a very different visual environment from that encountered by people living in the tropics. It is well known that lighting alters our perception and naming of colours.

Since the naming of colours is so subjective, we must build Machine Vision systems so that they are able to learn what colours to expect in a given application. This is where Pattern Recognition comes into its own.

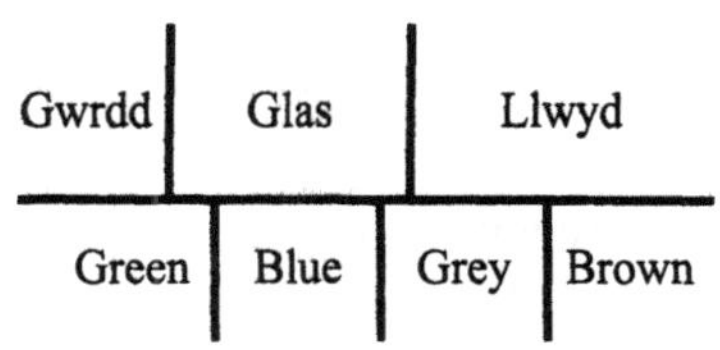

Figure 1.10. Differences in the naming of certain colours in Welsh and English. A Welsh speaker will refer to the colour of grass and blue sky using the same word.

Colour Sensors

By themselves, most video cameras are incapable of discriminating reliably between healthy and diseased tissue, or providing a reliable basis for deciding when fruit on the tree is ready for picking. However, more accurate instruments for measuring the spectral composition of light (from a point source) are available and can be used to calibrate a video camera. Assuming that we have given proper consideration to calibrating the camera, guaranteeing that the light source is stable and cleaning all optical surfaces, there is no reason why a low-noise solid-state camera cannot be used for colour analysis. The major problem for colour recognition lies elsewhere.

RGB Representation

A colour video camera measures the so-called RGB components of the light falling at each point in its retina. It does this by placing monochrome photosensors behind three differently coloured filters. These selectively attenuate different parts of the spectrum. R, G and B have traditionally been associated with detecting "red", "green" and "blue" although this idea must not be carried too far. A camera that outputs an HSI video signal (measuring Hue Saturation and Intensity) uses an RGB sensor and performs the RGB-HSI transformation within its internal electronics.

Let $[r(i,j),g(i,j),b(i,j)]$ denote the measured RGB values at the point (i,j) in a digital image. The set of colours in the image can be represented by S_3 where

$$S_3 = \{[r(i,j), g(i,j), b(i,j)]/1 \leq i, j \leq \mathrm{N}\} \tag{1.2}$$

By mapping the image into this particular three-dimensional measurement space (called RGB-space), we forsake all knowledge of the position of each colour vector. However, by extending this definition slightly, we can preserve this information. The set of points

$$S_5 = \{[r(i,j), g(i,j), b(i,j), i, j]/1 \leq i, j \leq N\} \tag{1.3}$$

lies in a five-dimensional space and preserves both position and colour information. For most purposes, S_5 is usually too cumbersome to be useful in practice. The representation embodied in S_3 is normally preferred, when we want to recognise colours independently of where they lie within an image.

We can combine the output of a colour recogniser with the position co-ordinates, (i,j), to obtain essentially the same information as is implicit in S_5. The following is a hybrid representation which combines colour labelling (not "raw" RGB values) with geometric position (i,j) within an image.

$$S_{\text{hybrid}} = \{[colour\_name, i, j]/1 \leq i, j \leq N\} \tag{1.4}$$

This representation is more economical than S_5 yet contains more information than S_3. Moreover, S_{hybrid} can be stored and processed as a monochrome image, in which "intensity" (i.e., an integer) represents symbolic colour names. (The result is an image like that obtained when "painting by numbers".) It is often convenient to display such an image in pseudo-colour, since similar values of the integers defining the values of *colour_name* may represent quite different physical colours. The design and use of a colour recogniser will be explained in Chapter 3, which also shows the innate differences between colour variation in natural products and human artefacts.

1.3.7 Transient Phenomena

I heard the following story from a colleague, who is a highly experienced and well-respected vision systems designer. He was commissioned to design a machine to detect patches of mould on the surface of (edible) nuts. The customer supplied him with a large bag of nuts, enabling him to study the problem in detail. He conducted a detailed feasibility study, then designed and built an inspection system. Of course, this took him several weeks. Then, he contacted the client, to arrange a demonstration, prior, to delivering and installing the system in a food-processing factory. A person from the client company arrived for the demonstration, carrying another large bag of nuts. Following good engineering practice, both parties wanted to test the inspection system on previously unseen samples. However, it did not respond well in these tests and failed to satisfy the client's expectations. There was a very simple reason for this disappointment: the new sample of nuts had fresh mould, which was *green*, whereas the original samples had older *brown* mould. The client had omitted to tell the vision engineer that the green mould quickly changes colour to brown. This fact was so obvious to the client that he had not thought it important enough to tell the vision engineer. Since brown and green mould patches can be identified by eye, it was "obvious" to the client that both could be detected easily by a Machine Vision system. It is a universal truth that, whenever the client tries to tell the vision engineer how to do his/her job, trouble will follow.

The lesson for us is that materials of biological origin change with time. The surfaces of industrial artefacts can also change their appearance as a result of oxidation, drying, as well as other chemical, physical and biological processes. The visible changes in biological materials can at times be very significant, yet in certain situations they can be very subtle. Dehydration will cause foliage to change shape quickly as it wilts, but early stages of decay are manifest as very slight changes of colour. It is essential that Machine Vision systems be designed on exactly the same type of product/material samples as will be seen by the final, ("target") system. As we have just seen, this obvious principle is not always obeyed.

Transient phenomena occur in a wide variety of applications, particularly in the food and pharmaceutical industries, agriculture and horticulture. On-line data capture is essential, even for initial (feasibility) studies, whenever the appearance is susceptible to significant changes, over a period of a year or less. In this event, we must build an image-acquisition rig over the production line, even for a feasibility study; off-line analysis will not suffice.

1.3.8 Very Complex Objects

It is nonsensical to search for optimal lighting methods for very complicated objects, because features that are enhanced by one lighting method may be obscured by another, and vice versa. Few objects can match the complexity and unpredictability of form of a living plant. Imagine that we want to view its leaves, perhaps searching for spots indicating fungal disease. Lighting from the sides inevitably casts shadows, leaving "internal" features poorly illuminated. Even omni-directional lighting will fail to produce good images within a dense thicket (Figure 1.11). Coaxial illumination and viewing produces no shadows whatsoever but is prone to creating highlights, due to glinting (Figure 1.12). This can be problematical if the object surface is shiny, perhaps through it being wet or greasy. Some relief can be provided using a circular polariser (Figure 1.13), or crossed linear polarisers (Figure 1.14). In many applications, however, one standard lighting method can be used, even though it is not optimal for any particular task. Such an arrangement is used in a wide variety of applications where the shape of an object is too complicated to allow optimal lighting methods to be devised. On the other hand, where object shape is more predictable a formal approach to lighting system design is advisable.

Machine Vision has the potential to do something that natural vision finds impossible: combining several views of the same object/scene, taken under different lighting conditions [7]. So far, this exciting idea has not been exploited to its full potential. Intelligent lighting, perhaps generating different colours from the same set of "point" sources (multi-colour LEDs) is now a real possibility. Imagine a lighting rig, consisting of an array of multi-colour LEDs, each one individually controlled. By switching the LEDs and combining the images that result, a variety of interesting effects can be achieved that have no counterpart in human vision. Although this has very real potential for inspecting objects that have a highly complicated structure, much more fundamental research needs to be done.

1.3.9 Uncooperative Objects

By definition, inanimate objects are neutral in neither cooperating nor obstructing visual examination. On the other hand, animals such as pigs, chickens and fish will avoid strange and seemingly threatening equipment, particularly if it makes an unusual noise, has a strange smell, brightly coloured or flashing lights or even superficially resembles the animal's natural predator. These can be major obstacles to gaining an animal's acceptance of a vision system. Moreover, it must be borne in mind that an animal does not see the world as we do. It may be sensitive to a different part of the electro-magnetic spectrum. It may perceive two LEDs as being the eyes of a potential predator. Animals can sometimes be encouraged to move past a camera in a semi-controlled manner by the judicious use of guide rails (pigs, sheep, etc.), or glass panels/tubes (insects, fish, etc.). Some animals may also need encouragement, in the form of some prospect of a reward (food), to move at all. Needless to say, the inspection equipment must be very robust to work safely in close proximity to animals and it must be able to continue working unattended for long periods in extremely dirty working conditions. Designing a suitable image acquisition system for this type of application is far removed from a vision engineer's normal experience, developing algorithms, writing software and designing hardware. However, it is just as important. Failure to pay attention to these matters will significantly alter the vision system's prospects of success.

1.3.10 Texture

Texture is one of the most difficult areas for Machine Vision engineers, because there is no clear objective definition of what it actually is. Moreover, it is often extremely difficult to collect samples of "good" and "bad" texture, particularly in those situations where texture is merely an indicator of another more significant parameter. For example, the texture of the cut surface of a loaf is known to be related to the "in mouth" feel of the bread. The nature of the texture varies across the slice. (Indeed, bakers often deliberately cause this to happen, by rolling the dough to maximise the surface brightness of white bread when it is cut in the normal manner.) The sensation of biting an apple can be predicted, to some extent, by the cellular-level texture, although the exact nature of the relationship is unclear. The adhesion of paint to a steel surface is known to be affected by the surface texture of the metal. The texture of a coated surface is an indicator of its weathering/wear properties. The surface texture of many types of engineering products, most notably cars, white goods, woven fabrics and knitwear, is critical to customer acceptance.

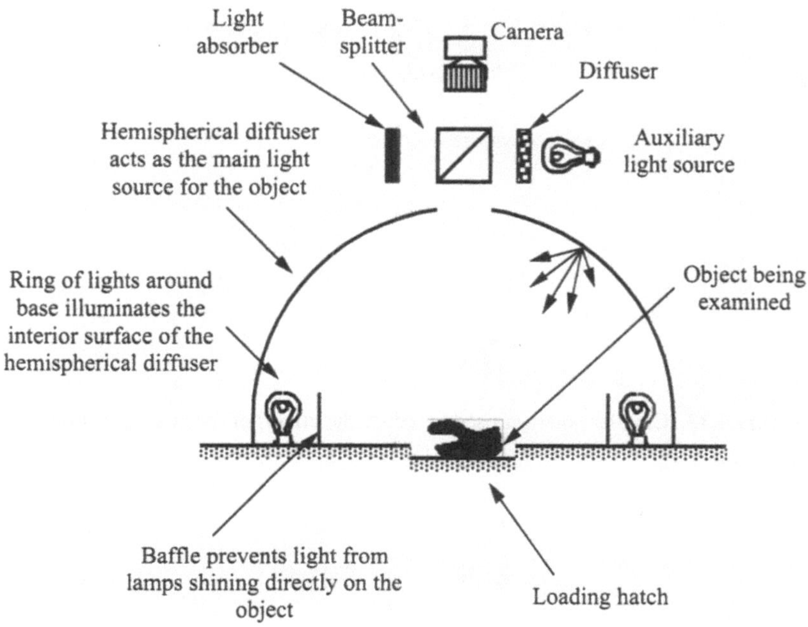

Figure 1.11. Omni-directional lighting. The auxiliary light source and beam-splitter are used to compensate for the hole cut in the hemispherical diffuser.

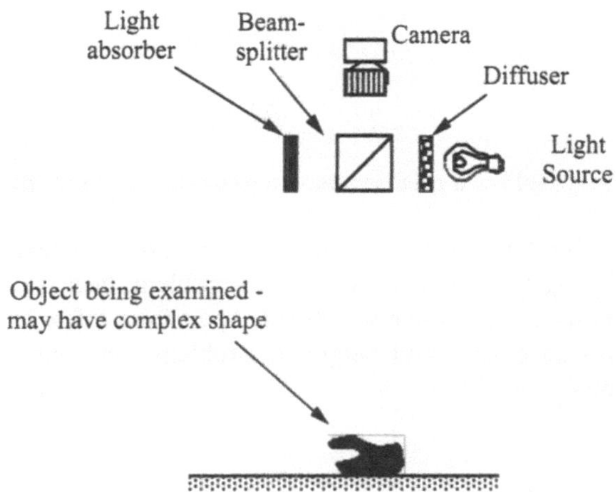

Figure 1.12. Coaxial illumination and viewing.

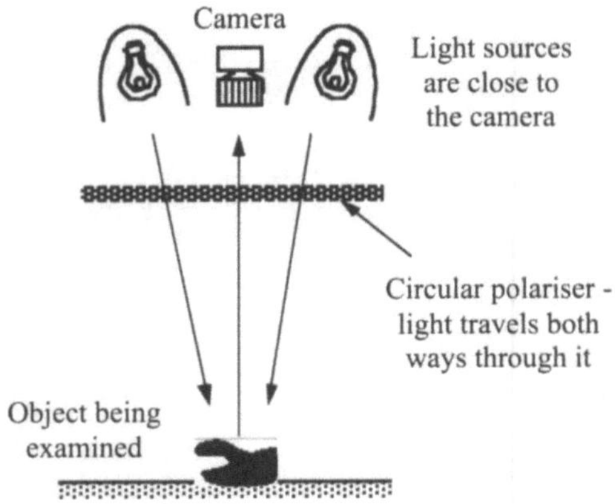

Figure 1.13. Circular polariser used to reduce the effects of specular reflection.

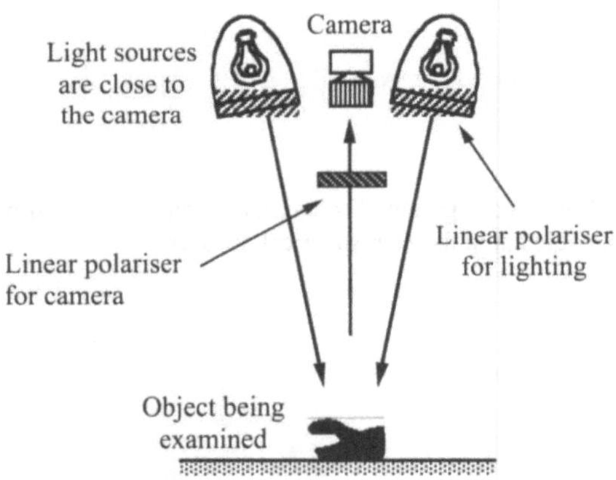

Figure 1.14. Crossed linear polarisers used to reduce the effects of specular reflection.

It is not difficult to derive measures for representing texture (Chapter 2). However, analysing the data causes a severe problem, because we do not know how to interpret the numbers that we obtain. Apart from a few exceptions, texture analysis remains an elusive and enigmatic problem that continues to challenge vision engineers.

1.4 Systems Issues

We conclude this brief introduction to our subject by making some pertinent remarks about the practical issues that must be considered to ensure that an inspection system works properly and reliably over a long period of time. The most important points to remember are:

- In direct contradiction to the title of this chapter, natural products are not all alike; they vary very considerably.
- It is important that the inspection system does not damage, pollute or infect the object being examined.
- It is important that the object being examined does not damage the inspection system.
- A series of general principles relating to the design of industrial vision systems has been published earlier [1]. Nearly all of these apply with equal validity to systems for examining natural products.

The criteria by which an inspection system are judged are the same, whether it be applied to natural products or engineering artefacts:

- It must be easy to use; a degree in engineering or computer science should not be necessary to operate it.
- The system must be acceptable to the work-force. This requires careful attention to a wide variety of factors, including the user interface and other points in this list.
- It must perform at the speed required. (For example, it must be able to keep up with production in a food-processing plant.)
- It must perform the necessary function, with a minimum of human intervention. Stringent tests are required to ensure that the system functions properly. Judgement will be made on a statistical basis, against previously agreed criteria, as set out in the specification.
- As far as possible, it should be possible to reuse the system to inspect new product lines.
- The system must be robust and reliable.
- Since cost is always of great importance, the system should be able to pay for itself in a few months. The food industry is notable for operating on very low profit margins, so this is often difficult to achieve.
- The system must be safe and hygienic. Since hygiene is critical in the food and pharmaceutical industries, we deal with this point separately.

Animals are liable to cause particular problems for a vision system, because they create a lot of pollution. They are liable to create a considerable amount of air-borne pollution in the form of droplets, when they urinate, defæcate, sneeze, cough or simply bellow. They also increase the amount of dust in the atmosphere by kicking up straw or feedstuff, as well as by shedding hair. Animals are likely to lick a lamp, or camera, if it is within reach. Making sure that the vision system is protected from pollution of this kind is vitally important, if it is to have a long and useful working life. There are many techniques for doing this, some of which are summarised in Figure 1.15. Of course, for safety's sake, the animal must also be protected from the vision system. Animal pens are hostile places for vision systems, so protection for both livestock and machine is essential.

Of course, inanimate objects also cause pollution, which can render a vision system useless. Biological materials are even worse than most standard engineering materials in causing air-borne pollution. Flour, and other fine powders abound in many food factories, as do mists, fumes, smoke, occasional splashes, etc.

All of these can settle on lenses, mirrors, lamp surfaces, and other optical components, causing malfunction of an inspection system. (This and other comments relate to the close proximity of the objects being inspected. The working environment occupied by factory workers is controlled by law and should therefore be relatively clean.)

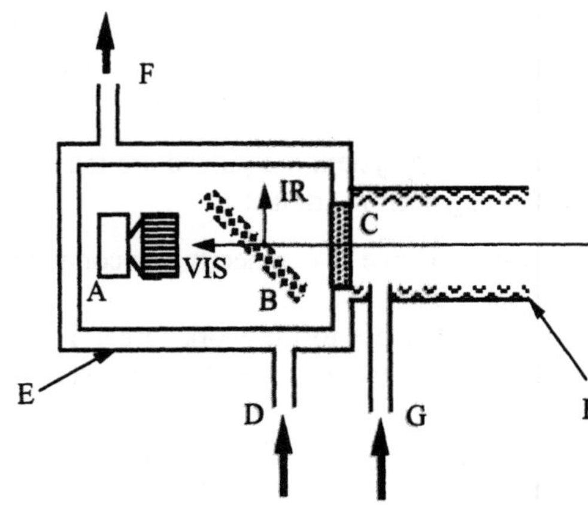

Figure 1.15. Protecting the camera: A – camera; B – hot mirror (reflects IR, transmits VIS); C – protective optical quality window; D – input for cold air/water; E - cooling jacket; F – outlet for cooling air/water; G – input for air purge (to keep protective window clean); H – tube acts as dust trap (rifled bore eliminates unwanted reflections).

One of the biggest hazards for a vision system in a food, toiletry, or pharmaceutical factory is *hygiene*. The need to keep all factory equipment scrupulously clean, including all exposed components of the vision system, means that equipment will be washed down on a regular basis. Not only must the optical-lighting-camera sub-system survive being hosed down, it must be easy for factory personnel to dismantle it, so that any traps where debris could accumulate are cleaned. In Chapter 16, we will learn about an application in which provision for easy cleaning had to be planned from the outset.

1.5 References

[1] Batchelor B.G., Whelan P.F. (1997) *Intelligent Vision Systems for Industry*, Springer Verlag, ISBN 3-540-19969 1.

[2] Batchelor B.G.,Waltz F.M. (2001) *Intelligent Machine Vision: Techniques, Implementation and Applications*, Springer-Verlag, ISBN 3-540-76224-8

[3] Graves M. (2000) X-ray machine vision for on-line food inspection, PhD Thesis, Cardiff University, Wales

[4] Graves M., Smith A., Batchelor B.G. (1998) Approaches to foreign body detection in foods, *Trends in Food Science and Technology* 9, No. 1, pp. 21–27.

[5] Batchelor B.G., Charlier J.R. (November 1998) *Machine Vision is Not Computer Vision*, Keynote paper, Proc. SPIE Conf., Machine Vision Systems for Inspection and Metrology VII, Boston, MA, Vol 3521, pp. 2–13, ISBN 0-8194-2982-1.

[6] Davies E.R. (2000) *Image Processing for the Food Industry*, World Scientific, Singapore, ISBN 981-02-4022-8.

[7] Batchelor B.G. (November 1994) *HyperCard lighting advisor*, Proc. Conf. on Machine Vision Applications, Architectures and Systems III, Boston, MA, pub. SPIE, Bellingham, WA, U.S.A., Vol. 2347, ISBN 0-8194-1682-7, pp.180–188. Also URL: http://bruce.cs.cf.ac.uk/bruce/index.html

Chapter 2

Editorial Introduction

A Machine Vision system is like a chain: it is only as strong as its weakest link. To most people, the ability to handle images inside a computer, or dedicated electronic hardware, is a complete mystery, although the basic concepts are apparently quite straightforward. It is usually far easier to understand *what* an image processing operator does than *how* it does it. When designing/choosing a Machine Vision system, understanding the image processing at this level is usually adequate, although, in Chapter 15, Professor Roy Davies points out the shortcomings of this attitude. One of the major problems facing vision engineers is that everyone thinks that they are an expert on vision. Would-be customers certainly do! It is common for a vision engineer to be given gratuitous and utterly inappropriate advice about how to analyse an image by clients. For some inexplicable reason, many people feel that they must contribute in this way, even though they have no knowledge whatsoever about the details of any of the operators described in this chapter. It must be appreciated that machines do not see as human beings and it is fruitless to try to design them to do so. There are several reasons for this:

a. We do not know enough about human vision to design algorithms properly on this basis.
b. The basic computational "atoms" available in electronics and networks of neurons are completely different in nature.
c. Despite their superficial simplicity, image processing operators possess a variety of subtle nuances and interactions, which together make their use far from straightforward.

The term *image processing* has two distinct uses. On the one hand, it has a specific meaning when it relates to the manipulation of pictures by transforming one image into another. It is also used in a generic sense, to encompass this meaning, as well as feature identification, location and measurement. These are all viewed as *low-level* functions and are discussed in this chapter. These are the principal methods used within most present-day industrial vision systems. *High-level* reasoning about images is discussed in Chapter 3.

Chapter 2

Basic Machine Vision Techniques

B.G. Batchelor

This chapter introduces "low-level" vision processing operators, which lack the ability to perform the quintessential functions of intelligence: logical deduction, searching, classification and learning. Many intermediate-level operations can be implemented by concatenating these basic commands in simple macros. The operators described here are assigned mnemonic names, indicated in square brackets, and are used in Section 3.3.2, in combination with the AI language Prolog, to produce smart vision systems.

2.1 Representations of Images

We shall first consider the representation of *Monochrome* (grey-scale) images. Let i and j denote two integers where $1 \le i \le m$ and $1 \le j \le n$. In addition, let $f(i,j)$ denote an integer function such that $0 \le f(i,j) \le W$. (W denotes the white level in a grey-scale image.) An array F will be called a *digital image*.

$$F = \begin{vmatrix} f(1,1), & f(1,2), & \cdots & f(1,n) \\ f(2,1), & f(2,2), & \cdots & f(2,n) \\ . & . & \cdots & . \\ . & . & \cdots & . \\ f(m,1), & f(m,2), & \cdots & f(m,n) \end{vmatrix}$$

An address (i,j) defines a position in F, called a *pixel*, *pel* or *picture element*. The elements of F denote the intensities within a number of small rectangular regions within a real (i.e., optical) image. (Figure 2.1). Strictly speaking, $f(i,j)$ measures the intensity at a single point but if the corresponding rectangular region is small enough, the approximation will be accurate enough for most purposes. The array F contains a total of $m.n$ elements and this product is called the *spatial resolution* of F. We may *arbitrarily* assign intensities according to the following scheme:

$f(i,j) = 0$	black	
$0 < f(i,j) \leq 0.33W$	dark grey	
$0.33W < f(i,j) \leq 0.67W$	mid-grey	
$0.67W < f(i,j) < W$	light grey	
$f(i,j) = W$	white	

Let us consider how much data is required to represent a grey-scale image in this form. Each pixel requires the storage of $\log_2(1 + W)$ bits. This assumes that (1 + W) is an integer power of two. If it is not, then $\log_2(1 + W)$ must be rounded up to the next integer. This can be represented using the ceiling function, $< \ldots >$. Thus, a grey-scale image requires the storage of $< \log_2(1 + W) >$ bits. Since there are $m.n$ pixels, the total data storage for the entire digital image F is equal to $m.n.< \log_2(1 + W) >$ bits. If $m = n \geq 128$, and $W \geq 64$, we can obtain a good image of a human face. Many of the industrial image processing systems in use nowadays manipulate images in which $m = n = 512$ and $W = 255$. This leads to a storage requirement of 256 Kbytes/image. A *binary image* is one in which only two intensity levels, black (0) and white (1), are permitted. This requires the storage of $m.n$ bits/image.

An impression of colour can be conveyed to the eye by superimposing *four* separate imprints. (Cyan, magenta, yellow and black inks are often used in printing.) Ciné film operates in a similar way, except that when different colours of light, rather than ink, are added together, *three* components (red, green and blue) suffice. Television operates in a similar way to film; the signal from a colour television camera may be represented using three components: $R = \{r(i,j)\}$; $G = \{g(i,j)\}$; $B = \{b(i,j)\}$, where R, G and B are defined in a similar way to F. The vector $\{r(i,j), g(i,j), b(i,j)\}$ defines the intensity and colour at the point (i,j) in the colour image. Colour image analysis is discussed in more detail in Chapter 3. Multispectral images can also be represented using several monochrome images. The total amount of data required to code a colour image with r components is equal to $m.n.r.< \log_2(1 + W) >$ bits, where W is simply the maximum signal level on each of the channels.

Ciné film and television will be referred to, in order to explain how moving scenes may be represented in digital form. A ciné film is, in effect, a time-sampled representation of the original moving scene. Each frame in the film is a standard colour, or monochrome image, and can be coded as such. Thus, a monochrome ciné film may be represented digitally as a sequence of two-dimensional arrays $[F_1, F_2, F_3, F_4, \ldots]$. Each F_i is an $m.n$ array of integers as we defined above, when discussing the coding of grey-scale images. If the film is in colour, then each of the F_i has three components. In the general case, when we have a sequence of r-component colour images to code, we require $m.n.p.r.< \log_2(1+W) >$bits/image sequence, where the spatial resolution is $m.n$ pixels, each spectral channel permits $(1 + W)$ intensity levels, there are r spectral channels and p is the total number of "stills" in the image sequence.

We have considered only those image representations, which are relevant to the understanding of simple image processing and analysis functions. Many alternative methods of coding images are possible but these are not relevant to this discussion.

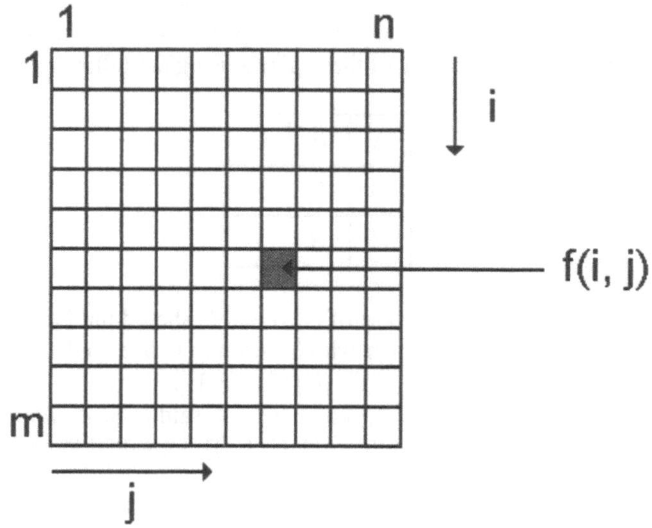

Figure 2.1. A digital image consisting of an array of $m.n$ pixels. The pixel in the ith row and the jth column has an intensity equal to $f(i,j)$.

2.2 Elementary Image Processing Functions

The following notation will be used throughout this section, in which we shall concentrate upon grey-scale images, unless otherwise stated:

- i and j are row and column address variables and lie within the ranges: $1 \le i \le$ m and $1 \le j \le$ n. (Figure 2.1)
- $A = \{a(i,j)\}$, $B = \{b(i,j)\}$ and $C = \{c(i,j)\}$.
- W denotes the white level.
- $g(X)$ is a function of a single independent variable X.
- $h(X,Y)$ is a function of two independent variables, X and Y.
- The assignment operator '$\leftarrow$' will be used to define an operation that is performed upon one data element. In order to indicate that an operation is to be performed upon all pixels within an image, the assignment operator '$\Leftarrow$' will be used.
- k, k1, k2, k3 are constants.
- $N(i,j)$ is that set of pixels arranged around the pixel (i,j) in the following way:

$(i$-1,j-1)	$(i$-1,$j)$	$(i$-1,j+1)

$(i,j\text{-}1)$	(i,j)	$(i,j+1)$
$(i+1,j\text{-}1)$	$(i+1,j)$	$(i+1,j+1)$

Notice that $N(i,j)$ forms a 3 × 3 set of pixels and is referred to as the 3 × 3 *neighbourhood* of (i,j). In order to simplify some of the definitions, we shall refer to the intensities of these pixels using the following notation:

A	B	C
D	E	F
G	H	I

Ambiguities over the dual use of A, B and C should not be troublesome; as the context will make it clear which meaning is intended. The points $\{(i\text{-}1, j\text{-}1), (i\text{-}1, j), (i\text{-}1, j+1), (i, j\text{-}1), (i, j+1), (i+1, j\text{-}1), (i+1, j), (i+1, j+1)\}$ are called the *8-neighbours* of (i, j) and are also said to be *8-connected* to (i, j). The points $\{(i\text{-}1, j), (i, j\text{-}1), (i, j+1), (i+1, j)\}$ are called the *4-neighbours* of (i, j) and are said to be *4-connected* to (i, j).

2.2.1 Monadic Point-by-point Operators

These operators have a characteristic equation of the form:

$$c(i,j) \Leftarrow g(a(i,j)) \text{ or } E \Leftarrow g(E)$$

Such an operation is performed for all (i,j) in the range $[1,m].[1,n]$. (Figure 2.2). Several examples will now be described.

Intensity shift [acn]

$$c(i,j) \Leftarrow \begin{vmatrix} 0 & a(i,j) + k < 0 \\ a(i,j) + k & 0 \le a(i,j) + k \le \mathrm{W} \\ W & W < a(i,j) + k \end{vmatrix}$$

k is a constant, set by the system user. Notice that this definition was carefully designed to maintain $c(i,j)$ within the same range as the input, viz. $[0,W]$. This is an example of a process referred to as *intensity normalisation*. Normalisation is important because it permits iterative processing by this and other operators in a machine having a limited precision for arithmetic (e.g., 8-bits). Normalisation will be used frequently throughout this chapter.

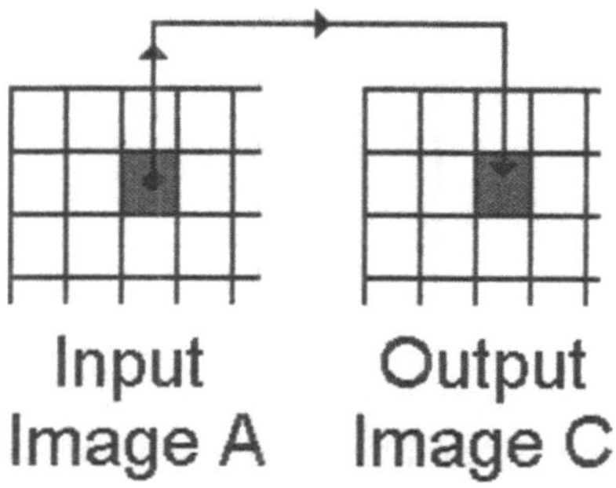

Figure 2.2. Monadic point-by-point operator. The (i,j)th pixel in the input image has intensity $a(i,j)$. This value is used to calculate $c(i,j)$, the intensity of the corresponding pixel in the output image.

Intensity multiply [mcn]

$$c(i,j) \Leftarrow \begin{vmatrix} 0 & a(i,j)\,.\,k < 0 \\ a(i,j)\,.\,k & 0 \le a(i,j)\,.\,k \le W \\ W & W < a(i,j)\,.\,k \end{vmatrix}$$

Logarithm [log]

$$c(i,j) \Leftarrow \begin{vmatrix} 0 & a(i,j) = 0 \\ W\left[\frac{\text{Log}(a(i,j))}{\text{Log}(W)}\right] & \text{otherwise} \end{vmatrix}$$

This definition arbitrarily replaces the infinite value of log(0) by zero, and thereby avoids a difficult rescaling problem.

Antilogarithm (exponential) [exp]

$$c(i,j) \Leftarrow W.\ \exp(a(i,j))/\exp(W)$$

Negate [neg]

$$c(i,j) \Leftarrow W - a(i,j)$$

Threshold [thr]

$$c(i,j) \Leftarrow \begin{cases} W & k1 \leq a(i,j) \leq k2 \\ 0 & \text{otherwise} \end{cases}$$

This is an important function, which converts a grey-scale image to a binary format. Unfortunately, it is often difficult, or even impossible to find satisfactory values for the parameters *k1* and *k2*.

Highlight [hil]

$$c(i,j) \Leftarrow \begin{cases} k3 & k1 \leq a(i,j) \leq k2 \\ a(i,j) & \text{otherwise} \end{cases}$$

Squaring [sqr]

$$c(i,j) \Leftarrow [a(i,j)\,]^2/W$$

2.2.2 Dyadic Point-by-point Operators

Dyadic operators have a characteristic equation of the form:

$$c(i,j) \Leftarrow h(a(i,j), b(i,j))$$

There are two input images: $A = \{a(i,j)\}$ and $B = \{b(i,j)\}$ (Figure 2.3), while the output image is $C = \{c(i,j)\}$. It is important to realise that $c(i,j)$ depends upon only $a(i,j)$ and $b(i,j)$. Here are some examples of dyadic operators.

Add [add]

$$c(i,j) \Leftarrow [a(i,j) + b(i,j)]/2.$$

Subtract [sub]

$$c(i,j) \Leftarrow [(a(i,j) - b(i,j)) + W\,]/2$$

Multiply [mul]

$$c(i,j) \Leftarrow [a(i,j).b(i,j)]/W$$

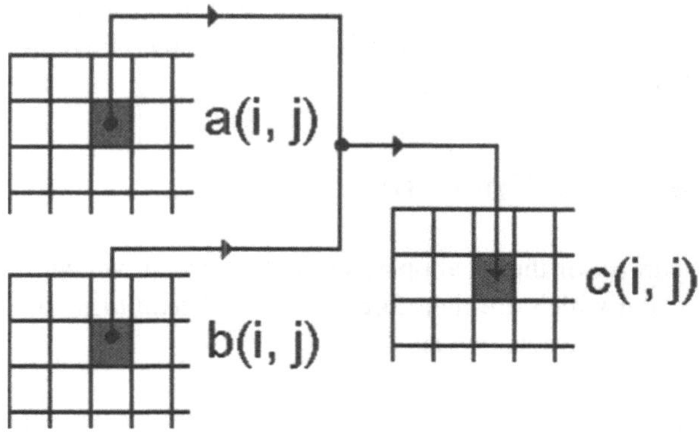

Figure 2.3. Dyadic point-by-point operator. The intensities of the (i,j)th pixels in the two input images (i.e., $a(i,j)$ and $b(i,j)$) are combined to calculate the intensity, $c(i,j)$, at the corresponding address in the output image.

Maximum [max]

$$c(i,j) \Leftarrow MAX\,[a(i,j),\, b(i,j)]$$

When the maximum operator is applied to a pair of binary images, the *union* (OR function) of their white areas is computed. This function may also be used to *superimpose* white writing onto a grey-scale image.

Minimum [min]

$$c(i,j) \Leftarrow MIN\,[a(i,j),\, b(i,j)]$$

When A and B are both binary, the *intersection* (AND function) of their white areas is calculated.

2.2.3 Local Operators

Figure 2.4 illustrates the principle of the operation of local operators. Notice that the intensities of several pixels are combined together, in order to calculate the intensity of just one pixel. Amongst the simplest of the local operators are those which use a set of nine pixels arranged in a 3 × 3 square. These have a characteristic equation of the following form:

$$c(i,j) \Leftarrow g(a(i\text{-}1,j\text{-}1),\, a(i\text{-}1,j),\, a(i\text{-}1,j+1),\, a(i,j\text{-}1),\, a(i,j),\, a(i,j+1),\\ a(i+1,j\text{-}1),\, a(i+1,j),\, a(i+1,j+1))$$

where $g(.)$ is a function of 9 variables. This is an example of a local operator, which uses a *3 × 3 processing window*. (That is, it computes the value for one pixel on the basis of the intensities within a region containing 3 × 3 pixels. Other local operators employ larger windows and we shall discuss these briefly later.) In the simplified notation which we introduced earlier, the above definition reduces to: $E \Leftarrow g(A, B, C, D, E, F, G, H, I)$.

2.2.4 Linear Local Operators

An important sub-set of the local operators is that group, which performs a linear, weighted sum, and which are therefore known as *linear locdl operators*. For this group, the characteristic equation is:

$$E \Leftarrow k1.(A.W1 + B.W2 + C.W3 + D.W4 + E.W5 + F.W6 + G.W7 + H.W8 + I.W9) + k2$$

where $W1$, $W2$,...,$W9$ are weights, which may be positive, negative or zero. Values for the normalisation constants, $k1$ and $k2$ are given later. The matrix illustrated below is termed the *weight matrix* and is important, because it determines the properties of the linear local operator.

$W1$	$W2$	$W3$
$W4$	$W5$	$W6$
$W7$	$W8$	$W9$

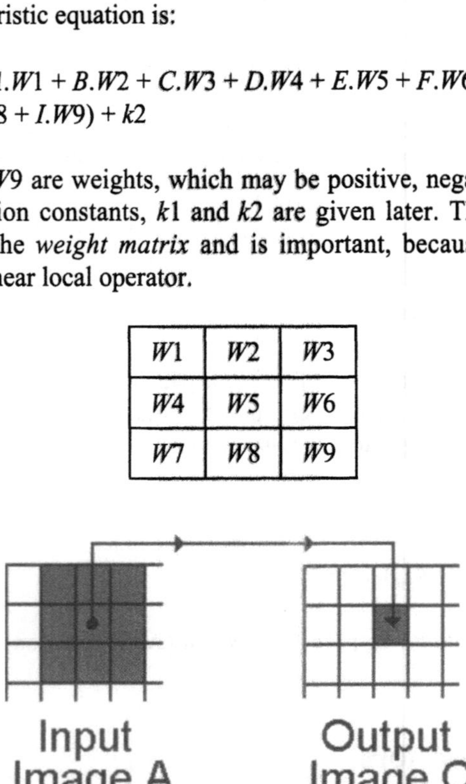

Figure 2.4. Local operator. In this instance, the intensities of nine pixels arranged in a 3 × 3 window are combined together. Local operators may be defined which uses other, possibly larger windows. The window may, or may not, be square and the calculation may involve linear or non-linear processes.

The following rules summarise the behaviour of this type of operator. (They exclude the case where *all* the weights and normalisation constants are zero, since this would result in a null image.):

i. If all weights are either positive or zero, the operator will *blur* the input image. Blurring is referred to as *low-pass filtering*. Subtracting a blurred image from the original results in a highlighting of those points where the intensity is changing rapidly and is termed *high-pass filtering*.

ii. If $W1 = W2 = W3 = W7 = W8 = W9 = 0$, and $W4, W5, W6 > 0$, then the operator blurs along the rows of the image; horizontal features, such as edges and streaks, are not affected.
iii. If $W1 = W4 = W7 = W3 = W6 = W9 = 0$, and $W2, W5, W8 > 0$, then the operator blurs along the columns of the image; vertical features are not affected.
iv. If $W2 = W3 = W4 = W6 = W7 = W8 = 0$, and $W1, W5, W9 > 0$, then the operator blurs along the diagonal (top-left to bottom-right). There is no smearing along the orthogonal diagonal.
v. If the weight matrix can be reduced to a matrix product of the form $P.Q$, where

$P =$

0	0	0
V4	V5	V6
0	0	0

and

$Q =$

0	V1	0
0	V2	0
0	V3	0

the operator is said to be of the "separable" type. The importance of this is that it is possible to apply two simpler operators in succession, with weight matrices P and Q, in order to obtain the same effect as that produced by the separable operator.
vi. The successive application of linear local operators which use windows containing 3×3 pixels produces the same results as linear local operators with larger windows. For example, applying that operator which uses the following weight matrix

1	1	1
1	1	1
1	1	1

twice in succession results in a similar image as that obtained from the 5×5 operator with the following weight matrix. (For the sake of simplicity, normalisation has been ignored here.)

1	2	3	2	1
2	4	6	4	2
3	6	9	6	3
2	4	6	4	2
1	2	3	2	1

Applying the same 3 × 3 operator thrice is equivalent to using the following 7 × 7 operator

1	3	6	7	6	3	1
3	9	18	21	18	9	3
6	18	36	42	36	18	6
7	21	42	49	42	21	7
6	18	36	42	36	18	6
3	9	18	21	18	9	3
1	3	6	7	6	3	1

Notice that all of these operators are also separable. Hence it would be possible to replace the last-mentioned 7 × 7 operator with four simpler operators: 3 × 1, 3 × 1, 1 × 3 and 1 × 3, applied in any order. It is not always possible to replace a large-window operator with a succession of 3 × 3 operators. This becomes obvious when one considers, for example, that a 7 × 7 operator uses 49 weights and that three 3 × 3 operators provide only 27 degrees of freedom. Separation is often possible, however, when the larger operator has a weight matrix with some redundancy, for example when it is symmetrical.

vii. In order to perform normalisation, the following values are used for $k1$ and $k2$.

$$k1 \leftarrow 1/\sum_{p,q} |W_{p,q}|$$

$$k2 \leftarrow [1 - \sum_{p,q} W_{p,q} / \sum_{p,q} |W_{p,q}|].W/2$$

viii. A filter using the following weight matrix performs a *local averaging function* over an 11 × 11 window [*raf(11,11)*].

1	1	1	1	1	1	1	1	1	1	1
1	1	1	1	1	1	1	1	1	1	1
1	1	1	1	1	1	1	1	1	1	1
1	1	1	1	1	1	1	1	1	1	1
1	1	1	1	1	1	1	1	1	1	1
1	1	1	1	1	1	1	1	1	1	1
1	1	1	1	1	1	1	1	1	1	1
1	1	1	1	1	1	1	1	1	1	1
1	1	1	1	1	1	1	1	1	1	1
1	1	1	1	1	1	1	1	1	1	1
1	1	1	1	1	1	1	1	1	1	1

This produces quite a severe two-directional blurring effect. Subtracting the effects of a blurring operation from the original image generates a picture in which spots, streaks and intensity steps are all emphasised. On the other hand, large areas of constant or slowly changing intensity become uniformly grey. This process is called *high-pass filtering*, and produces an effect similar to unsharp masking, which is familiar to photographers.

2.2.5 Non-linear Local Operators

Largest intensity neighbourhood function [lnb]

$$E \Leftarrow \mathbf{MAX}(A, B, C, D, E, F, G, H, I)$$

This operator has the effect of spreading bright regions and contracting dark ones.

Edge detector [command sequence: lnb, sub]

$$E \Leftarrow \mathbf{MAX}(A, B, C, D, E, F, G, H, I) - E$$

This operator is able to highlight edges (i.e., points where the intensity is changing rapidly).

Median filter [mdf(5)]

$$E \Leftarrow \mathbf{FIFTH\_LARGEST}\ (A, B, C, D, E, F, G, H, I)$$

This filter is particularly useful for reducing the level of noise in an image. (Noise is generated from a range of sources, such as video cameras and x-ray detectors, and can be a nuisance if it is not eliminated by hardware or software filtering.)

Crack detector*[1] *[lnb, lnb, neg, lnb, lnb, neg]

This operator is equivalent to applying the above sequence of operations and then subtracting the result from the original image. This detector is able to detect thin dark streaks and small dark spots in a grey-scale image; it ignores other features, such as bright spots and streaks, edges (intensity steps) and broad dark streaks.

Roberts edge detector [red]

The Roberts gradient is calculated using a 2 × 2 mask. This will determine the edge gradient in two diagonal directions (i.e., the cross-differences).

$$E \Leftarrow \sqrt{(A - E)^2 + (B - D)^2}$$

The following approximation to the Roberts gradient magnitude is called the Modified Roberts operator. This is simpler and faster to implement and it more precisely defines the operator *red*. It is defined as

$$E \Leftarrow \{ \left| A - E \right| + \left| B - D \right| \}/2$$

Sobel edge detector [sed]

This popular operator highlights the edges in an image; points where the intensity gradient is high are indicated by bright pixels in the output image. The Sobel edge detector uses a 3_3 mask to determine the edge gradient.

$$E \Leftarrow \sqrt{[(A + 2.B + C) - (G + 2.H + I)]^2 + [(A + 2.D + G) - (C + 2.F + I)]^2}$$

The following approximation is simpler to implement in software and hardware and more precisely defines the operator *sed*:

$$E \Leftarrow \{| (A + 2.B + C) - (G + 2.H + I) | + | (A + 2.D + G) - (C + 2.F + I) |\}/6$$

Figure 2.5 shows a comparison of the Roberts and Sobel edge detector operators when applied to a sample monochrome image. Note that, while the Roberts operator produces thinner edges, these edges tend to break up in regions of high curvature. The primary disadvantage of the Roberts operator is its high sensitivity to noise, since fewer pixels are used in the calculation of the edge gradient. There is also a slight shift in the image, when the Roberts edge detector is used. The Sobel edge detector does not produce such a shift.

---

[1] This is an example of an operator that can be described far better using computer notation rather than mathematical notation.

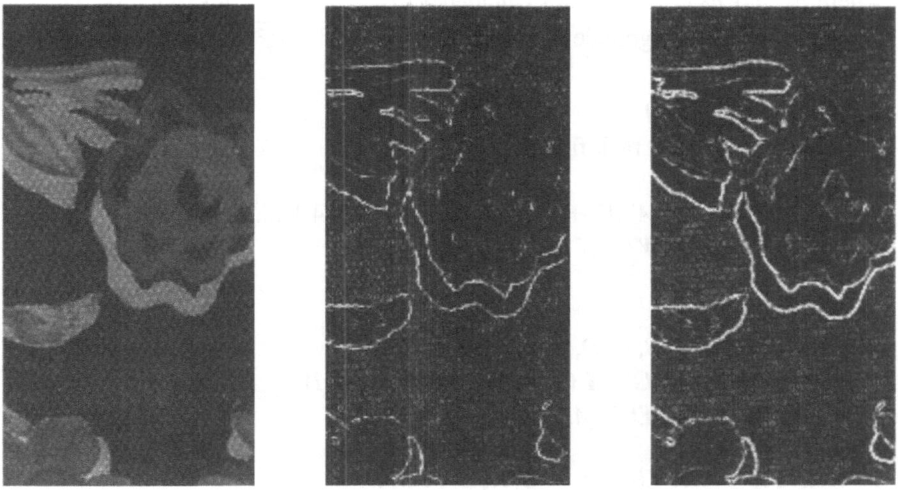

(a) (b) (c)

Figure 2.5. Edge detection applied to an image derived from a piece of dress fabric: (a) original image; (b) Roberts gradient; (c) Sobel gradient

Prewitt edge detector

The Prewitt edge-detector is similar to the Sobel operator, but is more sensitive to noise as it does not possess the same inherent smoothing. This operator uses the two 3 × 3 weight matrices shown below to determine the edge gradient,

-1	-1	-1
0	0	0
1	1	1

P_1

-1	0	1
-1	0	1
-1	0	1

P_2

where P_1 and P_2 are the values calculated from each mask respectively. The Prewitt gradient magnitude is defined as $E \Leftarrow \sqrt{P_1^2 + P_2^2}$

Frei and Chen edge detector

This operator uses the two 3 × 3 masks shown below to determine the edge gradient,

-1	-√2	-1
0	0	0
1	√2	1

F_1

-1	0	1
-√2	0	√2
-1	0	1

F_2

where F_1 and F_2 are the values calculated from each mask respectively. The Frei and Chen gradient magnitude is defined as $E \Leftarrow \sqrt{F_1^2 + F_2^2}$

Rank filters [mdf, rid]

The generalised 3 × 3 rank filter is:

$$c(i,j) \Leftarrow k1.(A'.W1 + B'.W2 + C'.W3 + D'.W4 + E'.W5 + F'.W6 + G'.W7 + H'.W8 + I'.W9) + k2$$

where

$A' =$ **LARGEST** $(A, B, C, D, E, F, G, H, I)$
$B' =$ **SECOND_LARGEST** $(A, B, C, D, E, F, G, H, I)$
$C' =$ **THIRD_LARGEST** $(A, B, C, D, E, F, G, H, I)$
...
$I' =$ **NINTH_LARGEST** $(A, B, C, D, E, F, G, H, I)$

and $k1$ and $k2$ are the normalisation constants defines previously. With the appropriate choice of weights ($W1$, $W2$,...,$W9$), the rank filter can be used for a range of operations including edge detection, noise reduction, edge sharping and image enhancement.

Direction codes [dbn]

This function can be used to detect the *direction* of the intensity gradient. A direction code function DIR_CODE is defined thus:

$$\text{DIR\_CODE}(A,B,C,D,F,G,H,I) \Leftarrow \begin{cases} 1 & \text{if } A \geq \text{MAX}(B,C,D,F,G,H,I) \\ 2 & \text{if } B \geq \text{MAX}(A,C,D,F,G,H,I) \\ 3 & \text{if } C \geq \text{MAX}(A,B,D,F,G,H,I) \\ 4 & \text{if } D \geq \text{MAX}(A,B,C,F,G,H,I) \\ 5 & \text{if } F \geq \text{MAX}(A,B,C,D,G,H,I) \\ 6 & \text{if } G \geq \text{MAX}(A,B,C,D,F,H,I) \\ 7 & \text{if } H \geq \text{MAX}(A,B,C,D,F,G,I) \\ 8 & \text{if } I \geq \text{MAX}(A,B,C,D,F,G,H) \end{cases}$$

Using this definition the operator *dbn* may be defined as:

$$E \Leftarrow \text{DIR\_CODE}(A,B,C,D,F,G,H,I)$$

2.2.6 N-tuple Operators

The N-tuple operators are closely related to the local operators and have a large number of linear and non-linear variations. N-tuple operators may be regarded as generalised versions of local operators. In order to understand the N-tuple operators, let us first consider a *linear* local operator, which uses a large processing window, (say *r.s* pixels) with most of its weights equal to zero. Only *N* of the weights are non-zero, where $N << r.s$. This is an N-tuple filter (Figure 2.6.). The N-tuple filters are usually designed to detect specific patterns. In this role, they are able to locate a simple feature, such as a corner, annulus, the numeral "2", in any position etc. However, they are sensitive to changes of orientation and scale. The N-tuple can be regarded as a sloppy template, which is convolved with the input image.

Non-linear tuple operators may be defined in a fairly obvious way. For example, we may define operators which compute the average, maximum, minimum or median values of the intensities of the *N* pixels covered by the N-tuple. An important class of such functions is the *morphological operators*. (Sections 2.4 and 2.5.) Figure 2.7 illustrates the recognition of the numeral '**2**' using an N-tuple. Notice how the goodness of fit varies with the shift, tilt, size, and font. Another character ('**Z**' in this case) may give a score that is close to that obtained from a '**2**', thus making these two characters difficult to distinguish reliably.

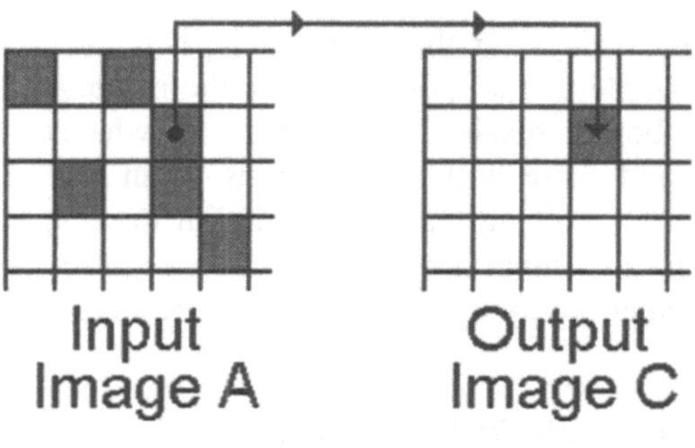

Figure 2.6. An N-tuple filter operates much like a local operator. The only difference is that the pixels whose intensities are combined together do not form a compact set. A linear N-tuple filter can be regarded as being equivalent to a local operator which uses a large window and in which many of the weights are zero.

Figure 2.7. Recognising a numeral '2' using an N-tuple.

2.2.7 Edge Effects

All local operators and N-tuple filters are susceptible to producing peculiar effects around the edges of an image. The reason is simply that, in order to calculate the intensity of a point near the edge of an image, we require information about pixels outside the image, which of course are simply not present. In order to make some attempt at calculating values for the edge pixels, it is necessary to make some assumptions, for example that all points outside the image are black, or have the same values as the border pixels. This strategy, or whatever one we adopt, is perfectly arbitrary and there will be occasions when the edge effects are so pronounced that there is nothing that we can do but to remove them by masking [*edg*]. Edge effects are important because they require us to make special provisions for them when we try to patch several low-resolution images together.

2.2.8 Intensity Histogram *[hpi, hgi, hge, hgc]*

The intensity histogram is defined in the following way:

a. let

$$s(p,i,j) \leftarrow \begin{cases} 1 & a(i,j) = p \\ 0 & \text{otherwise} \end{cases}$$

b. let $h(p)$ be defined thus: $h(p) \leftarrow \sum_{i,j} s(p,i,j)$

It is not, in fact, necessary to store each of the $s(p,i,j)$, since the calculation of the histogram can be performed as a serial process in which the estimate of $h(p)$ is updated iteratively, as we scan through the input image. The *cumulative histogram*, $H(p)$, can be calculated using the following recursive relation:

$$H(p) = H(p\text{-}1) + h(p), \text{ where } H(0) = h(0).$$

Both the cumulative and the standard histograms have a great many uses, as will become apparent later. It is possible to calculate various intensity levels which indicate the occupancy of the intensity range [pct]. For example, it is a simple matter to determine that intensity level, $p(k)$, which when used as a threshold parameter ensures that a proportion k of the output image is black, $p(k)$ can be calculate using the fact that $H(p(k)) = m.n.k$. The *mean intensity* [avg] is equal to:

$$\sum_p (h(p).p\,)/(m.n)$$

while the *maximum intensity* [gli] is equal to $\text{MAX}(p \mid h(p) > 0)$ and the *minimum intensity* is equal to $\text{MIN}(p \mid h(p) > 0)$.

One of the principal uses of the histogram is in the selection of threshold parameters. It is useful to plot $h(p)$ as a function of p. It is often found from this graph that a suitable position for the threshold can be related directly to the position of the "foot of the hill" or to a "valley" in the histogram.

An important operator for image enhancement is given by the transformation:

$$c(i,j) \Leftarrow [W.H(a(i,j))]/(m.n)$$

This has the interesting property that the histogram of the output image $\{c(i,j)\}$ is flat, giving rise to the name *histogram equalisation* [heq] for this operation. Notice that histogram equalisation is a *data-dependent* monadic, point-by-point operator.

An operation known as *"local area histogram equalisation"* relies upon the application of histogram equalisation within a small window. The number of pixels in a small window that are darker than the central pixel is counted. This number defines the intensity at the equivalent point in the output image. This is a powerful filtering technique, which is particularly useful in texture analysis applications. (Section 2.7.).

2.3 Binary Images

For the purposes of this description of binary image processing, it will be convenient to assume that $a(i,j)$ and $b(i,j)$ can assume only two values: 0 (black) and 1(white). The operator "+" denotes the Boolean **OR** operation, "•" represents the **AND** operation and where '⊗' denotes the Boolean **Exclusive OR** operation. Let $\#(i,j)$ denote the number of white points addressed by $N(i,j)$, including (i,j) itself.

Inverse [not]

$$c(i,j) \Leftarrow \mathbf{NOT}(a(i,j))$$

AND white regions [and]

$$c(i,j) \Leftarrow a(i,j) \bullet b(i,j)$$

OR [ior, max]

$$c(i,j) \Leftarrow a(i,j) + b(i,j)$$

Exclusive OR [xor] (Find differences between white regions.)

$$c(i,j) \Leftarrow a(i,j) \otimes b(i,j)$$

Expand white areas [exw]

$$c(i,j) \Leftarrow a(i\text{-}1,j\text{-}1) + a(i\text{-}1,j) + a(i\text{-}1,j+1) + a(i,j\text{-}1) + a(i,j) + a(i,j+1) \\ + a(i+1,j\text{-}1) + a(i+1,j) + a(i+1,j+1)$$

Notice that this is closely related to the local operator *lnb* defined earlier. This equation may be expressed in the simplified notation: $E \Leftarrow A + B + C + D + E + F + G + H + I$

Shrink white areas [skw]

$$c(i,j) \Leftarrow a(i\text{-}1,j\text{-}1) \bullet a(i\text{-}1,j) \bullet a(i\text{-}1,j+1) \bullet a(i,j\text{-}1) \bullet a(i,j) \bullet a(i,j+1) \\ \bullet a(i+1,j\text{-}1) \bullet a(i+1,j) \bullet a(i+1,j+1)$$

or more simply $c(i,j) \Leftarrow A \bullet B \bullet C \bullet D \bullet E \bullet F \bullet G \bullet H \bullet I$

Edge detector [bed]

$$c(i,j) \Leftarrow E \bullet \mathbf{NOT}(A \bullet B \bullet C \bullet D \bullet F \bullet G \bullet H \bullet I)$$

Remove isolated white points [wrm]

$$c(i,j) \Leftarrow \begin{cases} 1 & a(i,j) \bullet (\#(i,j) > 1) \\ 0 & \text{otherwise} \end{cases}$$

Count white neighbours [cnw]

$c(i,j) \Leftarrow \#(a(i,j) = 1).$

Where *#(Z)* is the number of times *Z* occurs. Notice that $\{c(i,j)\}$ is a grey-scale image.

Connectivity detector [cny].
Consider the following pattern:

1	0	1
1	X	1
1	0	1

If *X*=1, then all of the 1s are 8-connected to each other. Alternatively, if X=0, then they are not connected. In this sense, the point marked *X* is critical for connectivity. This is also the case in the following examples:

1	0	0
0	*X*	1
0	0	0

1	1	0
0	*X*	0
0	0	1

0	0	1
1	*X*	0
1	0	1

However, those points marked *X* below are not critical for connectivity, since setting *X*=0 rather than 1 has no effect on the connectivity of the 1s.

1	1	1
1	*X*	1
0	0	1

0	1	1
1	*X*	0
1	1	1

0	1	1
1	*X*	0
0	1	1

A connectivity detector shades the output image with 1s to indicate the position of those points which are critical for connectivity and which were white in the input image. Black points, and those which are not critical for connectivity, are mapped to black in the output image.

Euler number [eul]
The Euler number is defined as the number of connected components (blobs) minus the number of holes in a binary image. The Euler number represents a simple method of counting blobs in a binary image, provided they have no holes in them. Alternatively, it can be used to count holes in a given object, providing they have no "islands" in them. The reason why this approach is used to count blobs, despite the fact that it may seem a little awkward to use, is that the Euler number is

very easy and fast to calculate. It is also a useful means of classifying shapes in an image. The Euler number can be computed by using three local operators. Let us define three numbers *N*1, *N*2 and *N*3, where Nα indicates the number of times that one of the patterns in the pattern set α (α = 1, 2 or 3) occur in the input image.

0	0
0	1

0	0
1	0

1	0
0	0

0	1
0	0

Pattern set 1 (*N*1)

0	1
1	0

1	0
0	1

Pattern set 2 (*N*2)

1	1
1	0

1	1
0	1

0	1
1	1

1	0
1	1

Pattern set 3 (*N*3)

The *8-connected* Euler number, where holes and blobs are defined in terms of 8-connected figures, is defined as: (*N*1-2.*N*2-*N*3)/4. It is possible to calculate the *4-connected* Euler number using a slightly different formula, but this parameter can give results which seem to be anomalous when we compare them to the observed number of holes and blobs.

Filling holes [blb]

Consider a white blob-like figure containing a hole (lake), against a black background. The application of the hole-filling operator will cause all of the holes to be *filled-in*; by setting all pixels in the holes to white. This operator will not alter the outer edge of the figure.

Region labelling [ndo]

Consider an image containing a number of separate blob-like figures. A region-labelling operator will shade the output image so that each blob is given a separate intensity value. We could shade the blobs according to the order in which they are found, during a conventional raster scan of the input image. Alternatively, the blobs could be shaded according to their areas; the biggest blobs becoming the brightest. This is a very useful operator, since it allows objects to be separated and analysed individually (Figure 2.8). Small blobs can also be eliminated from an image using this operator. Region labelling can also be used to count the number of distinct binary blobs in an image. Unlike the Euler number, counting based on region labelling is not effected by the presence of holes.

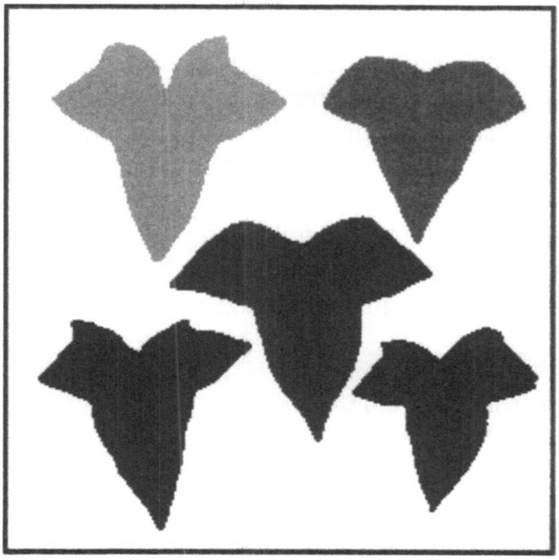

Figure 2.8. Shading blobs in a binary according to the order in which they are found during a raster scan (left to right; top to bottom).

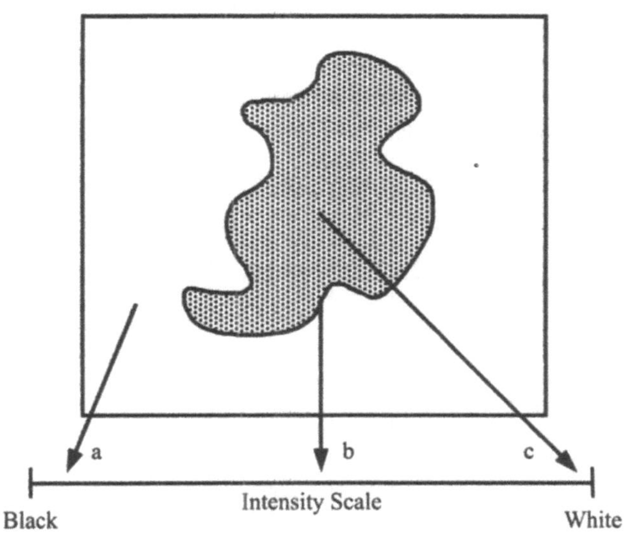

Figure 2.9. Using a grey-scale blurring filter to remove noise from a binary image: a. background points are mapped to black; b. edge points are mapped to the central part of the intensity range. Thresholding at mid-grey has the effect of smoothing the edge of large blobs; c. central areas of large white blobs are mapped to white.

Other methods of detecting/removing small spots

A binary image can be represented in terms of a grey-scale image in which only two grey levels, 0 and W, are allowed. The result of the application of a conventional low-pass (blurring) filter to such an image is a grey-scale image in which there is a larger number of possible intensity values. Pixels which were well inside large white areas in the input image are mapped to very bright pixels in the output image. Pixels which were well inside black areas are mapped to very dark pixels in the output image. However, pixels which were inside small white spots in the input image are mapped to mid-grey intensity levels (Figure 2.9). Pixels on the edge of large white areas are also mapped to mid-grey intensity levels. However, if there is a cluster of small spots, which are closely spaced together, some of them may also disappear.

Based on these observations, the following procedure has been developed. It has been found to be effective in distinguishing between small spots and, at the same time, achieving a certain amount of edge smoothing of the large bright blobs which remain:

```
raf(11,11),     % Low-pass filter using a 11x11 local operator
thr(128),       % Threshold at mid-grey
```

This technique is generally easier and faster to implement than the blob shading technique described previously. Although it may not achieve the desired result exactly, it can be performed at high speed.

An N-tuple filter having the weight matrix illustrated below can be combined with simple thresholding to distinguish between large and small spots. Assume that there are several small white spots within the input image and that they are spaced well apart. All pixels within a spot which can be contained within a circle of radius three pixels will be mapped to white by this particular filter. Pixels within a larger spot will become darker than this. The image is then thresholded at white to separate the large and small spots.

			-1	-1	-1			
		-1				-1		
	-1						-1	
-1								-1
-1				20				-1
-1								-1
	-1						-1	
		-1				-1		
			-1	-1	-1			

Grass-fire transform and skeleton [gfa, mdl, mid]

Consider a binary image containing a single white blob (Figure 2.10). Imagine that a fire is lit at all points around the blob's outer edge and the edges of any holes it may contain. The fire will burn inwards, until at some instant, advancing fire lines meet. When this occurs, the fire becomes extinguished locally. An output image is generated and is shaded in proportion to the time it takes for the fire to reach each point. Background pixels are mapped to black.

The importance of this transform, referred to as the *grass-fire* transform, lies in the fact that it indicates distances to the nearest edge point in the image [1]. It is therefore possible to distinguish thin and fat limbs of a white blob. Those points at which the fire lines meet are known as *quench* points. The set of quench points form a "match-stick" figure, usually referred to as a *skeleton or medial axis transform*. These figures can also be generated in a number of different ways [2] (Figure 2.11).

One such approach is described as *onion-peeling*. Consider a single white blob and a "bug" which walks around the blob's outer edge, removing one pixel at a time. No edge pixel is removed if by doing so we would break the blob into two disconnected parts. In addition, no white pixel is removed if there is only one white pixel amongst its 8-neighbours. This simple procedure leads to an undesirable effect in those instances when the input blob has holes in it; the skeleton which it produces has small loops in it which fit around the holes like a tightened noose. More sophisticated algorithms have been devised which avoid this problem.

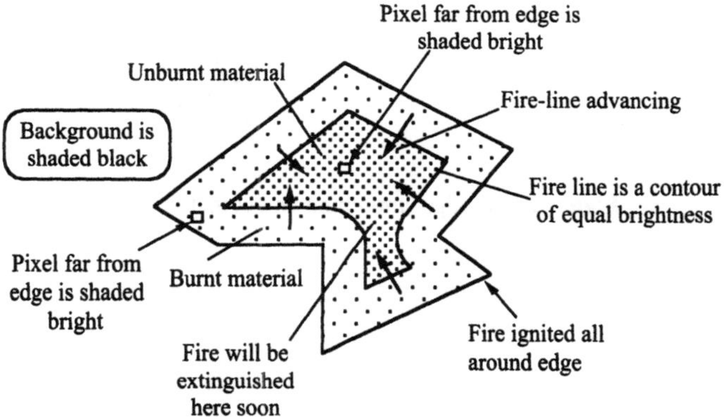

Figure 2.10. Grass-fire transform.

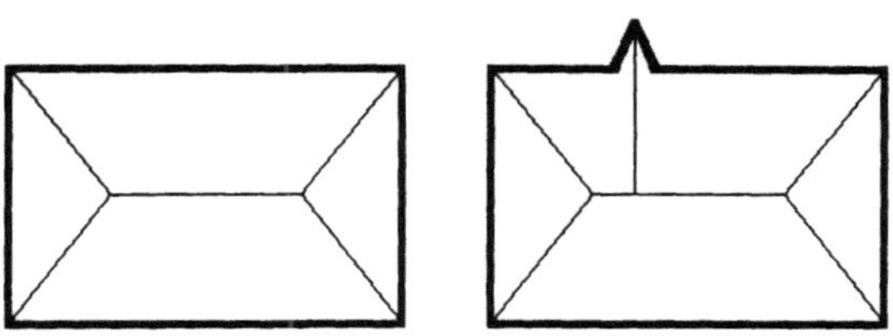

Figure 2.11. Application of the Medial Axis Transform.

Edge smoothing and corner detection

Consider three points B1, B2 and B3, which are placed close together on the edge of a single blob in a binary image (Figure 2.12). The perimeter distance between B1 and B2 is equal to that between B2 and B3. Define the point P to be that at the centre of the line joining B1 and B3. As the three points now move around the edge of the blob, keeping the spacing between them constant, the locus of P traces a smoother path than that followed by B2 as it moves around the edge. This forms the basis of a simple edge smoothing procedure. A related algorithm, for corner detection, shades the edge according to the distance between P and B2. This results in an image in which the corners are highlighted, while the smoother parts of the image are much darker.

Many other methods of edge smoothing are possible. For example, we may map white pixels which have fewer than, say, three white 8-neighbours to black. This has the effect of eliminating "hair" around the edge of a blob-like figure. One of the techniques described previously for eliminating small spots offers another possibility. A third option is to use the processing sequence: [*exw*, *skw*, *skw*, *exw*], where *exw* represents expand white areas and *skw* denotes shrink white areas.

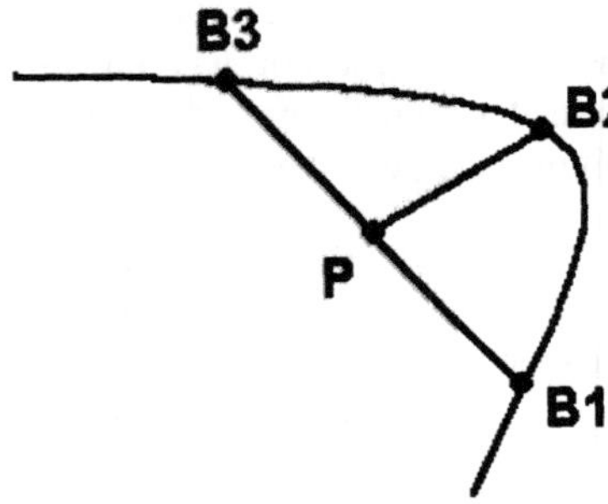

Figure 2.12. Edge smoothing and corner detection.

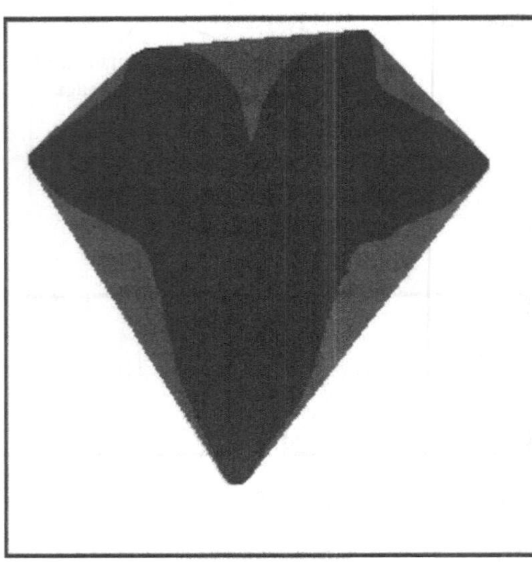

Figure 2.13. Convex hull of an ivy leaf. The lightly shaded region indicates the shape's convex deficiency.

Convex hull [chu]
Consider a single blob in a binary image. The convex hull is that area enclosed within the smallest convex polygon which will enclose the shape (Figure 2.13). This can also be described as the region enclosed within an elastic string, stretched around the blob. The area enclosed by the convex hull, but not within the original blob is called the *convex deficiency*, which may consist of a number of disconnected parts, and includes any holes and indentations. If we regard the blob as being like an *island*, we can understand the logic of referring to the former as *lakes* and the latter as *bays*.

2.3.1 Measurements on Binary Images

To simplify the following explanation, we will confine ourselves to the analysis of a binary image containing a single blob. The area of the blob can be measured by the total number of object (white) pixels in the image. However, we must first define two different types of edge points, in order to measure an object's perimeter.

The *4-adjacency* convention (Figure 2.14) only allows the four main compass points to be used as direction indicators, while *8-adjacency* uses all eight possible directions. If 4-adjacency convention is applied to the image segment given in Figure 2.14c, then none of the four segments (two horizontal and two vertical) will appear as touching, i.e., they are not connected. Using the 8-adjacency convention, the segments are now connected, but we have the ambiguity that the inside of the shape is connected to the outside. Neither convention is satisfactory, but since 8-adjacency allows diagonally connected pixels to be represented, it leads to a more faithful perimeter measurement.

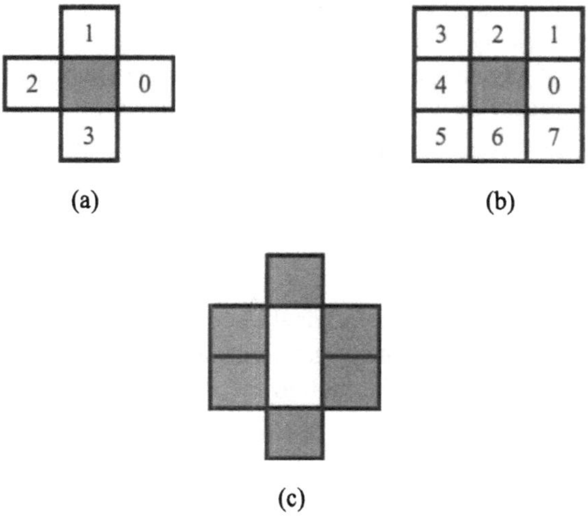

Figure 2.14. Chain code: (a) 4-adjacency coding convention; (b) 8-adjacency coding convention; (c) image segment.

Assuming that the 8-adjacency convention is used, we can generate a coded description of the blob's edge. This is referred to as the *chain code* or *Freeman code* [*fcc*]. As we trace around the edge of the blob, we generate a number, 0–7, to indicate which of the eight possible directions we have taken (i.e., from the centre, shaded pixel in Figure 2.14(b)). Let N_o indicate how many *odd-numbered* code values are produced as we code the blob's edge, and N_e represent the number of *even-numbered* values found. The *perimeter* of the blob is given *approximately* by the formula: $N_e + \sqrt{2}.N_o$

This formula will normally suffice for use in those situations where the perimeter of a smooth object is to be measured. The *centroid* of a blob [*cgr*] determines its position within the image and can be calculated using the formulae:

$$I \leftarrow \sum_j \sum_i (a(i,j).i)/N_{i,j} \qquad \text{and} \qquad J \leftarrow \sum_j \sum_i (a(i,j).j)/N_{i,j}$$

where $N_{i,j} \leftarrow \sum_j \sum_i a(i,j)$

Although we are considering images in which the $a(i,j)$ are equal to 0 (black) or 1 (white), it is convenient to use $a(i,j)$ as an ordinary arithmetic variable as well.

2.3.2 Shape Descriptors

The following are just a few of the numerous shape descriptors that have been proposed:

a. the distance of the furthest point on the edge of the blob from the centroid;
b. the distance of the closest point on the edge of the blob from the centroid;
c. the number of protuberances, as defined by that circle whose radius is equal to the average of the parameters measured in a. and b.;
d. the distances of points on the edge of the blob from the centroid, as a function of angular position. This describes the silhouette in terms of polar co-ordinates. (This is not a single-valued function.);
e. Circularity = Area/Perimeter^2. This will tend to zero for irregular shapes with ragged boundaries, and has a maximum value (=1/4π) for a circle;
f. the number of holes. (Use *eul* and *ndo* to count them.);
g. the number of bays;
h. Euler number;
i. the ratio of the areas of the original blob and that of its convex hull;
j. the ratio of the areas of the original blob and that of its circumcircle;
k. the ratio of the area of the blob to the square of the total limb-length of its skeleton;
l. distances between joints and limb ends of the skeleton;
m. the ratio of the projections onto the major and minor axes.

2.4 Binary Mathematical Morphology

The basic concept involved in mathematical morphology is simple: an image is probed with a template shape, called a structuring element, to find where the structuring element fits, or does not fit within a given image [3] (Figure 2.15). By marking the locations where the template shape fits, structural information, can be gleaned about the image. The structuring elements used in practice are usually geometrically simpler than the image they act on, although this is not always the case. Common structuring elements include points, point pairs, vectors, lines, squares, octagons, discs, rhombi and rings. Since shape is a prime carrier of information in machine vision applications, mathematical morphology has an important role to play in industrial systems [4].

The language of binary morphology is derived from that of set theory [5]. General mathematical morphology is normally discussed in terms of Euclidean *N*-space, but in digital image analysis we are only interested in a discrete or digitised equivalent in two-space. The following analysis is therefore restricted to binary images, in a digital two-dimensional integer space, Z^2. The image set (or scene) under analysis will be denoted by *A*, with elements $a = (a1, a2)$. The shape parameter, or structuring element, that will be applied to scene *A* will be denoted by *B*, with elements $b = (b1, b2)$. The primary morphological operations that we will examine are dilation, erosion, opening and closing.

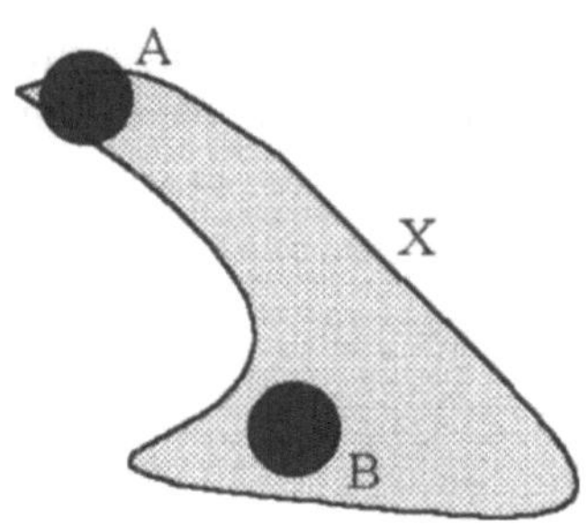

Figure 2.15. A structuring element fitting, *B*, and not fitting, *A*, into a given image scene *X* [3].

Dilation

Dilation (also referred to as filling and growing) is the expansion of an image set *A* by a structuring element *B*. It is formally viewed as the combination of the two sets using vector addition of the set elements. The dilation of an image set *A* by a structuring element *B*, will be denoted *A* d *B*, and can be represented as the union of translates of the structuring element *B* [5]:

$$A \,\mathrm{d}\, B = \bigcup_{a \in A} B_a$$

where $\bigcup$ represents the union of a set of points and the translation of B by point a is given by, $B_a = \{c \in Z^2 \mid c = b + a \text{ for some } b \in B\}$. This is best explained by visualising a structuring element B moving over an image A in a raster fashion. Whenever the origin of the structuring element touches one of the image pixels in A, then the entire structuring element is placed at that location. For example, in Figure 2.16 the grid image is dilated by a cross-shaped structuring element, contained within a 3 × 3 pixel grid.

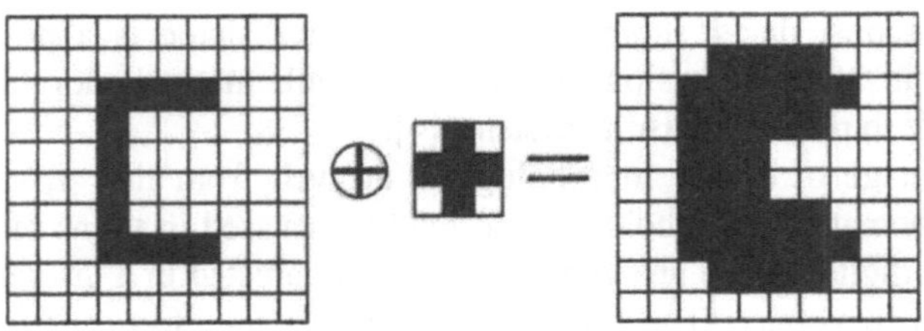

Figure 2.16. Dilation of a grid image by a cross structuring element.

Erosion

Erosion is the dual morphological operation of dilation and is equivalent to the shrinking (or reduction) of the image set A by a structuring element B. This is a morphological transformation, which combines two sets using vector subtraction of set elements [5]. The erosion of an image set A by a structuring element B, denoted $A \text{ c } B$, can be represented as the intersection of the negative translates:

$$A \text{ c } B = \bigcap_{b \in B} A_{-b}$$

where $\bigcap$ represents the intersection of a set of points. Erosion of the image A by B is the set of all points for which B translated to a point x is contained in A. This consists of sliding the structuring element B across the image A, and where B is fully contained in A (by placing the origin of the structuring element at the point x) then x belongs to the eroded image $A \text{ c } B$. For example, in Figure 2.17 the grid image is eroded by a cross-shaped structuring element, contained within a 3 × 3 pixel grid.

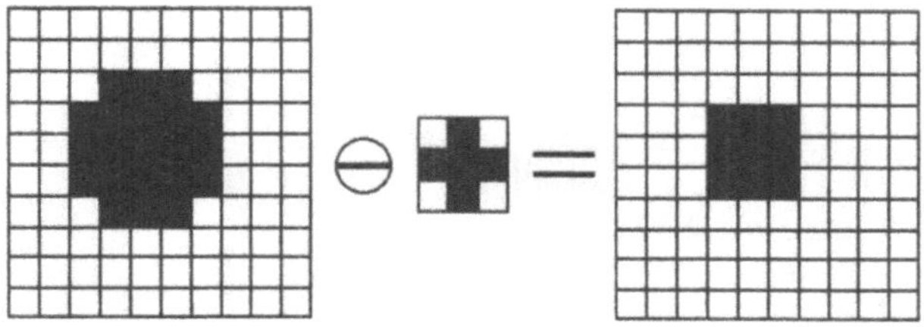

Figure 2.17. Erosion of a grid image by a cross structuring element

A duality relationship exists between certain morphological operators, such as erosion and dilation. This means that the equivalent of such an operation can be performed by its dual on the complement (negative) image and by taking the complement of the result [6]. Although duals, erosion and dilation operations are not inverses of each other. Rather they are related by the following duality relationships:

$$\left(A \,\mathrm{c}\, B\right)^C = A^C \mathrm{d}\, \overline{B} \quad \text{and} \quad \left(A \,\mathrm{d}\, B\right)^C = A^C \,\mathrm{c}\, \overline{B}$$

Where A^C refers to the complement of the image set A and $\overline{B} = \{ x \mid \text{for some } b \in B, x = -b \}$ refers to the reflection of B about the origin. (Serra [7,8] refers to this as the *transpose* of the structuring element.)

2.4.1 Opening and Closing Operations

Erosion and dilation tend to be used in pairs to extract, or impose, structure on an image. The most commonly found erosion-dilation pairings occur in the *opening* and *closing* transformations.

Opening

Opening is a combination of erosion and dilation operations that have the effect of removing isolated spots in the image set A that are smaller than the structuring element B and those sections of the image set A narrower than B. This is also viewed as a geometric *rounding* operation (Figure 2.18). The opening of the image set A by the structuring element B, is denoted $A \bigcirc B$, and is defined as $(A \text{ c } B) \text{ d } B$.

(a) (b) (c)

Figure 2.18. Application of a 3 × 3 square structuring element to a binary image of a small plant: (a) original image; (b) result of morphological opening; (c) result of morphological closing.

Closing

Closing is the dual morphological operation of opening. This transformation has the effect of filling in holes and blocking narrow valleys in the image set A, when a structuring element B (of similar size to the holes and valleys) is applied (Figure 2.18). The closing of the image set A by the structuring element B, is denoted $A \bullet B$, and is defined as $(A \, \mathrm{d} \, B) \, \mathrm{c} \, B$.

One important property that is shared by both the opening and closing operations is *idempotency*. This means that successful reapplication of the operations will not change the previously transformed image [4]. Therefore, $A \circ B = (A \circ B) \circ B$ and $A \bullet B = (A \bullet B) \bullet B$.

Unfortunately, the application of morphological techniques to industrial tasks, which involves complex operations on "real-world" images, can be difficult to implement. Practical imaging applications tend to have structuring elements that are unpredictable in shape and size. In practice, the ability to manipulate arbitrary structuring elements usually relies on their decomposition into component parts.

2.4.2 Structuring Element Decomposition

Some vision systems [9,10,11] can perform basic morphological operations very quickly in a parallel and/or pipelined manner. Implementations that involve such special-purpose hardware tend to be expensive, although there are some notable exceptions [12]. Unfortunately, some of these systems impose restrictions on the shape and size of the structuring elements that can be handled. Therefore, one of the key problems involved in the application of morphological techniques to industrial image analysis is the generation and/or decomposition of large structuring elements. Two main strategies are used to tackle this problem.

The first technique is called *dilation* or *serial decomposition*. This decomposes certain large structuring elements into a sequence of successive erosion and dilation operations, each step operating on the preceding result. Unfortunately, the decomposition of large structuring elements into smaller ones is not always possible. Also, those decompositions that are possible are not always easy to identify and implement.

If a large structuring element B can be decomposed into a chain of dilation operations, $B = B_1 \mathrm{d} B_2 \mathrm{d} \ldots\ldots \mathrm{d} B_N$ (Figure 2.19), then the dilation of the image set A by B is given by:

$$A \mathrm{d} B = A \mathrm{d} (B_1 \mathrm{d} B_2 \mathrm{d} \ldots \mathrm{d} B_N) = (((A \mathrm{d} B_1) \mathrm{d} B_2) \ldots) \mathrm{d} B_N.$$

Similarly, using the so-called chain rule [13], which states that $A \mathrm{c} (B \mathrm{d} C) = (A \mathrm{c} B) \mathrm{c} C$, the erosion of A by B is given by:

$$A \mathrm{c} B = A \mathrm{c} (B_1 \mathrm{d} B_2 \mathrm{d} \ldots \mathrm{d} B_N) = (((A \mathrm{c} B_1) \mathrm{c} B_2) \ldots) \mathrm{c} B_N$$

A second approach to the decomposition problem is based on "breaking up" the structuring element, B, into a union of smaller components, $B_1 \ldots, B_N$. We can think of this approach as 'tiling' of the structuring element by sub-structuring elements (Figure 2.20). Since the 'tiles' do not need to be contiguous or aligned, any shape can be specified without the need for serial decomposition of the structuring element, although the computational cost of this approach is proportional to the area of the structuring element [11]. This is referred to as *union* or *parallel decomposition*. Therefore, with B decomposed into a union of smaller structuring elements, $B = B_1 \cup B_2 \cup \ldots \cup B_N$, then the dilation of an image A by the structuring element B can be rewritten as:

$$A \mathrm{d} B = A \mathrm{d} (B_1 \cup B_2 \cup \ldots \cup B_N)$$
$$= (A \mathrm{d} B_1) \cup (A \mathrm{d} B_2) \cup \ldots \cup (A \mathrm{d} B_N)$$

Likewise, the erosion of A by the structuring element B can be rewritten as:

$$A \mathrm{c} B = A \mathrm{c} (B_1 \cup B_2 \cup \ldots \cup B_N)$$
$$= (A \mathrm{c} B_1) \cap (A \mathrm{c} B_2) \cap \ldots \cap (A \mathrm{c} B_N)$$

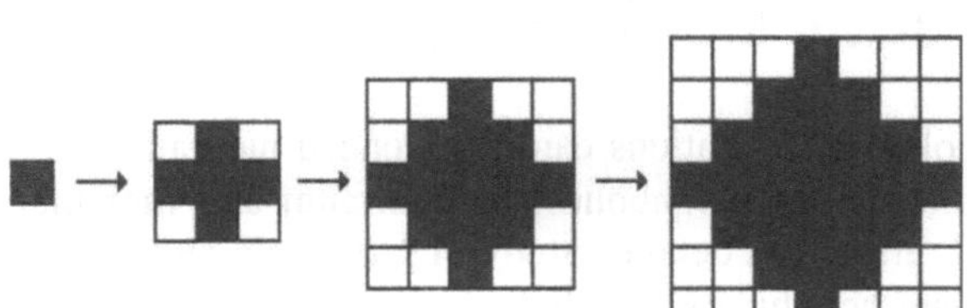

(a) (b) (c) (d)

Figure 2.19. Construction of a 7 × 7 structuring element by successive dilation of a 3 × 3 structuring element: (a) initial pixel; (b) 3 × 3 structuring element and the result of the first dilation; (c) result of the second dilation; (d) result of the third dilation [11]

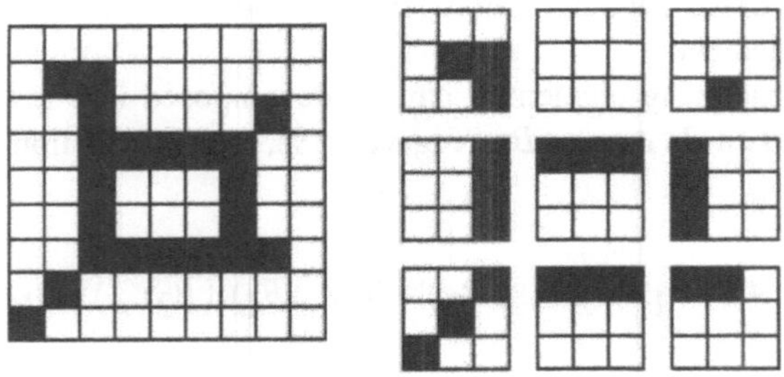

(a) (b)

Figure 2.20. Tiling of a 9 × 9 arbitrary structuring element: (a) the initial 9 × 9 structuring element; (b) tiling with nine 3 × 3 sub-structuring elements [11].

This makes use of the fact that $A \mathrm{c} (B \cup C) = (A \mathrm{c} B) \cap (A \mathrm{c} C)$ [4]. Due to the nature of this decomposition procedure, it is well suited to implementation on parallel computer architectures.

Waltz [11] compared these structural element decomposition techniques, and showed that the serial approach has a 9:4 speed advantage over its parallel equivalent. (This was based on an arbitrarily specified 9 × 9 pixel-structuring element, when implemented on a commercially available vision system.) However, the parallel approach has a 9:4 advantage in the number of degrees of freedom. (Every possible 9_9-structuring element can be achieved with the *parallel decomposition*, but only a small subset can be realised with the serial approach.) Although slower than the serial approach, it has the advantage that there is no need for serial decomposition of the structuring element.

Classical parallel and serial methods mainly involve the numerous scanning of image pixels and are therefore inefficient when implemented on conventional computers. This is so, because the number of scans depends on the total number of pixels (or edge pixels) in the shape to be processed by the morphological operator. Although the parallel approach is suited to some customised (parallel) architectures, the ability to implement such parallel approaches on *serial* machines is discussed by Vincent [14].

2.5 Grey-scale Morphology

Binary morphological operations can be extended naturally to process grey-scale imagery, by the use of neighbourhood minimum and maximum functions [4]. Heijmans [15], presents a detailed study of grey-scale morphological operators, in which he outlines how binary morphological operators and thresholding techniques can be used to build a large class of useful grey-scale morphological operators. Sternberg [16], discusses the application of such morphological techniques to industrial inspection tasks.

In Figure 2.21, a one-dimensional morphological filter, operates on an analogue signal (equivalent to a grey-scale image). The input signal is represented by the thin curve and the output by the thick black curve. In this simple example, the structuring element has an approximately parabolic form. In order to calculate a value for the output signal, the structuring element is pushed upwards, from below the input curve. The height of the top of the structuring element is noted. This process is then repeated, by sliding the structuring element sideways. Notice how this particular operator attenuates the intensity peak but follows the input signal quite accurately everywhere else. Subtracting the output signal from the input would produce a result in which the intensity peak is emphasised and all other variations would be reduced.

The effect of the basic morphological operators on two-dimensional grey-scale images can also be explained in these terms. Imagine the grey-scale image as a landscape, in which each pixel can be viewed in 3-D. The extra height dimension represents the grey-scale value of a pixel. We generate new images by passing the structuring element above/below this landscape. (See Figure 2.21.)

Grey-scale dilation

This is computed as the maximum of translations of the grey surface. Grey-level dilation of image A by the structuring element B produces an image C defined by:

$$C(r,c) = \mathrm{Max}_{(i,j)}\{A(r-i, c-j) + B(i,j)\} = (A \mathrm{~d~} B)(r,c)$$

where A, B and C are grey level images. Commonly used grey-level structuring elements include rods, discs, cones and hemispheres. This operation is commonly used to smooth small negative contrast grey-level regions in an image.

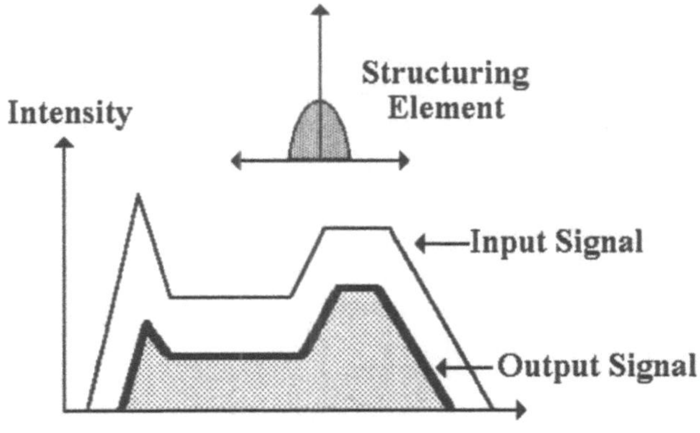

Figure 2.21. A 1-dimensional morphological filter, operating on an analogue signal.

Grey-scale erosion

The grey value of the erosion at any point is the *maximum* value for which the structuring element centred at that point still fits entirely within the foreground under the surface. This is computed by taking the *minimum* of the grey surface translated by all the points of the structuring element (Figure 2.21). Grey-level erosion of image A by the structuring element B produces an image C defined by:

$$C(r,c) = \text{Min}_{(i,j)}\{A(r+i,\ c+j) - B(i,j)\} = (A \text{ c } B)(r,c)$$

This operation is commonly used to smooth small positive contrast grey level regions in an image.

Grey-scale opening

This operation is defined as the grey level erosion of the image followed by the grey-level dilation of the eroded image. That is, it will cut down the peaks in the grey-level topography to the highest level for which the elements fit under the surface.

Grey-scale closing

This operation is defined as the grey-level dilation of the image followed by the grey-level erosion of the dilated image. Closing fills in the valleys to the maximum level for which the element fails to fit above the surface. For a more detailed discussion on binary and grey-scale mathematical morphology, see Haralick and Shapiro [17] and Dougherty [3].

2.6 Global Image Transforms

An important class of image processing operators is characterised by an equation of the form $B \Leftarrow f(A)$, where $A = \{a(i,j)\}$ and $B = \{b(p,q)\}$. Each element in the output picture, B, is calculated using all or, at least a large proportion of the pixels in A. The output image B may well look quite different from the input image A. Examples of this class of operators are: lateral shift, rotation, warping, Cartesian to polar co-ordinate conversion, Fourier and Hough transforms.

Integrate intensities along image rows [rin]

This operator is rarely of great value when used on its own, but can be used with other operators to good effect, for example detecting horizontal streaks and edges. The operator is defined recursively:

$$b(i,j) \Leftarrow b(i,j\text{-}1) + a(i,j)/n \text{ where } b(0,0) = 0$$

Row maximum [rox]

This function is often used to detect local intensity minima: $c(i,j) \Leftarrow \text{MAX}(a(i,j), c(i,j\text{-}1))$

Geometric transforms

Algorithms exist by which images can be shifted [*psh*], rotated [*tur*], undergo axis conversion [*ctr, rtc*], magnified [*pex* and *psq*] and warped. The reader should note that certain operations, such as rotating a digital image, can cause some difficulties because pixels in the input image are not mapped exactly to pixels in the output image. This can cause smooth edges to appear stepped. To avoid this effect, interpolation may be used, but this has the unfortunate effect of blurring edges. (See [18] for more details.)

The utility of axis transformations is evident when we are confronted with the examination of circular objects, or those displaying a series of concentric arcs, or streaks radiating from a fixed point. Inspecting such objects is often made very much easier, if we first convert from *Cartesian* to *polar* co-ordinates. Warping is also useful in a variety of situations. For example, it is possible to compensate for *barrel*, or *pincushion* distortion in a camera. Geometric distortions introduced by a wide-angle lens, or trapezoidal distortion due to viewing the scene from an oblique angle can also be corrected. Another possibility is to convert simple curves of known shape into straight lines, in order to make subsequent analysis easier.

2.6.1 Hough Transform

The Hough transform provides a powerful and robust technique for detecting lines, circles, ellipses, parabolae, and other curves of pre-defined shape, in a binary image. Let us begin our discussion of this fascinating topic, by describing the simplest version, the basic Hough Transform, which is intended to detect straight lines. Actually, our objective is to locate *nearly linear* arrangements of disconnected white spots and *"broken" lines*. Consider that a straight line in the input image is defined by the equation $r = x.\cos\phi + y.\sin\phi$, where r and ϕ are two unknown parameters, whose values are to be found. Clearly, if this line intersects the point (x_i, y_i), then $r = x_i.\cos\phi + y_i.\sin\phi$ can be solved for many different values of (r, ϕ). So, each white point (x_i, y_i) in the input image may be associated with a *set* of (r, ϕ) values. Actually, this set of points forms a *sinusoidal curve* in (r, ϕ) space. (The latter is called the *Hough Transform (HT)* image.) Since each point in the input image generates such a sinusoidal curve, the whole of that image creates a multitude of overlapping sinusoids, in the HT image. In many instances, a large number of sinusoidal curves are found to converge on the same spot in the HT image. The (r, ϕ) address of such a point indicates the slope, ϕ, and position, r, of a straight line that can be drawn through a large number of white spots in the input image.

The implementation of the Hough transform for line detection begins by using a two-dimensional accumulator array, $A(r, \phi)$, to represent quantised (r, ϕ) space. (Clearly, an important choice to be made is the step size for quantising r and ϕ. However, we shall not dwell on such details here.) Assuming that all the elements of $A(r, \phi)$ are initialised to zero, the Hough Transform is found by computing a set $S(x_i, y_i)$ of (r, ϕ) pairs satisfying the equation $r = x_i.\cos\phi + y_i.\sin\phi$. Then, for all (r, ϕ) in $S(x_i, y_i)$, we increment $A(r, \phi)$ by one. This process is then repeated for all values of i such that the point (x_i, y_i) in the input image is white. We repeat that bright spots in the HT image indicate "linear" sets of spots in the input image. Thus, line detection is transformed to the rather simpler task of finding local maxima in the accumulator array, $A(r, \phi)$. The co-ordinates (r, ϕ) of such a local maximum give the parameters of the equation of the corresponding line in the input image. The HT image can be displayed, processed and analysed just like any other image, using the operators that are now familiar to us.

The robustness of the HT techniques arises from the fact that, if part of the line is missing, the corresponding peak in the HT image is simply darker. This occurs because fewer sinusoidal curves converge on that spot and the corresponding accumulator cell is incremented less often. However, unless the line is almost completely obliterated, this new darker spot can also be detected. In practice, we find that "near straight lines" are transformed into a cluster of points. There is also a spreading of the intensity peaks in the HT image, due to noise and quantisation effects. In this event, we may conveniently threshold the HT image and then find the centroid of the resulting spot, to calculate the parameters of the straight line in the input image. Pitas [19] and Davies [20] give a more detailed description of this algorithm. Figure 2.22 illustrates how this approach can be used to find a line in a noisy binary image.

(a)

(b)

(c)

Figure 2.22. Hough transform: (a) original image; (b) Hough transform; (c) inverse Hough transform applied to a single white pixel located at the point of maximum intensity in (b). Notice how accurately this process locates the line in the input image, despite the presence of a high level of noise.

The Hough transform can also be generalised to detect groups of points lying on a curve. In practice, this may not be a trivial task, since the complexity increases very rapidly with the number of parameters needed to define the curve. For circle detection, we define a circle parametrically as:

$$r^2 = (x - a)^2 + (y - b)^2$$

where (a, b) determines the co-ordinates of the centre of the circle and r is its radius. This requires a three-dimensional parameter space, which cannot, of course, be represented and processed as a single image. For an arbitrary curve, with no simple equation to describe its boundary, a look-up table is used to define the relationship between the boundary co-ordinates an orientation and the Hough transform parameters. (See [21] for more details.)

2.6.2 Two-dimensional Discrete Fourier Transform

We have just seen how the transformation of an image into a different domain can sometimes make the analysis task easier. Another important operation to which this remark applies is the Fourier Transform. Since we are discussing the processing of images, we shall discuss the *two-dimensional Discrete Fourier Transform.* This operation allows spatial periodicities in the intensity within an image to be investigated, in order to find, amongst other features, the dominant frequencies. The two-dimensional Discrete Fourier Transform of an N.N image f(x,y) is defined as follows: [22]

$$F(u,v) = \frac{1}{N}\sum_{x=0}^{N-1}\sum_{y=0}^{N-1} f(x,y)\exp[-j2\pi(ux+vy)/N]$$

where $0 \le u,v \le N-1$. The inverse transform of $F(u,v)$ is defined as:

$$f(x,y) = \frac{1}{N}\sum_{u=0}^{N-1}\sum_{v=0}^{N-1} F(u,v)\exp[j2\pi(ux+vy)/N]$$

where $0 \le x,y \le N-1$.

Several algorithms have been developed to calculate the two-dimensional Discrete Fourier Transform. The simplest makes use of the observation that this is a *separable transform* which can be computed as a sequence of two one-dimensional transforms. Therefore, we can generate the two-dimensional transform by calculating the one-dimensional Discrete Fourier Transform along the image rows and then repeating this on the resulting image but, this time, operating on the columns [22]. This reduces the computational overhead when compared to direct two-dimensional implementations. The sequence of operations is as follows:

$$f(x,y) \quad \rightarrow \text{Row Transform} \rightarrow \quad F_1(x,v) \quad \rightarrow \text{Column Transform} \rightarrow \quad F_2(u,v)$$

Although this is still computationally slow compared to other many shape measurements, the Fourier transform is quite powerful. It allows the input to be represented in the frequency domain, which can be displayed as a *pair* of images. (It is not possible to represent both amplitude and phase using a single monochrome image.) Once the processing within the frequency domain is complete, the inverse transform can be used to generate a new image in the original, so-called, *spatial* domain.

The Fourier power, or amplitude, spectrum plays an important role in image processing and analysis. This can be displayed, processed and analysed as an intensity image. Since the Fourier transform of a real function produces a complex function: $F(u,v) = R(u,v) + i.I(u,v)$, the frequency spectrum of the image is the magnitude function

$$|F(u,v)| = \sqrt{R^2(u,v) + I^2(u,v)}$$

and the power spectrum (spectral density) is defined as $P(u,v) = |F(u,v)|^2$.

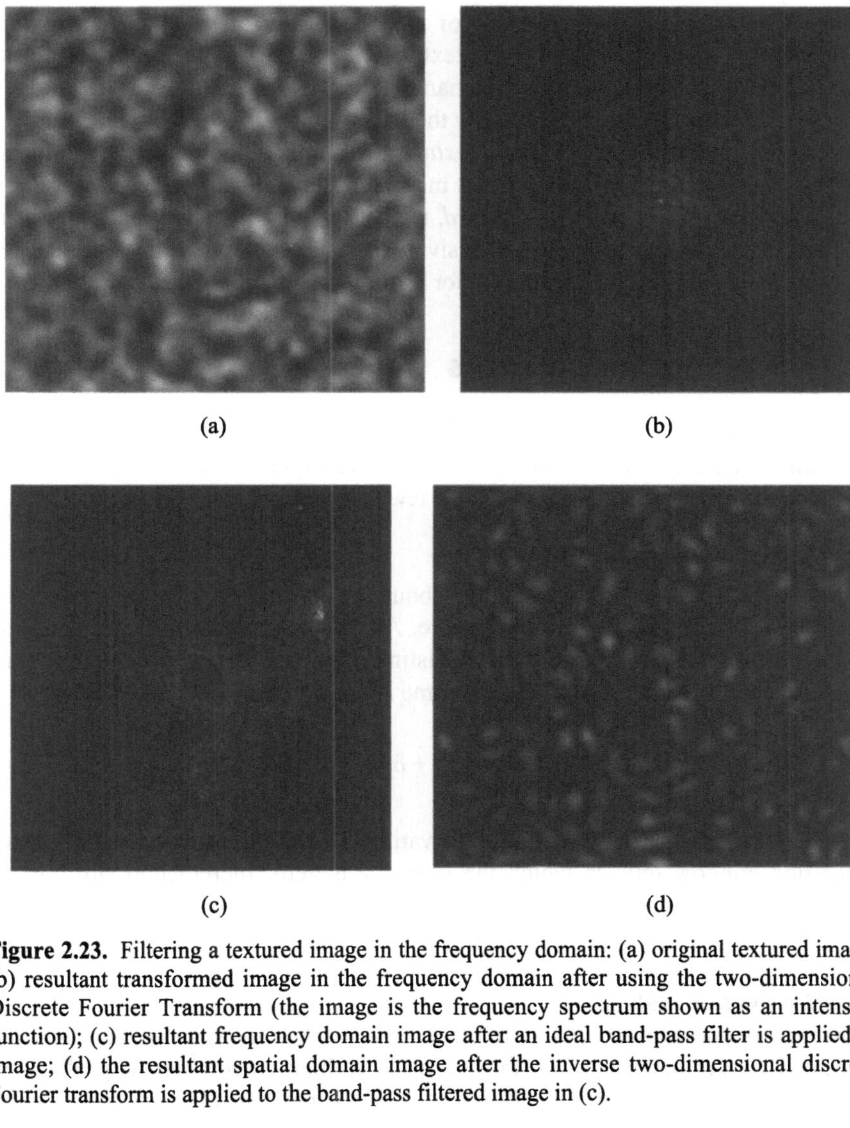

Figure 2.23. Filtering a textured image in the frequency domain: (a) original textured image; (b) resultant transformed image in the frequency domain after using the two-dimensional Discrete Fourier Transform (the image is the frequency spectrum shown as an intensity function); (c) resultant frequency domain image after an ideal band-pass filter is applied to image; (d) the resultant spatial domain image after the inverse two-dimensional discrete Fourier transform is applied to the band-pass filtered image in (c).

Figure 2.23 illustrates how certain textured features can be highlighted using the two-dimensional Discrete Fourier Transform. The image is transformed into the frequency domain and an ideal band-pass filter (with a circular symmetry) is applied. This has the effect of limiting the frequency information in the image. When the inverse transform is calculated, the resultant textured image has a different frequency content which can then be analysed. For more details on the Fourier transform and its implementations, see [19] and [22].

2.7 Texture Analysis

Texture is observed in the patterns of a wide variety of synthetic and natural surfaces (e.g., wood, metal, paint and textiles). If an area of a textured image has a large *intensity* variation then the dominant feature of that area would be *texture*. If this area has little variation in *intensity* then the dominant feature within the area is *tone*. This is known as the *tone-texture concept*. Although a precise formal definition of texture does not exist, it may be described subjectively using terms such as *coarse, fine, smooth, granulated, rippled, regular, irregular* and *linear,* and of course these features are used extensively in manual region segmentation. There are two main classification techniques for texture: *statistical* and *structural.*

2.7.1 Statistical Approaches

The statistical approach is well suited to the analysis and classification of random or natural textures. A number of different techniques have been developed to describe and analyse such textures [23], a few of which are outlined below.

Auto-correlation Function (ACF)

Auto-correlation derives information about the basic 2-D tonal pattern that is repeated to yield a given periodic texture. Although useful at times, the ACF has severe limitations. It cannot always distinguish between textures, since many subjectively different textures have the same ACF, which is defined as follows:

$$A(\delta x, \delta y) = \left(\sum_{i,j} [\, I(i,j).I(i+\delta x, j+\delta y) \,] \right) / \sum_{i,j} [\, I(i,j) \,]^2$$

where $\{I(i,j)\}$ is the image matrix. The variables (i,j) are restricted to lie within a specified window outside which the intensity is zero. Incremental shifts of the image are given by $(\delta x, \delta y)$. It is worth noting that the ACF and the power spectral density are Fourier transforms of each other.

Fourier spectral analysis

The Fourier spectrum is well suited to describing the directionality and period of repeated texture patterns, since they give rise to high-energy narrow peaks in the power spectrum (Section 2.6 and Figure 2.23). Typical Fourier descriptors of the power spectrum include: the location of the highest peak, mean, and variance and the difference in frequency between the mean and the highest value of the spectrum. This approach to texture analysis is often used in aerial/satellite and medical image analysis. The main disadvantage of this approach is that the procedures are not invariant even, under monotonic transforms of its intensity.

Edge density

This is a simple technique in which an edge detector or high-pass filter is applied to the textured image. The result is then thresholded and the edge density is measured by the average number of edge pixels per unit area. Two-dimensional, or directional filters/edge detectors may be used as appropriate.

Histogram features

This useful approach to texture analysis is based on the intensity histogram of all or part of an image. Common histogram features include: *moments*, *entropy dispersion*, *mean* (an estimate of the average intensity level), *variance* (this second moment is a measure of the dispersion of the region intensity), *mean square value* or *average energy*, *skewness* (the third moment which gives an indication of the histogram's symmetry) and *kurtosis* (cluster prominence or "peakness"). For example, a narrow histogram indicates a low-contrast region, while two peaks with a well-defined valley between them indicates a region that can readily be separated by simple thresholding.

Texture analysis, based solely on the grey-scale histogram, suffers from the limitation that it provides no information about the relative position of pixels to each other. Consider two binary images, where each image has 50% black and 50% white pixels. One of the images might be a checkerboard pattern, while the second one may consist of a salt and pepper noise pattern. These images generate exactly the same grey-level histogram. Therefore, we cannot distinguish them using first-order (histogram) statistics alone. This leads us naturally to the examination of the co-occurrence approach to texture measurement.

2.7.2 Co-occurrence Matrix Approach

The co-occurrence matrix technique is based on the study of *second-order grey-level spatial dependency statistics*. This involves the study of the grey-level spatial interdependence of pixels and their spatial distribution in a local area. Second-order statistics describe the way grey levels tend to occur together, in pairs and therefore provide a description of the type of texture present. A *two-dimensional* histogram of the spatial dependency of the various grey-level picture elements within a textured image is created. While this technique is quite powerful, it does not describe the shape of the primitive patterns making up the given texture.

The co-occurrence matrix is based on the estimation of the second-order joint conditional probability density function, $f(p,q,d,a)$, for angular displacements, a, equal to 0, 45, 90 and 135 degrees. Let $f(p,q,d,a)$ be the probability of going from one pixel with grey level p to another with grey level q, given that the distance between them is d and the direction of travel between them is given by the angle a. (For N_g grey levels, the size of the co-occurrence matrix will be $N_g.N_g$.) For example, assuming the intensity distribution shown in the sub-image given below, we can generate the co-occurrence matrix for $d = 1$ and a is taken as 0 degrees.

2	3	3	3
1	1	0	0
1	1	0	0
0	0	2	2
2	2	3	3

Sub-image with four grey-levels

Grey Scale	0	1	2	3
0	6	2	1	0
1	2	4	0	0
2	1	0	4	2
3	0	0	2	6

Co-occurrence matrix {f(p,q,1,0)} for the sub-image

A co-occurrence distribution that changes rapidly with distance, *d*, indicates a fine texture. Since the co-occurrence matrix also depends on the image intensity range, it is common practice to normalise the textured image's grey scale prior to generating the co-occurrence matrix. This ensures that first-order statistics have standard values and avoids confusing the effects of first- and second-order statistics of the image.

A number of texture measures (also referred to as texture attributes) have been developed to describe the co-occurrence matrix numerically and allow meaningful comparisons between various textures [23] (Figure 2.24). Although these attributes are computationally intensive, they are simple to implement. Some sample texture attributes for the co-occurrence matrix are given below.

Energy

Energy, or angular second moment, is a measure of the homogeneity of a texture. It is defined thus,

$$Energy = \Sigma_p \Sigma_q [\, f(p,q,d,a)\,]^2$$

In a uniform image, the co-occurrence matrix will have few entries of large magnitude. In this case the *Energy* attribute will be large.

Entropy

Entropy is a measure of the complexity of a texture and is defined thus:

$$Entropy = -\Sigma_p \Sigma_q [\, f(p,q,d,a).\log(f(p,q,d,a)) \,]$$

It is commonly found that what a person judges to be a complex image tends to have a higher *Entropy* value than a simple one.

Inertia

Inertia is the measurement of the moment of inertia of the co-occurrence matrix about its main diagonal. This is also referred as the *contrast* of the textured image. This attribute gives an indication of the amount of local variation of intensity present in an image.

$$Inertia = \Sigma_p \Sigma_q [\, (p-q)^2 . f(p,q,d,a) \,]$$

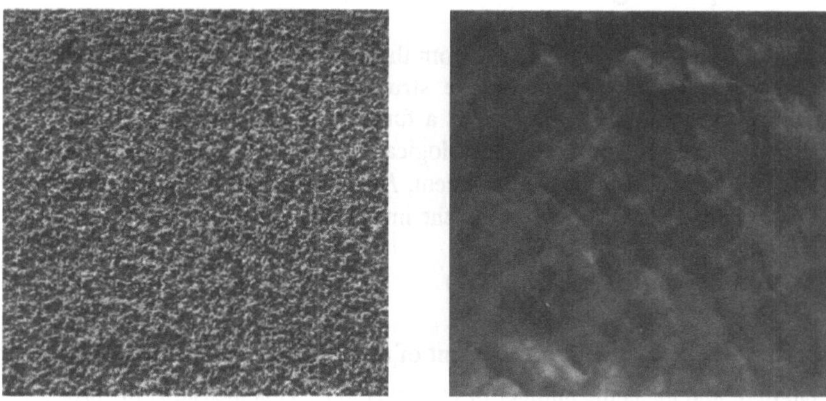

(a) (b)

	Sand		Paper	
	$f(p,q,1,0)$	$f(p,q,1,90)$	$f(p,q,1,0)$	$f(p,q,1,90)$
Energy ($\times 10^6$)	1.63	1.7	3.49	3.42
Inertia ($\times 10^8$)	5.4	6.5	0.181	0.304

(c)

Figure 2.24. Co-occurrence based texture analysis: (a) sand texture; (b) paper texture; (c) texture attributes.

2.7.3 Structural Approaches

Certain textures are deterministic in that they consist of identical *texels* (basic texture element), which are placed in a repeating pattern according to some well-defined but unknown placement rules. To begin the analysis, a texel is isolated by identifying a group of pixels having certain invariant properties, which repeat in the given image. A texel may be defined by its: grey level, shape, or homogeneity of some local property, such as size or orientation. Texel spatial relationships may be expressed in terms of adjacency, closest distance and periodicities.

This approach has a similarity to language; with both image elements and grammar, we can generate a syntactic model. A texture is labelled *strong* if it is defined by deterministic placement rules, while a *weak* texture is one in which the texels are placed at random. Measures for placement rules include: edge density, run lengths of maximally connected pixels and the number of pixels per unit area showing grey levels that are locally maxima or minima relative to their neighbours

2.7.4 Morphological Texture Analysis

Textural properties can be obtained from the erosion process (Sections 2.4 and 2.5) by appropriately parameterising the structuring element and determining the number of elements of the erosion as a function of the parameters value [3]. The number of white pixels of the morphological opening operation as a function of the size parameter of the structuring element, H, can determine the size distribution of the *grains* in an image. Granularity of the image F is defined as:

$$G(d) = 1 - (\#[F \circ H_d]/\#F)$$

Where H_d is a disc structuring element of diameter d or a line structuring element of length d, and $\#F$ is the number of elements in F. This measures the proportion of pixels participating in grains smaller than d.

2.8 Implementation Considerations

Of course, all of the image processing and analysis operators that have been mentioned above can be implemented using a conventional programming language, such as C or C++. However, it is important to realise that many of the algorithms are time-consuming when realised in this way. The monadic, dyadic and local operators can all be implemented in time $K.m.n$ seconds, where K is a constant that is different for each function and (m,n) define the image resolution. However, some of the global operators require $O(m^2.n^2)$ time. With these points in mind, we see that a low-cost, slow but very versatile image processing system can be assembled, simply by embedding a *frame-grabber* into a conventional desk-top computer. (A *frame-grabber* is a device for digitising video images and displaying computer-processed/generated images on a monitor.)

The monadic operators can be implemented using a look-up table, which can be realised simply in a ROM or RAM. The dyadic operators can be implemented using a straightforward *Arithmetic and Logic Unit* (ALU), which is a standard item of digital electronic hardware. The linear local operators can be implemented, nowadays, using specialised integrated circuits. One manufacturer sells a circuit board which can implement an 8 × 8 linear local operator in real time on a standard video signal. Several companies market a broad range of image processing modules that can be plugged together, to form a very fast image processing system that can be tailored to the needs of a given application. Specialised architectures have been devised for image processing. Among the most successful are parallel processors, which may process one row of an image at a time (vector processor), or the whole image (array processor). Competing with these are *systolic array*, *neural networks* and field programmable gate arrays. See Dougherty and Laplante [24] for a discussion on the considerations that need to be examined in the development of real-time imaging systems.

2.8.1 Morphological System Implementation

While no single image processing operation is so important that all others can be ignored, it is interesting to consider the implementation of the morphological operators, since it reflects the range of hardware and software techniques that can be applied to achieve high speed.

There are two classical approaches to the implementation of morphological techniques on computer architectures, parallel and sequential (serial) methods (Section 2.4). Morphological operations with 3 × 3 pixel structuring elements, are easily implemented by array architectures, such as CLIP [9]. Other system implementations include Sternberg [10]. Waltz [11,12] describes examples of a near real-time implementation of binary morphological processing using large (up to 50 × 50 pixels), arbitrary structuring elements, based on commercially available image processing boards. The success of this approach, referred to as SKIPSM (*Seperated-Kernal Image Processing using Finite State Machines*), was achieved by reformulating the algorithm in such a way that it permitted high-speed hardware implementation. Similar algorithmic methods allow fast implementation of these operators in software.

2.9 Commercial Devices

In this section, we discuss generic types of computing sub-systems for machine vision, rather than giving details of existing commercial products, since any review of current technology would become out of date quite quickly. The discussion will concentrate on the computing aspects of machine vision systems, rather than the remaining systems issues, such a lighting and optics. Also, there are numerous trade magazines that deal with commercial machine vision products.

For the purposes of this book, we have classified commercial systems into three main categories:

- plug-in board-based systems (frame-stores, dedicated function);
- self-contained vision systems;
- turnkey systems.

2.9.1 Plug-in Boards: Frame-grabbers

The imaging engine in many low-cost machine vision systems consists of a host computer working in conjunction with single or multiple plug-in boards. The most common example of these systems consists of a personal computer, or workstation, and a frame-grabber card, which allows an image to be captured from a standard CCD camera (array image format) and displayed. Many of the current, extensive range of frame-store cards also offer on-board processing. Plug-in accelerator cards which enable certain functions to be implemented in real time are available as daughter boards for many frame-stores. Some frame-stores have slow-scan capabilities and the ability to interface to line-scan cameras. When used in conjunction with the current range of high-speed personal computers, such a vision system is an attractive option for small to medium applications of low complexity. Even personal computers now offer direct video input via USB or IEEE 1394 ("firewire") ports, without the need for an additional plug-in frame-grabber cards.

Such systems offer a number of significant advantages, most important of which is their relatively low cost. Another significant advantage is their ease of use and familiarity. This is especially the case when used in conjunction with standard personal computers, which have become commonplace both in the home and the workplace. The fact that the host computer for the imaging system is a widely available commercial product also widens the base for software applications and maximises the use of the frame-grabber. Many of the software packages available today use 'point and click' interaction with the user, making it easy for him to investigate image processing ideas. (Unfortunately, the majority of these packages are for image processing, rather than image analysis.) The majority of the plug-in frame-store boards can be programmed using commonly used high-level languages, such as C, C++ or Java! This is important, since the use of standard programming languages can have a major impact on program development costs.

A common disadvantage with frame-grabber cards is that they rely on the power of the host computer to do all of the required imaging tasks. Since the host computer is generally not tuned for imaging applications, the system operation may be too slow, despite the constantly increasing the performance of commercial computers. So, for many high-speed industrial applications, such systems are not suitable. Many machine vision integrators would not consider personal computer systems as robust enough for industrial applications. The use of industrial PCs in conjunction with a wide range of dedicated processing and interface cards counters this argument to a certain extent. Despite these disadvantages, the use of frame-grabber plug-in cards offers a low-cost introduction to machine vision, and is suitable for educating, training, system design and other less-demanding applications.

2.9.2 Plug-in Boards: Dedicated Function

For greater speed and ability, engineers often turn to plug-in boards that have a specific functionality, such as real-time edge detection, binary correlation, and convolution. Typically the host computer for such boards would be a VME rack fitted with a CPU card. Quite often, such special-purpose boards are pipelined. That is, they perform different operations on the image, in a sequential manner that allows a new image to be captured while the previous image is still undergoing processing. The main advantage of such systems is their speed and the ability to increase the systems image throughput rate by the addition of extra plug-in boards. The disadvantage of such systems is that they can be difficult to program and quite often require programmers with highly specialist skills. There is also a significant cost factor involved in the capital equipment, along with the application development costs.

While the majority of dedicated plug-in boards for pipelined systems are tuned to deal with array CCD cameras, newer systems have appeared on the market that are specifically designed for a line-scan camera.

2.9.3 Self-contained Systems

Some system manufactures have taken the option of designing specific machine vision engines that are not tuned for a specific application, but rather designed for their general functionality. Such systems may be totally self-contained and ready to install in an industrial environment. That is, they contain the imaging optics, camera, imaging engine and interfaces for various mechanical actuators and sensors. They differ from turnkey systems in that the software is supplied with the self-contained system has yet to be moulded into a form that would solve the vision application. Such systems have significant advantages, the main one being speed. The majority of self-contained systems are custom designed, although they may contain some plug-in boards and are tuned to provide whatever functionality is required by the application. The self-contained nature of the mechanical and image acquisition and display interfaces is also a significant benefit when installing vision systems. However, it can be difficult to add further functionality at a later date without upgrading the system.

2.9.4 Turnkey Systems

Turnkey vision systems are self-contained machine vision systems, designed for a specific industrial use. While some such systems are custom designed, many turnkey systems contain commercially available plug-in cards. Turnkey systems tend to be designed for a specific market niche, such as can-end inspection, high-speed print recognition and colour print registration. So, not only is the hardware tuned to deal with high-speed image analysis applications, it is also optimised for a specific imaging task. While the other systems discussed usually require significant development to produce a final solution for an imaging application, turnkey systems are fully developed, although they need to be integrated into the industrial environment. This should not be taken lightly, as this can often be a difficult task. It may not be possible to find a turnkey system for a specific application.

While we have avoided the discussion of any specific commercial devices, there are a number of valuable information sources available, some of these are provided by commercial organisations but some of the most valuable are free! One resource that is well worth considering is the "Machine Vision Resources" website operated by P.F. Whelan [25]. This is a machine vision database that gives details of a large number of machine vision vendors and their products and services.

2.9.5 Software

As was mentioned earlier, there is a large number of image processing, and analysis, packages available, for a wide range of computing platforms. Several of these packages are freely available over the Internet. Some of these packages are tightly tied to a given vision system, while others are compiled for a number of host computers and operating systems. The majority of the software packages have interactive imaging tools that allow ideas to be tested prior to coding for the efficient operation. For more information on the hardware and software aspects of real-time imaging, including a survey of commonly used languages, see [24].

2.10 Further Remarks

The low-level image processing operators described in this chapter are always used in combination with one another, since none can, on its own, provide the kind of quantitative information needed to solve practical applications. Very often, high-level operators can be expressed simply, in terms of sequences of the basic procedures described above. (The high-level procedure may be implemented in this way, or by reformulating it and coding it directly, to improve computational efficiency.) However, it is sometimes necessary to combine the basic operators with sophisticated decision-making, search and control techniques, based on Artificial Intelligence and Pattern Recognition. These are the subjects of the next chapter, where we emphasise that product variation almost invariably requires that we employ more intelligent algorithms.

2.11 References

[1] Borgefors, G. (1986) Distance transformations in digital images, Computer Vision, Graphics and Image Processing, vol. 34, pp. 344–371.

[2] Gonzalez, R.C. and Wintz, P. (1987) *Digital Image Processing*, Addison Wesley, Reading, MA.
[3] Dougherty, E.R (1992) *An Introduction to Morphological Image Processing*, Tutorial Textvol. TT9, SPIE Press.
[4] Haralick, R.M. (1987) Image analysis using mathematical morphology, *IEEE Trans. Pattern Analysis and Machine Intelligence*, vol. 9, no. 4, pp. 532–550.
[5] Haralick, R.M. and Shapiro L.G. (1992) *Computer and Robot Vision: Volumes I and II,* Addison Wesley, Reading MA.
[6] Vogt, R.C. (1989) *Automatic Generation of Morphological Set Recognition Algorithms*, Springer-Verlag.
[7] Serra, J. (1982) *Image Analysis and Mathematical Morphology Vol. 1*, Academic Press, New York.
[8] Serra, J. (1988) *Image Analysis and Mathematical Morphology Vol. 2, Theoretical Advances*, Academic Press, New York.
[9] Duff, M.J.B, Watson, D.M., Fountain, T.M., and Shaw, G.K. (1973) A cellular logic array for image processing, *Pattern Recognition,* volume/issue/pp,.
[10] Sternberg, S.R. (1978) Parallel architectures for image processing, Proc. IEEE Conf. Int.Computer Software and Applications, Chicago, pp. 712–717.
[11] Waltz, F.M., Hack, R., and Batchelor, B.G. (1998) Fast, efficient algorithms for 3×3 rankedfilters using finite-state machines, Proc. SPIE Conf. on Machine Vision Systems for Inspection and Metrology VII, Vol. 3521, Paper No. 31, Boston.
[12] Waltz, F.M. and Garnaoui, H.H. (1994) Application of SKIPSM to binary morphology, Proc. SPIE Conf. on Machine Vision Applications, Architectures, and Systems Integration III, Vol. 2347, Paper No. 37, Boston.
[13] Zhuang, X. and Haralick, R.M. (1986) Morphological structuring element decomposition, *Computer Vision, Graphics and Image Processing*, vol. 35, pp. 370–382.
[14] Vincent, L. (1991) *Morphological transformations of binary imnages with arbitrary structuring elements*, *Signal Processing*, vol. 22, pp. 3–23.
[15] Heijmans, H.J.A.M. (1991) Theoretical aspects of grey-scale morphology, *IEEE Trans. Pattern Analysis and Machine Intelligence*, vol. 13, no. 6, pp. 568–582.
[16] Sternberg, S.R. (1986) Grey-scale morphology, *Computer Vision, Graphics and Imag eProcessing*, vol. 35, pp. 333–355.
[17] Haralick, R.M. (1979) Statistical and structural approaches to texture, *Proc. IEEE*, vol. 67, no. 5, p. 768–804.
[18] Batchelor, B.G. (1991) *Intelligent Image Processing in Prolog*, Springer-Verlag, Berlin, ISBN 3-540-19647-1.
[19] Pitas, I (1993) *Digital Image Processing Algorithms*, Prentice-Hall, Englewood Cliffs NJ.
[20] Davies, E.R (1990) *Machine Vision: Theory, Algorithms, Practicalities*, Academic Press, London.
[21] Sonka, M., Hlavac, V., and Boyle, R. (1993) *Image Processing, Analysis and Machine Vision,* Chapman and Hall.

[22] Gonzalez and Wintz, P. (1987) *Digital Image Processing*, Addison Wesley, Reading MA.
[23] Haralick, R.M. (1979) Statistical and structural approaches to texture, *Proc. IEEE*, vol. 67,no. 5, pp. 768–804.
[24] Dougherty, E.R. (1992) *An Introduction to Morphological Image Processing*, Tutorial Text vol. TT9, SPIE Press.
[25] www.eng.dcu.ie/~whelanp/resources/resources.html

Chapter 3

Editorial Introduction

Paradoxically, some of the so-called "high-level" operators for image understanding are actually concerned with apparently simple concepts. Their apparent simplicity can be deceiving. Consider, for example, the relationship "is to the left of". This is difficult to express in a computing language such as Java, C or Basic, although it is easy to define it in Prolog. Other abstract relationships of this type are "inside", "next to", "between", etc., and can also be represented conveniently in this language. When we are discussing natural products, abstract relationships between symbolic objects become important, because they allow us to escape from the confinements imposed by always working with precise numeric (e.g. geometric position) data. People do this all the time in everyday conversation. For example, we may note that a person has a mole on the left cheek. We do not necessarily want to spend a long time discussing what a mole is, what it looks like, or exactly where the limits of that person's left cheek are. Suppose that somebody says "The mole on Mary's left cheek is getting bigger." There are far more serious issues to consider than being pre-occupied with irrelevancies that would only obscure the sinister nature of this statement. There are no precise rules for identifying features like moles, eyes on potatoes, bird-pecks on apples, etc., so we may need to devise a vision system that can learn to do so. Anatomical features of an animal or plant are also ill-defined, and hence are often difficult to recognise. In situations like these, where precise terms cannot always be applied, we must resort to using techniques borrowed from Fuzzy Logic, Artificial Intelligence or Pattern Recognition

Chapter 3

Intelligent Image Processing

B.G. Batchelor

If it's green, we reject it.
If it's too ripe, we reject it.
If it's bruised, we reject it.
If it's diseased, we reject it.
If it's dirty, we reject it.
If it's just right, we squash it.

Advertisement, McDonald's Restaurants, 1992

3.1 Why We Need Intelligence

An introduction to the basic concepts and techniques of image processing is provided in the previous chapter. That alone would be adequate for a book describing engineering applications of Machine Vision but the inspection of natural products requires the use of techniques that are more flexible and intelligent. In this chapter, we shall therefore concentrate on additional methods needed to enhance the basic picture manipulation and measurement functions just described. We shall discuss computational techniques, borrowed from Pattern Recognition and Artificial Intelligence, that are appropriate for inspecting highly variable products. These two subjects, together with Computer Vision, all have an important contribution to make to Machine Vision when it is applied to natural product inspection. However, we must emphasise that, while we are happy to use "borrowed" techniques, we do not accept the ethos of any of these science-based subjects; our roots are very firmly established in engineering. In other words, pragmatism and systems-related issues outweigh theoretical ones that have little or no practical significance.

3.2 Pattern Recognition

Nowadays, *Pattern Recognition* is often mistakenly equated solely with the study of *Neural Networks*. In fact, this provides a limited view, since Pattern Recognition sometimes employs other types of decision-making and analysis procedures. For example, the Pattern Recognition methods that we describe for colour recognition could not legitimately be termed Neural Networks. The simple (traditional) model of a Pattern Recognition system is explained in Figure 3.1 and will be refined later.

3.2.1 Similarity and Distance

In Pattern Recognition, it is common practice to describe a pattern (e.g., an image, or feature within an image, an acoustic signal, a medical patient, the state of an engine, or machine tool) in terms of a fixed number of measurements, taken in parallel (Figure 3.1). This might be appropriate as a way of representing texture, colour, object shape, or some ill-defined feature in an image. Of course, we cannot always conveniently and efficiently describe image features in this way. For this reason, this approach is relatively uncommon in most existing industrial Machine Vision systems. However, this data format is widely used in Pattern Recognition systems for applications as widely varied as insurance-risk assessment, investment planning, differential diagnosis in medicine, electro-encephalography, analysing seismic events, speech recognition, fault detection in machine tools, biological classification, etc. It can also be used to recognise varieties of seeds, colours of fruit, shapes of leaves, etc.

Let us denote a set of N parallel measurements representing a pattern P by the N-dimensional vector

$$\mathbf{X}_P = (X_{P,1}, X_{P,2}, X_{P,3}, \ldots, X_{P,N}) \tag{3.1}$$

(Since this a vector, rather than a set, the order in which the elements $X_{P,1}$, $X_{P,2}$,... are written is critical.) It is reasonable to assume that, if two patterns, say P and Q are subjectively similar, then their corresponding vectors $\mathbf{X}_P$ and $\mathbf{X}_Q$ will also be similar. Of course, this presupposes that we have managed to find parameters that adequately describe the salient features of the pattern P. The *dissimilarity* between the vectors $\mathbf{X}_P$ and $\mathbf{X}_Q$ can be measured by one of the following distance metrics:

Euclidean distance

$$D_e(\mathbf{X}_P, \mathbf{X}_Q) = \sqrt{[\sum(X_{P,i} - X_{Q,i})^2]} \tag{3.2}$$

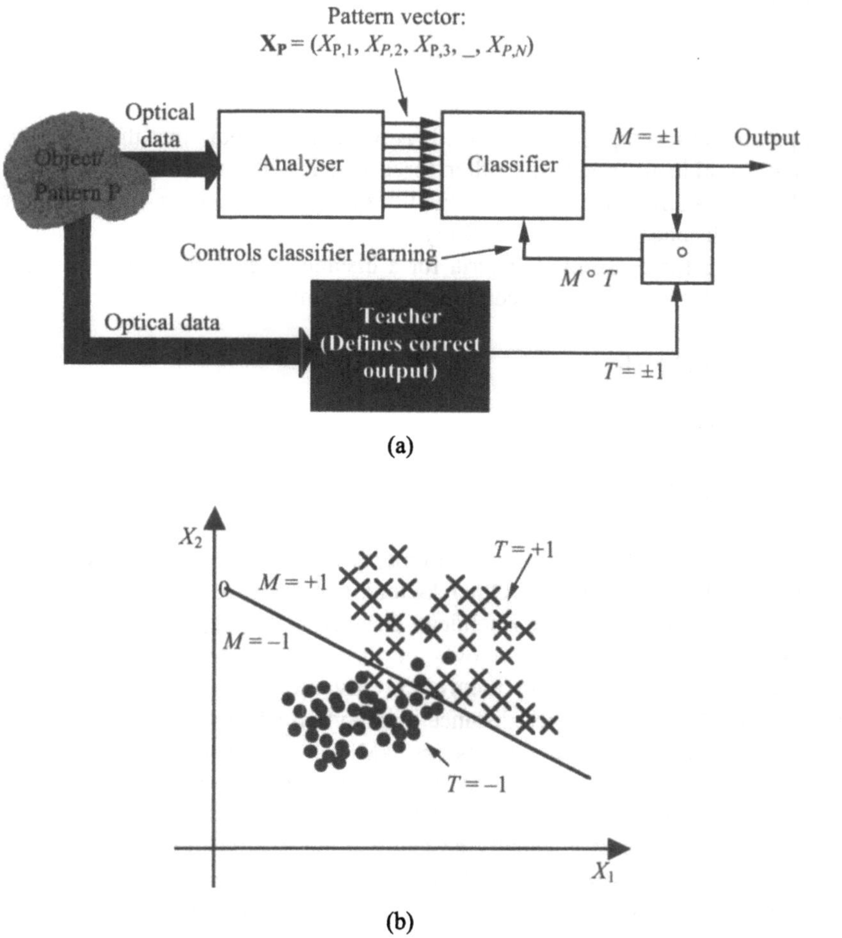

Figure 3.1. Traditional model of a Pattern Recognition system: (a) organisation of a system employing error-correction learning: (b) measurement space. In this simple case $N = 2$ and the machine decision (M) is computed by deciding whether a point falls above or below a straight line.

Square distance

$$D_s(\mathbf{X}_P, \mathbf{X}_Q) = \boldsymbol{MAX}(|(X_{P,i} - X_{Q,i})| \tag{3.3}$$

Manhatten (or City Block) distance

$$D_m(\mathbf{X}_P, \mathbf{X}_Q) = \sum |(X_{P,i} - X_{Q,i})| \tag{3.4}$$

Minkowski r-distance

$$D_r(\mathbf{X}_P, \mathbf{X}_Q) = [\sum(X_{P,i} - X_{Q,i})^2]^{1/r} \tag{3.5}$$

The parameter r is a positive integer. It is of interest to note that the Minkowski r-distance is able to model the other three metrics; it is identically equal to the Manhatten distance when $r = 1$; the Euclidean distance when $r = 2$; and approaches the Square distance as r tends to infinity.

There are three essential criteria for a distance metric that apply to all of the above formulae and any linear combination of them:

$$D(\mathbf{A}, \mathbf{A}) = 0 \tag{3.6}$$

$$D(\mathbf{X}_P, \mathbf{X}_Q) \geq 0 \tag{3.7}$$

$$D(\mathbf{A}, \mathbf{C}) \leq D(\mathbf{A}, \mathbf{B}) + D(\mathbf{B}, \mathbf{C}) \tag{3.8}$$

Which of these four distance measurements is best? In many cases, it does not matter which one we choose. Other factors, such as ease of implementation then become more important. The Euclidean distance is the most familiar to us, as it is the one that we can derive in two- or three-dimensional space, using a tape measure. It also has certain theoretical advantages over the others, in some situations [1]. The City Block distance is important when we drive or walk through New York City, where the streets are laid out on a grid. This and the Square distance were originally introduced into Pattern Recognition research in the 1960s, principally to avoid the difficulties that then existed in calculating the Euclidean distance. Nowadays, it is usually best to use the Euclidean distance to avoid the theoretical dangers associated with the City Block and Square distances.[1] Notice that, in many situations, it is not necessary to perform the square-root operation implicit in the Euclidean distance, $D_e(X,Y)$.

3.2.2 Compactness Hypothesis

The so-called *Compactness Hypothesis* relates similarity between patterns P and Q to the inverse of the distance between their corresponding vectors, $\mathbf{X}_P$ and $\mathbf{X}_Q$ [1]. Thus, in order to identify (classify) an unknown pattern **X**, we might reasonably compare it to a set of stored vectors, $S = \{\mathbf{X}^1, \mathbf{X}^2, \mathbf{X}^3, \ldots\}$, representing patterns $\{\mathbf{P}^1, \mathbf{P}^2, \mathbf{P}^3, \ldots\}$, which belong to classes $\{\mathbf{C}^1, \mathbf{C}^2, \mathbf{C}^3, \ldots\}$. A vector **X** obtained from a pattern that we wish to classify is then identified with the pattern class associated with the closest neighbour in *S*. That is, we find i such that $D(\mathbf{X},\mathbf{X}^i)$ is a minimum and then associate **X** with class $\mathbf{C}^i$. Such a classifier is called a *Maximum Similarity* or, more commonly, a *Nearest Neighbour Classifier* (Figure 3.2) [1].

[1] When the City Block, Square and Minkowski ($r \neq 2$) distances are used within a Nearest Neighbour Classifier, the decision surface is unstable.

(This is not a Neural Network in the strict sense but could be implemented by one.) The so-called *Compound Classifier* is closely related to the Nearest Neighbour Classifier and makes decisions by measuring distances from a (small) set of stored points. A wide variety of other methods for performing pattern classification, including *Neural Networks*, has been devised (Figure 3.3) [2]. They differ in the way that they combine the elements of the pattern vector. Conceptually, the Nearest Neighbour and Compound Classifiers are among the simplest. They are just as versatile as Neural Networks. Indeed, the Compound Classifier is more natural for certain types of application (Figures 3.4 and 3.5) [1].

3.2.3 Pattern Recognition Models

In this section, we shall see that the traditional model for Pattern Recognition is not always appropriate for the types of applications that interest us. Figure 3.5 illustrates various situations in which the original paradigm, expressed in Figure 3.1, is unsatisfactory.

Traditional Model

A classifier is simply a mathematical formula, for calculating a decision (M) given a vector of parallel measurements: $\mathbf{X} = (X_1, X_2, \ldots, X_N)$. In most applications, we do not know what function to use to compute M and we must derive an estimate by self-adaptive learning, based on feedback (Figure 3.1). Traditionally, the parameters of a classifier are adjusted if the decision (M) that it has just computed in response to an input $\mathbf{X} = (X_1, X_2, \ldots, X_N)$ differs from that of an abstract entity, called the *Teacher*. The Teacher is the ultimate authority that defines the correct classification (T) about $\mathbf{X}$ and might typically be

- a farmer showing the classifier a set of "ripe" and "unripe" items of fruit for harvesting;
- a human inspector watching a production line;
- a committee of people examining a batch of randomly chosen products from a manufacturing process;
- an analyst performing a "post-mortem" on a set of products that have been returned by customers;
- a post-production test machine.

The sole function of the Teacher is to act as the expert authority, helping us design a classifier by adjusting its internal parameters. Eventually, the Teacher must be removed, since it (or he or she) is too expensive, too slow, or does not have the capacity to be used all of the time. For example, an expert human inspector cannot work reliably for more than a few hours of each day. Nearly all work on Neural Networks and other self-adaptive classifiers has been based on this traditional model of Pattern Recognition, in which the Teacher labels samples of at least two pattern classes (Figure 3.5a).

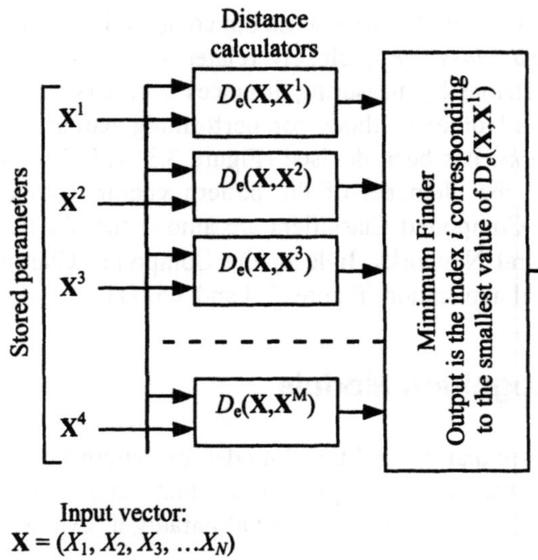

(a)

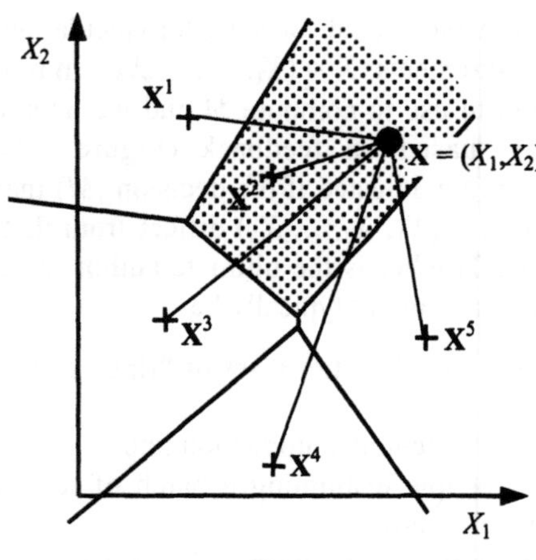

(b)

Figure 3.2. Nearest Neighbour Classifier: (a) implementing the algorithm; (b) the pattern represented by the vector $\mathbf{X} = (X_1,X_2)$ is identified with that class associated with the closest stored point: $\mathbf{X}^1$, $\mathbf{X}^2$, $\mathbf{X}^3$, ... (i.e., $\mathbf{C}^2$). All points in the shaded area are closer to $\mathbf{X}^2$ than they are to any of the other stored points: $\mathbf{X}^1$, $\mathbf{X}^3$, $\mathbf{X}^4$ or $\mathbf{X}^5$.

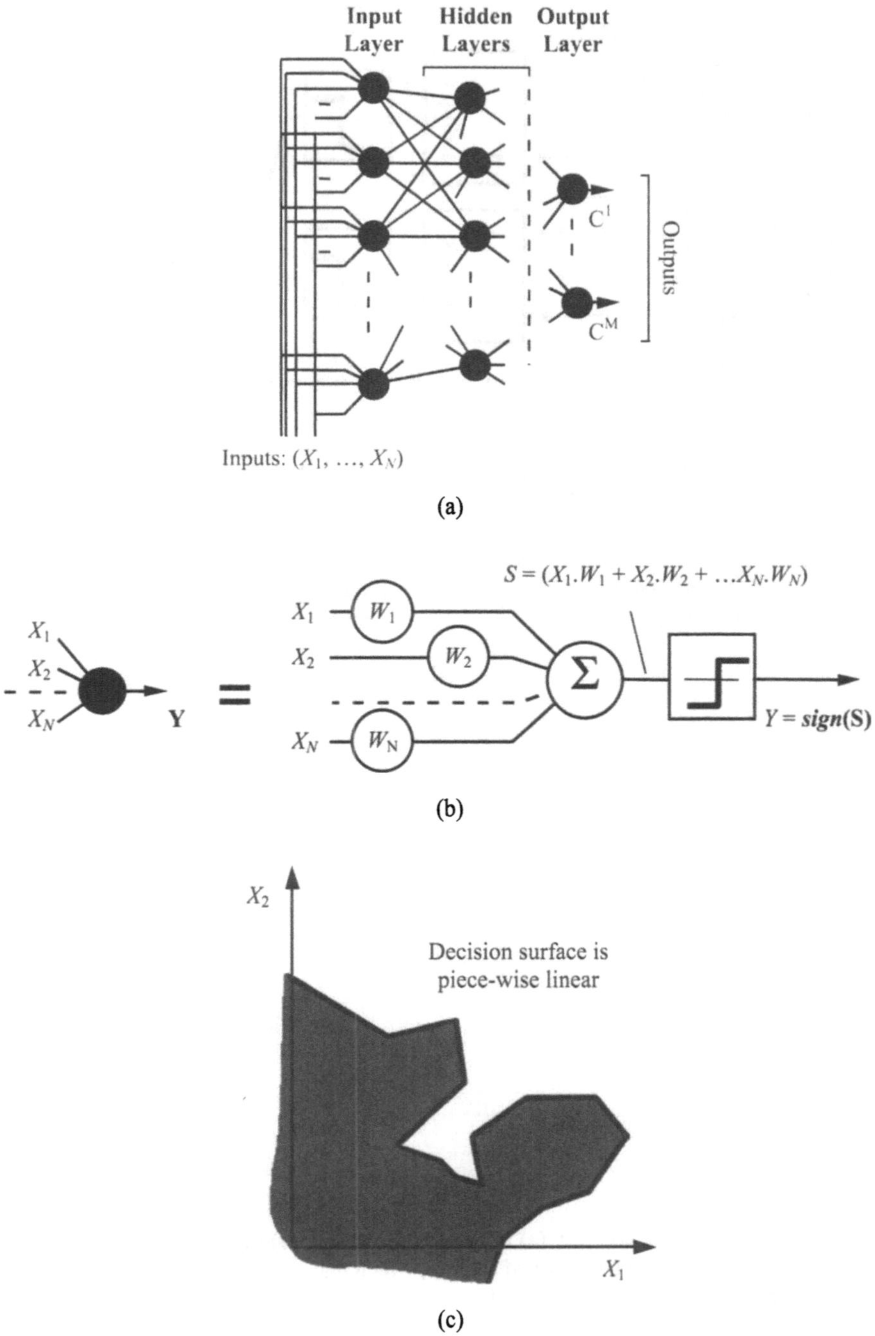

Figure 3.3. Neural Network: (a) impementation of the algorithm; (b) function of a single isolated "neuron"; (c) decision surface consists of a number of straight lines (2d), planes (3d) or hyperplanes (many dimensions). Notice that both the Nearest Neighbour Classifier and Neural Networks produce piece-wise linear decision surfaces.

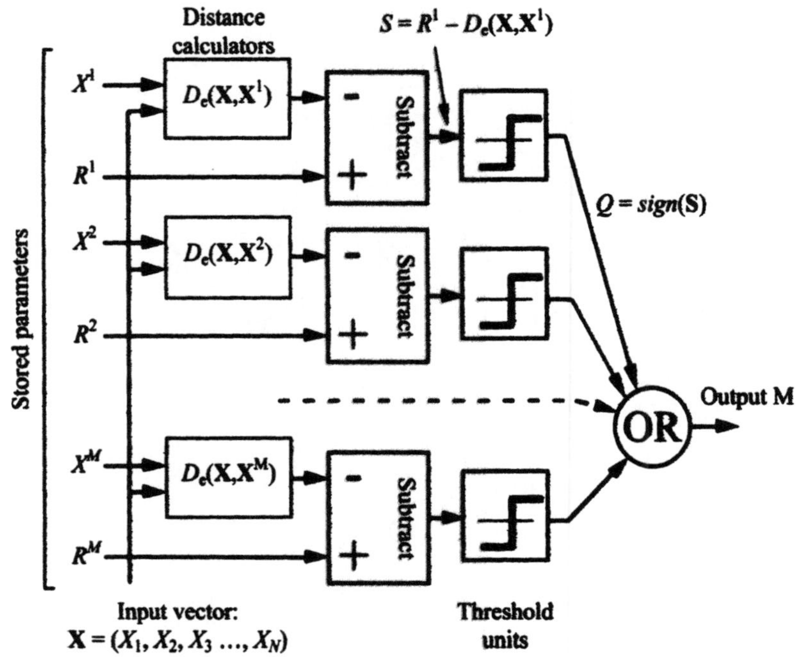

(a)

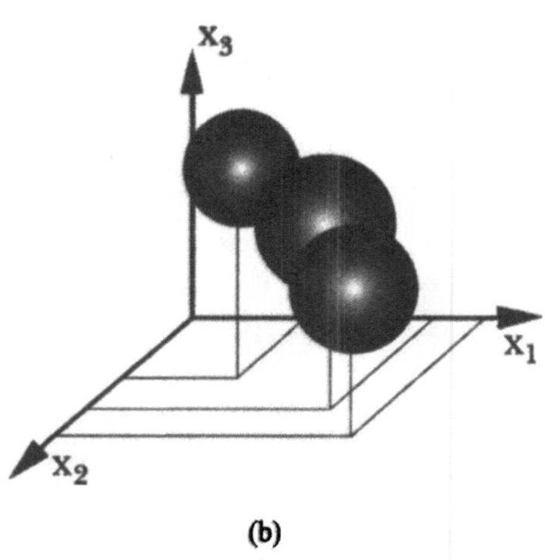

(b)

Figure 3.4. Compound Classifier: (a) internal organisation; (b) decision surface. In 3d, the decision surface consists of a number of spheres, which may or may not overlap. In 2d, they are circles and are hyperspheres in multi-dimensional space.

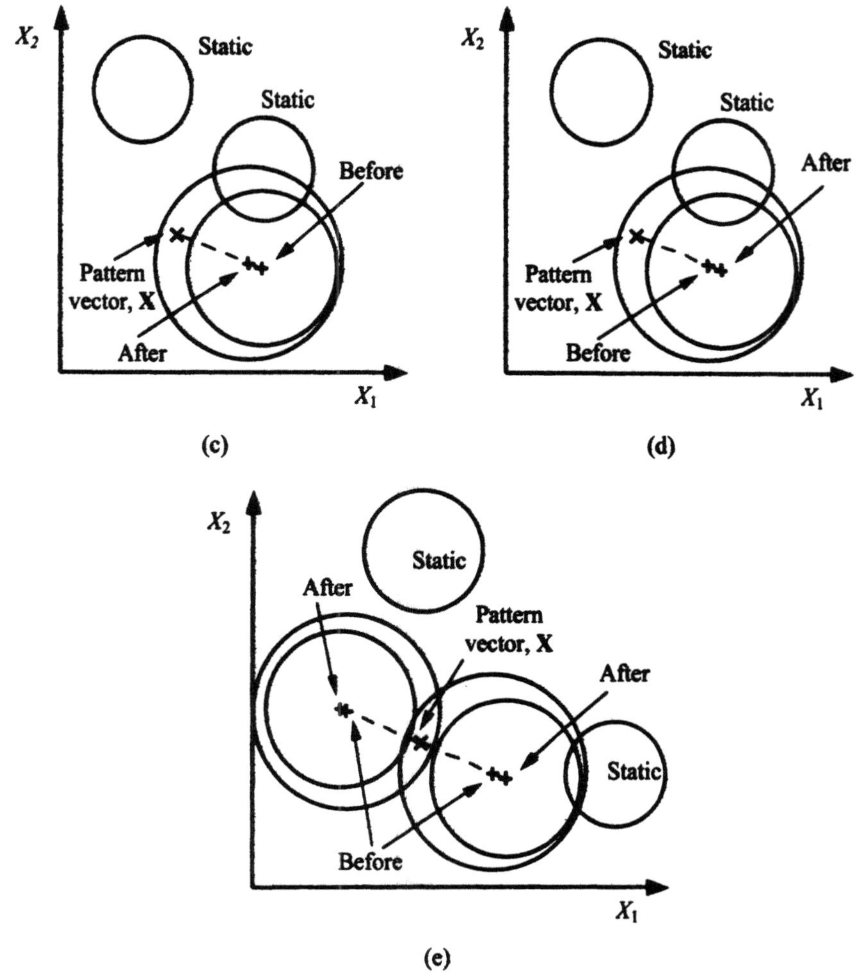

Figure 3.4 (continued) (c) learning rule for the situation in which the pattern vector **X** is outside all circles: the nearest circle is enlarged and is moved towards **X** (all other circles are unchanged). Here, and in the two other situations illustrated, the changes made to the circles are very small indeed; they have been greatly enlarged here for clarity; (d) learning rule for the situation in which the pattern vector **X** is inside just one circle: that circle is reduced in size and is moved away from **X** (all other circles are unchanged); (e) learning rule for the situation in which the pattern vector **X** is inside more than one circle: all circles enclosing **X** are reduced in size and are moved away from **X** (all other circles are unchanged).

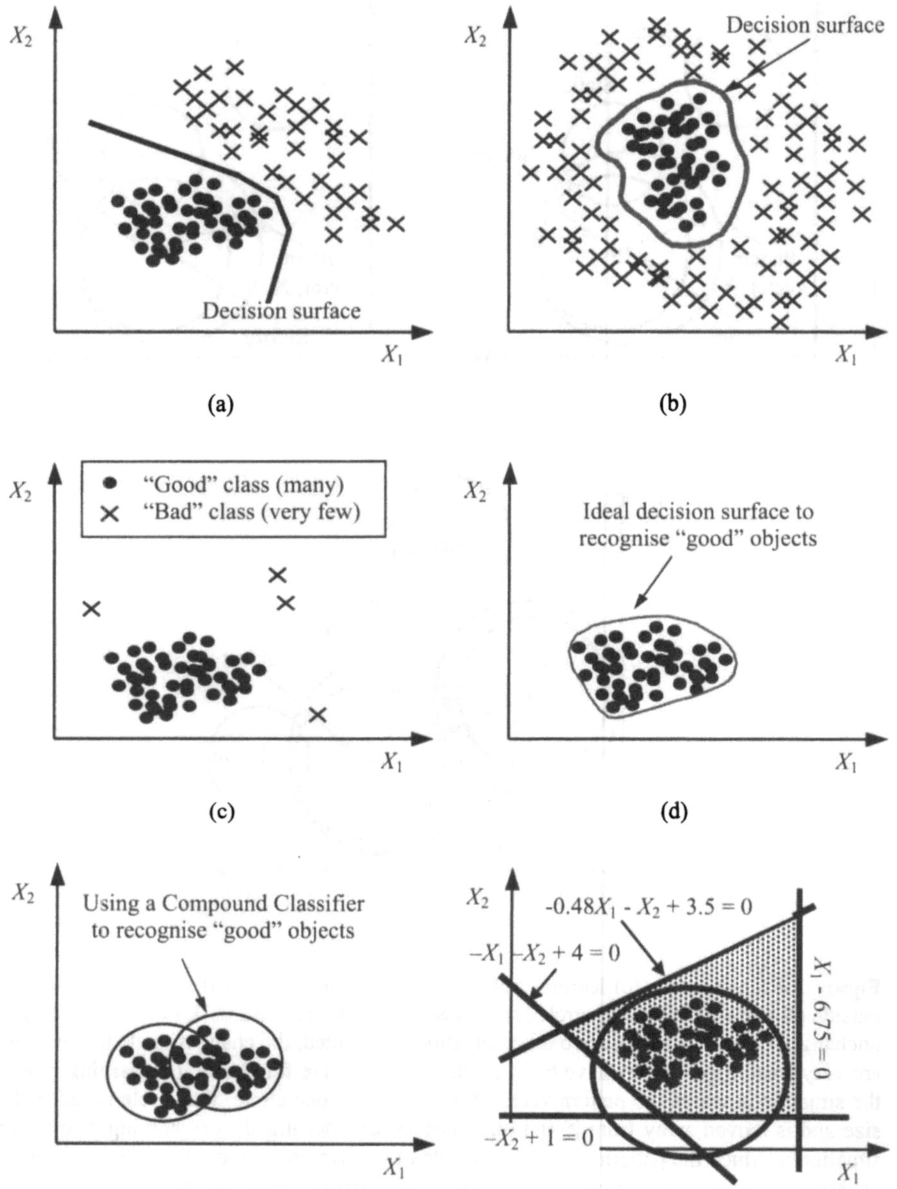

Figure 3.5. Pattern recognition models: (a) traditional model; (b) one pattern class ("good") lies in a compact region, while the "faulty" class surrounds it. (Typically occurs when homeostatic control breaks down. The Compound Classifier is better than Neural Networks in this situation.); (c) there are many samples of the "good", class and very few "faulty" ones; (d) a closed boundary is used to detect anomalies; (e) in practice, a Compound Classifier might be used, although the learning rules differ from those explained in Figure 3.4; (f) combining Pattern Recognition, Compound Classifier, in this case, and rule-based decision methods.

Learning on Single Class

To illustrate the shortcomings of the traditional model, we shall describe two situations in which we should like to design a classifier but cannot obtain a representative set of samples from more than one class.

First, consider an automated bakery. The manufacturing system *never* produces large quantities of "bad" loaves. (We shall ignore the problem of nomenclature in which bakers and production engineers never admit to making "bad" objects but will confess that they occasionally make "unsaleable" items. This and other euphemisms for "bad", "defective", or "faulty" are common-place and should not distract us.) A baking line is an expensive facility and any significant downtime leads to a large financial loss. As a result, the baker will immediately adjust the system, so that it quickly returns to making "good" loaves again; very few "bad" loaves are ever made. The reader might well question whether a sufficient number of "bad" products could be made as part of a research project. For reasons of cost this is most unlikely and is probably impossible in practice anyway. To understand why, let us suppose that there are N individual control parameters on a baking line. Then, a minimum of 2^N settings must be made to the manufacturing system, which may take several hours to settle after some controls are adjusted. (We need to set each control high/low individually.) Some parameters cannot even be altered at all. For example, the amount of moisture in a 10-tonne batch of flour cannot be altered at the whim of an engineer. The effect is that we can obtain very large quantities of the "good" class of products but almost none of the "bad" class, certainly not enough to match the huge appetite of learning systems for training data (Figure 3.5c).

Second, consider the task of inspecting apples. We cannot order a farmer to grow either "good" or "bad" fruit. Even if they could do so, farmers would not be willing to grow diseased or malformed fruit deliberately. Most of the harvest produce will be "good". Once again, we cannot obtain a truly representative set of "bad" apples. The situation is akin to that in medicine; there is a very large number of ways that we can be ill and we are unwilling/unable to be ill to order!

In both of these cases, we can reasonably expect to receive a large set of "good" produce to train a classifier but very few samples of "bad" ones. Pattern Recognition techniques have been developed for this type of situation. They fall into several categories: Unsupervised Learning, Probability Density Function Estimation and Learning on a Single Class (Figure 3.5e) [2].

Hybrid Model

Human beings have the ability to combine self-adaptive learning with rule-based reasoning. To demonstrate this, consider the problem of meeting a stranger, at a pre-arranged point. Prior to the first meeting, the person we are about to meet is described by a mutual acquaintance; recognition then relies on decision rules. At the first meeting, we learn a lot about our new friend's appearance and mannerisms, so that on subsequent occasions, we recognise him/her, by recalling this learned information. By then we may even have forgotten the initial recognition rules, given by our mutual friend. In any case, those rules have been augmented by learning.

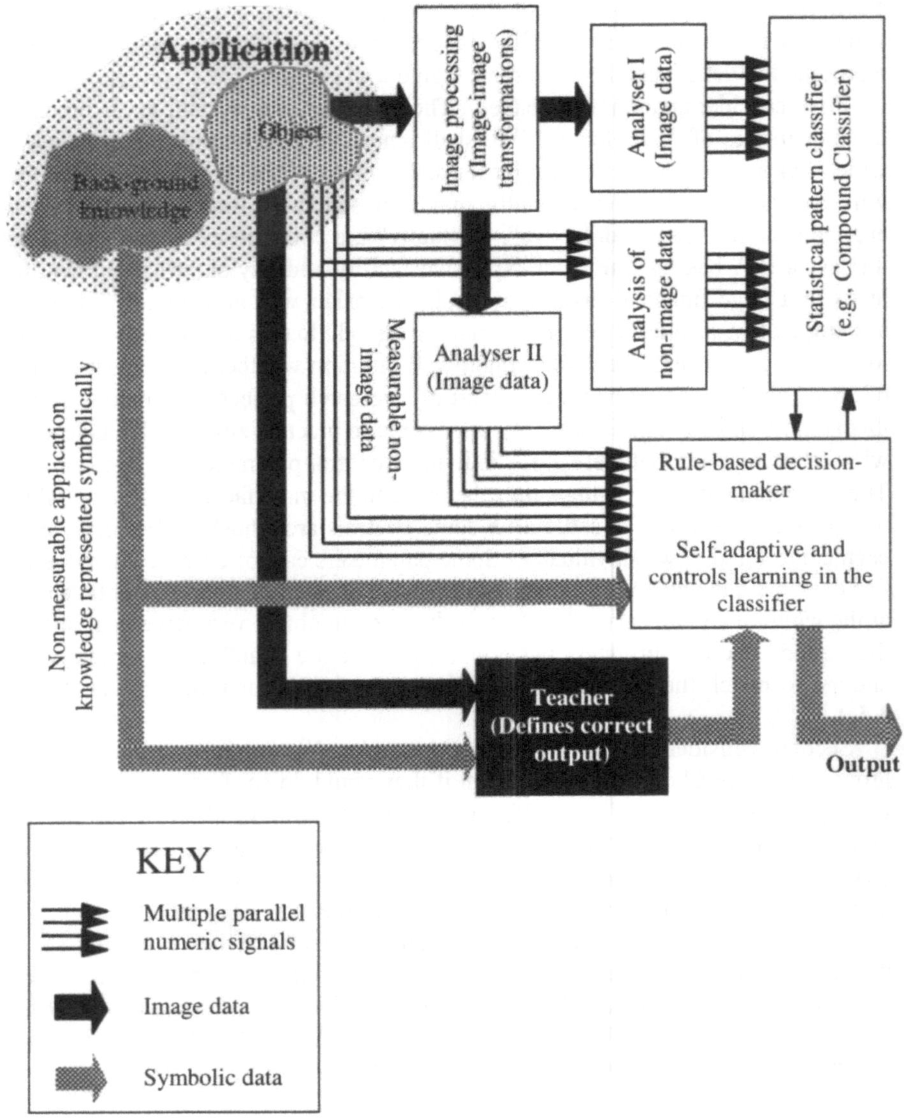

Figure 3.6. Combining Pattern Recognition and rule-based decision making. Notice the 2-way data link between the decision-maker and the classifier. The classifier can provide inputs to the rule-based system and the latter can control the training of the pattern classifier.

To understand how this can help us, let us consider the bakery and fruit-inspection applications again. We might reasonably expect to receive a large set of "good" loaves or apples *and* a set of rules that human inspectors *think* that they employ to grade them. Introspection is often far from accurate as a means of deriving decision rules but it may well provide the best criteria we can obtain in practice. The existence of a large training set (i.e., a collection of physical objects, such as loaves or apples) suggests the use of a self-adaptive learning system, while the rules just mentioned prompt us to contemplate using an Expert System (Figure 3.5f). With this in mind and for other reasons, a hybrid system is proposed for tasks that present both types of data. Figure 3.6 shows how multi-variate classification and rule-based reasoning might be combined. This has several interesting features: it allows a rule-based system to control a self-adaptive learning process. The former must possess meta-rules specifically for this purpose it allows an Expert System to base its decisions upon sub-decisions derived from the classifier. The Expert System can incorporate non-measurable application knowledge, specified in symbolic terms, by a human being, with both image and non-image data, derived mechanically from product samples. The hybrid system outlined in Figure 3.6 offers a possible solution to some of the problems caused by the special nature of the data that we have available. For far too long, researchers have relied on the traditional Pattern Recognition model (Figures 3.1 and 3.5a and b) without seriously questioning whether it is appropriate. In conceptual terms, the hybrid system has much to commend it. However, this is a speculative suggestion for future work and programming such a system may prove to be problematical.

3.3 Rule-based Systems

The hybrid system just described involving both Pattern Recognition and Rule-based decision-making sub-systems has an obvious appeal but it is simply an idea, put forward to help us analyse images from complex highly variable scenes. It remains untested. However, rule-based systems are already being used, for example, in examining solder joints [3]. In this section, we shall examine their use in a little more detail.

3.3.1 How Rules are Used

Many inspection applications involving natural products can be specified in terms of simple rules. Grading and inspecting vegetables and fruit are prime examples of this kind. Similar comments apply to many areas of food processing. For example, loaves may be examined by applying rules that define what is an acceptable shape. These may be expressed in terms of either two-dimensional slices, or three-dimensional height maps. Ready-made deserts, such as trifles, are specified by a series of rules, defining such parameters as the minimum/maximum amount of fruit. Similarly, limits are placed on the amount of custard and number of bubbles visible on the top surface. Pizzas are specified in a similar way, with limits being expressed for the amount and the distribution of each type of topping. Food products with an ill-defined form, such as croissants, cakes, meringues, and biscuits (cookies). Dollops of cream, icing (frosting) and even adhesive (applied to engineering components) can all be specified using a set of rules. In many applications, it is not possible to use self-adaptive learning, so a set of simple inspection rules specified by an experienced production engineer provides the best approach.

Many existing Machine Vision systems in engineering manufacturing effectively employ naïve rule-based decision criteria. For example, a porcelain plate may be judged to be "acceptable" if

Rule 1 *It contains a single crack-like feature not more than 1 mm long.*
Rule 2 *The total length of all crack/spot- like features is less than 1.5 mm.*
Rule 3 *There are no more than 20 spot-like defects (i.e., length of each is less than 0.1 mm).*
Rule 4 *There no more than 5 crack-like defects (i.e., length of each is more than 0.1 mm).*

As is evident here, individual rules are often quite trivial and may, for example, simply define the maximum length, or width of an object or feature. Sometimes, rules are contrived simply to avoid silly or dangerous situations that might arise as a result of an accident, or human negligence. For example, we might use a simple set of rules to detect a foreign object, such as a screwdriver, or a pair of spectacles, that has inadvertently been left on a food processing line. Simple rules, based only on length, colour and shape factor (ratio of the area to the square of perimeter), can readily distinguish objects that bear some superficial resemblance to one another and can therefore perform a useful "sanity check". For example, the following rule for recognising bananas, rejects cucumbers, courgettes, gherkins, carrots, lemons, oranges, tomatoes, parsnips, etc.

An object is a banana if
i. *it is yellow* and
ii. *it is curved* and
iii. *it is between 75 and 400mm long* and
iv. *it is between 15 and 60 mm wide.*

A computer program, which implements this rule, is listed in the next section. As evidence of the power of simple rule-based decision criteria, look around the room where you are sitting and try to find something, other than a banana, that conforms to this description. Rule-based criteria are used to define classes of fruit for the purposes of the General Agreement on Tariffs and Trade[2] (GATT) and to grade both fruit and vegetables. At the moment, human inspectors are more widely used for these inspection functions than vision systems. Rule-base criteria, based on size and shape, can also distinguish the seeds of certain varieties of cereals. This function is important, as there is a need for a machine that is able to check the honesty of suppliers of rice, wheat, other types of grain and even bananas!

3.3.2 Combining Rules and Image Processing

The idea of integrating image processing operators with the Artificial Intelligence language Prolog [4] was conceived in the mid-1980s and has since been

---

[2] A few years ago a popular UK national newspaper staged a campaign ridiculing the notion of classifying bananas, in an attempt to discredit certain political views. The relevance to the GATT was not mentioned!

implemented in a number of ways (Figure 3.7). Early systems used external hardware to implement the image processing functions, with Prolog running on a standard desktop computer [5]. The most powerful system of this type developed to date is called *PIP (Prolog Image Processing)* and combines image processing operators written in C with LPA MacProlog [6]. (Both Prolog and the image processing functions run on the same computer.) The latest system, called CIP (Cyber Image Processing) [7], is still under development and incorporates CKI Prolog, which was written by S. van Otterloo of the University of Utrecht, Netherlands [8]. This is a "freeware" implementation of the language and is written in Java. The image processing operators were written separately, also in Java, by students at Cardiff University, Wales, UK [9]. Hence, CIP employs only one computer language, whereas PIP uses two and is therefore more difficult to maintain. PIP currently has a command repertoire consisting of about 275 primitive image processing commands and CIP about 150.

Readers, who are familiar with Prolog, merely need to note that it is possible to implement a system like PIP, or CIP, by building on a single predicate *(ip/1* in what follows), within the standard language. Readers who have not encountered Prolog before are referred elsewhere for a more detailed description of this fascinating language [4,10].

Sample programs are given below, in order to demonstrate how Prolog can incorporate image processing operations.

Program 1: Move the Centroid to the Middle of the Picture.

In this example, we use the predicate *ip/1* to send individual commands to the image processor, from Prolog.

```
shift_to_image_centre :-
ip(enc),              % Enhance contrast.
ip(thr),              % Threshold at mid-grey
ip(cgr(X,Y)),         % Locate the centroid, [X,Y]
ip(imm(X0,Y0)),       % Locate the centre of the image, [X0,Y0]
X1 is X0 – X,         % Prolog-style arithmetic
Y1 is Y0 – Y,         % More Prolog-style arithmetic
ip(psh(X1,Y1)).       % Shift so that centroid is at middle of
                      image
```

Program 2: Revised Version, Using the # Operator

This time, we use the operator #, instead of *ip/1*. This is really just a cosmetic change but it can make both programming and subsequent reading of the program easier.

```
:- op(900,#,fx).      % # is prefix operator with precedence 900
# A :- ip(A).         % Defines what # does.
% Revised version of shift_to_image_centre/0
shift_to_image_centre :-
# enc,                % Enhance contrast
# thr,                % Threshold at mid grey
```

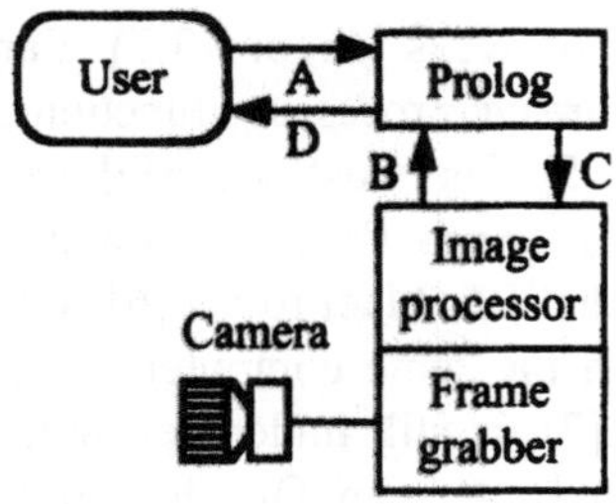

(a)

A	B	C	D
neg	neg	[0].	YES
thr (115, 134)	thr (115, 134)	[0].	YES
avg(A)	avg	[0,123].	A = 123 YES
cgr(X,Y)	cgr	[0,78,156].	X = 78 Y = 156 YES

(b)

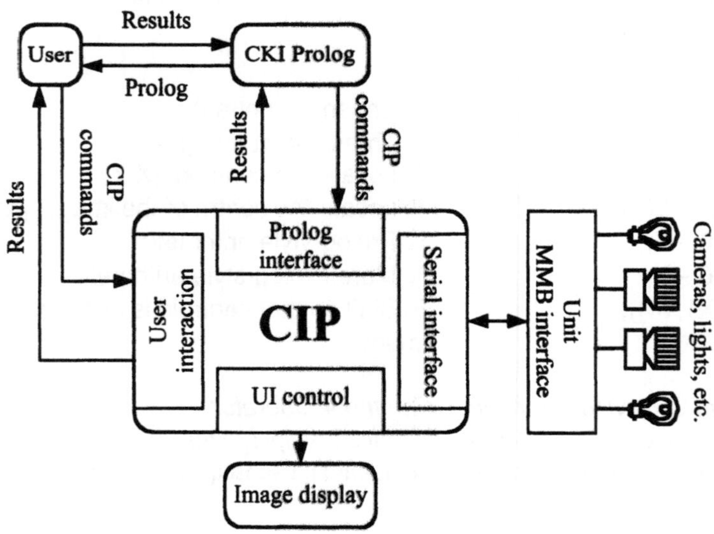

(c)

Figure 3.7. Using Prolog to control an image processor; (a) system block diagram; (b) signals transmitted through the system when various commands are issued by the user (these are PIP commands mnemonics); (c) CIP system architecture. MMB is a multi-function, low-speed interface unit. CKI Prolog was written by S. van Otterloo, University of Utrecht.

```
# cgr(X,Y),              % Locate the centroid, [X,Y]
# imm(X0,Y0),            % Locate the centre of the image, [X0,Y0]
X1 is X0 – X,            % Prolog-style arithmetic
Y1 is Y0 – Y,            % More Prolog-style arithmetic
# psh(X1,Y1).            % Shift so centroid is at middle of image
```

Program3: Crack Detector

In this example, commonly-used functions are defined in a library, to avoid using either *ip/1* or the # operator.

% Standard library definitions – available to all programs

% The library contains clauses like these, for each image procesing operator

```
lnb :- # lnb.
snb :- # snb.
wri(a) :- # wri(A).
rea(A) :- # rea(A).

% Operator: N•G satisfies a goal (G) N times - used in lieu of FOR loops:
:- op(1000, xfx, •).          % • is infix operator with precedence 1000

0•G :- !.                     % Terminates recursion. Stops when N = 0.

N•G :-
call(G),                      % Satisfy goal G (once)
M is N-1                      % Arithmetic
M•G, !.                       % Now, satisfy the goal (N-1) more times
% End of the library
% Crack detector (Morphological Closing operator)
crk(N) :-
wri(a) and                    % Save image in archive area #1
N•lnb and                     % Brightest neighbour; grey-scale dilation
N•snb and                     % Darkest neighbour; grey-scale erosion
rea(a) and                    % Recover image from archive area #1
sub.                          % Subtract images
```

This and the two preceding programs are simple linear sequences of image processing operations; there is no back-tracking or recursion

Program 4:Finding a Banana

The following program uses back-tracking to search for an object called "banana" that satisfies the rules given earlier. It fails, if no such object is found.

```
object(banana) :-
object(X),                    % Find object in given image. Call it X.
colour(X,yellow),             % Colour of X is yellow.
curved(X),                    % Object X is curved.
```

```
length(X,L),                    % Measure the length of X.
L ≥ 75, L ≤ 400,                % Check the length limits
width(X,W),                     % Measure the width of X
W ≥ 15, W ≤ 60.                 % Check the width limits
```

Colour recognition, implicit in the sub-goal *colour (X,yellow)* will be discussed in Section 3.4.3. For completeness, we list the following program, which determines whether or not an object in a binary image is curved.

```
curved(X) :-
isolate(X),                     % Isolate the object called X
blb,                            % Fill in the lakes, if there are any
((cvd,                          % Convex deficiency. Call this Image Z
big(2),                         % Find the second largest bay (B2)
cwp(A2));(A2 is 0)),            % Measure its area (A2)
swi,                            % Switch images - revert to Image Z
big(1),                         % Find the largest bay (B1)
cwp(A1),                        % Measure its area (A1)
((cvd,                          % Find the convex deficiency of B1
big(1),                         % Isolate its largest meta-bay (B3)
cwp(A3));(A3 is 0)),            % Measure its area (A3)
A1 > 5*A2,                      % Area of B1 ≥ 5 times area of B2
A1 ≥ 10*A3.                     % Area of B1 ≥ 10 time area of B3.
```

This is a heuristic procedure and is explained in Figure 3.8. Notice that, although this definition of "curved" may seem a little sloppy, it works most of the time – that is what good heuristics do! It will, for example, succeed when it is applied to the silhouette of a banana.

Program 5: Checking a Table Place Setting

This is a more sophisticated program. It examines a binary image, in order to verify that it conforms to an English-like description of a formal table place setting (Figure 3.9). It uses the relationships *left/2*, *right/2*, *above/2* and *below/2* to compare the positions of objects in two-dimensional space.

```
describe_table_scene :-
loa(original_image),                       % Load the original image
retractall(cutlery_item_found(_,_,_,_)),   % Forget all stored items
eab(identify_cutlery_item(_)),             % Identify each item in turn
loa(original_image),                       % Reload the original image
acceptable_place_setting.                  % Check layout is satisfactory
```

```
% Find an object of type Q, positioned at [X,Y] with orientation Z.
identify_cutlery_item([Q, [X,Y,Z]]) :-

wri,                            % Save image to temporary file on disc
cvd,                            % Convex deficiency
kgr(100),                       % Discard blobs with areas < 100 pixels
```

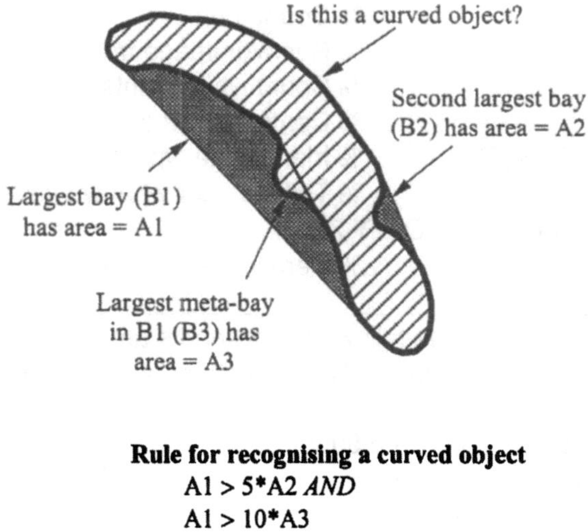

Figure 3.8. Heuristic procedure implemented by the predicate *curved/1* listed in the text.

Figure 3.9. Ideal table place setting. Such a scene can be described, in English, using a set of statements of the following form: *"There is a knife to the right of the mat."*; *"A dinner fork can be found on the left of the mat."*, etc. These can be used to program a vision system, which then checks that a real place setting is well laid.

```
ske,                          % Skeletonise the blob
cnw, min, thr(1,1),           % Identify the skeleton limb ends
cbl(B),                       % Count them
rea,                          % Recover image from temporary file
cwp(C),                       % Count white points (i.e., find area)
lmi(X,Y,Z),                   % Centroid, and orientation of principal axis
cutlery_item(Q,[C,A,B]),      % Identify the type of object found (Q)
tell_user(Q,A,B,C),           % Tell the user what we have found
assert(cutlery_item_found(Q,X,Y,Z)).
                              % Remember what we have found
/*______________________________________________________________
How to identify objects. The following arguments are used:
cutlery_item(Object_type, [Area, No_of_limb_ends, No_of_bays])
______________________________________________________________*/
cutlery_item(mat,[A,_,_]) :-  A ≥ 5000, !.      % Object is a mat
cutlery_item(knife,[_,0,_]) :- !.               % Object is a knife.
cutlery_item(fork,[_,_,4]) :- !.                % Object is a fork.

cutlery_item(spoon,[_,2,_]) :- !.               % Object is a spoon.
cutlery_item('not known',_).                    % Not of known type

% Defining spatial relationships between two objects, A and B
left(A,B) :-
location(A,Xa,Ya),                 % Centroid of object A is at [Xa,Ya]
location(B,Xb,Yb),                 % Centroid of object B is at [Xb,Yb]
Xa < Xb,                           % Check that Xa < Xb
about_same(Ya, Yb)                 % Are Ya and Yb about the same?

right(A,B) :- left(B,A).           % right/2 is the inverse of left/2

above(A,B) :-
location(A,Xa,Ya),                 % Centroid of object A is at [Xa,Ya]
location(B,Xb,Yb),                 % Centroid of object B is at [Xb,Yb]
Ya < Yb,                           % Check that Ya < Yb
about_same(Xa, Xb),
message(['I have found a ',A, ' above a ', B]).

below(A,B) :- above(B,A).

% Finding the position (X,Y) of an item of type A.
location(A,X,Y) :- cutlery_item_found(A,X,Y,_).

% An arbitrary, very naïve, way of checking that A and B are similar
about_same(A,B) :- abs(A-B) < 25.
```

```
/*_______________________________________________________________
The following rule specifies what constitutes an acceptable table place
setting. It can easily be reformulated to accept English-language
descriptions, using Definite Clause Grammars (DCGs). These form an
integral part of Prolog [10].
_______________________________________________________________*/
acceptable_place_setting :-
left(fork,mat),                 % There is a fork to the left of a mat
right(knife, mat),              % There is a knife to the right of a
                 mat.
left(knife,knife),
left(knife, spoon),
above(fork,mat),
below(fork,spoon),              % There is a fork to the left of a
                 spoon
message('The table place setting is acceptable.').

% What to do when we cannot prove that table place setting is acceptable

acceptable_place_setting :-
message('This is not an acceptable table place setting').
```

Why Prolog?

The reader may well wonder why Prolog, rather than a conventional language, such as C or Java, is used to control the image processing engine. In the short space available here, it is impossible to do full justice to this question. The reader is therefore referred elsewhere for a more complete discussion [5,11].

A hint of the expressional power of Prolog is seen in the last two examples. In Program 4, we employed heuristic definitions for *curved/2* and *colour/2* (yet to be described in detail) while in Program 5, we defined the "sloppy" spatial relationships: *left/2*, *right/2*, *above/2* and *below/2*. It is difficult to envisage how this could be done in a conventional computer language. Prolog's expressional power arises because it is a *Declarative Language* and is therefore able to manipulate abstract symbols (words) without explaining what they mean. For example, the relationship *left(A,B)* may refer to objects A and B in physical space, or in "political space".[3] Since we do not need to specify the nature of A and B, the definition given above for *left/2* will suffice for both. When programming in Prolog, we simply describe (a little bit of) the world. We then ask questions about it and rely on Prolog's built-in search engine to hunt for a solution. Conventional languages are said to be *imperative*, since programs consist of a series of commands. Implementing a search algorithm in a language such as C is tedious;

[3] Consider the following: Hitler ? Stalin If we are considering physical objects (words), *left(Hitler,Stalin)* is true, while it is clearly false, if political objects are compared. Our Prolog program can accommodate either.

especially when we do not know in advance precisely what kind of object we shall be looking for.

We conclude this section by considering a task, in which declarative programming excels. Suppose that a certain man (A) wishes to find a wife (B) for himself. The Prolog programming paradigm requires that A first specifies his "requirements" as illustrated in Table 3.1.

Writing a Prolog program to find a wife is very straightforward, as is evident from the program segment below. (To conserve space, only the top level is specified. The lower levels are also programmed easily.)

```
marriage_partner(A,B) :-
male(A),                          % This rule applies to men. A similar
                                  % rule should be added for women.
person(B),                        % Find a person called B
female(B),                        % Check that B is a female
age(B,C), C ≥ 38, C ≤ 42     ,    % Find how old B is and check limits
height(B,D), D ≥ 1500, D ≤ 1750, % Find height and check limits
hair_colour(B,blonde),            % Check that B is blonde
religion(B,protestant_christian), % Check B is Protestant Christian
personality(B, [kind, generous, loyal, truthful]), % Check all are true
hobbies(B, [swimming, theatre, antiques]). %  Check one or more is true
```

Table 3.1 Example of prolog programming paradigm requirements.

Feature	**Value**
Sex	Female
Age	38 – 42
Height	1500 – 1750
Race	Caucasian
Hair colour	Blonde
Religion	Protestant Christian
Personality (all must be true)	[Kind, generous, loyal, truthful]
Hobbies/interests (at least one must be true)	Swimming, theatre, antiques]

Note that we are combining data relating to "natural products" (people) that is derivable from images (e.g., height, hair colour), with non-image data (e.g., religion and personality).

It should be clear by now that Prolog with embedded image processing operators is well suited to tasks such as grading and sorting fruit, vegetables and decorative plants. There is another good reason for using Prolog: it is able to manipulate natural language [10]. Combined with a speech recognition system, this provides a powerful user-interface facility, for hands-free supervision of a robot vision system [5]. Natural language programming, albeit in a simple form, is also feasible. The level of complexity that we can reasonably expect to accommodate is roughly commensurate with that needed to program a visually guided robot to lay the table, or place tools in a tool-box More significant industrial applications are found in packing natural objects, stacking irregular packages, etc.

3.4 Colour Recognition

We turn our attention now to colour recognition, which we have already mentioned and used. Our goal in this section is to explain how colours can be recognised automatically, so that a person can then employ symbolic colour labels to program a vision system. He/she may then refer either to "standard" colours (e.g., yellow, turquoise, blue, etc.), or to application-specific colours ("tuna-can red", "margarine-tub red", "butter", "banana yellow", etc.). As we proceed, we shall apply the concepts and methods of Pattern Recognition that we introduced earlier.

3.4.1 RGB Representation

A colour video camera measures the so-called RGB components of the light falling at each point in its retina. It does this by placing monochrome photosensors behind three differently coloured filters. These selectively attenuate different parts of the spectrum, to give what are called the R, G, B colour channels. These have traditionally been associated with detecting "red", "green" and "blue", although this idea must not be carried too far. A camera that outputs an HSI video signal (measuring Hue Saturation and Intensity) uses an RGB sensor and performs the RGB-HSI transformation within its internal electronics.

Let $[r(i,j),g(i,j),b(i,j)]$ denote the measured RGB values at the point (i,j) in a colour image. The set of colours in the image can be represented by S_3 where

$$S_3 = \{[r(i,j), g(i,j), b(i,j)] / 1 \leq i,j \leq \{N\} \qquad (3.9)$$

By mapping the image into this particular three-dimensional measurement space (called RGB-space, with axes r, g and b), we forsake all knowledge of the position of each colour vector. However, by extending this, we can preserve this information. The set of points

$$S_5 = \{[r(i,j), g(i,j), b(i,j), i,j] / 1 \leq i,j \leq \{N\} \qquad (3.10)$$

lies in a 5-dimensional space and preserves both position and colour information. For most purposes, S_5 is too cumbersome to be useful in practice. The representation embodied in S_3 is normally preferred, when we want to recognise colours independently of where they lie within an image.

We can combine the output of a colour recogniser with the position co-ordinates, (i,j), to obtain essentially the same information as is implicit in S_5. The following is a hybrid representation which combines symbolic colour labelling (not "raw" RGB values) with geometric position (i,j) within an image.

$$S_{\text{hybrid}} = \{[colour\_name, i, j] / 1 \leq i, j \leq \{N\}\} \tag{3.11}$$

This representation is more economical than S_5 yet contains more information than S_3. Moreover, S_{hybrid} can be stored and processed as a monochrome image, in which "intensity" (i.e., an integer) represents symbolic colour names (Figure 3.10).

It is often convenient to display such an image in pseudo-colour, since similar values of the integers defining the values of *colour_name* may represent quite different physical colours.

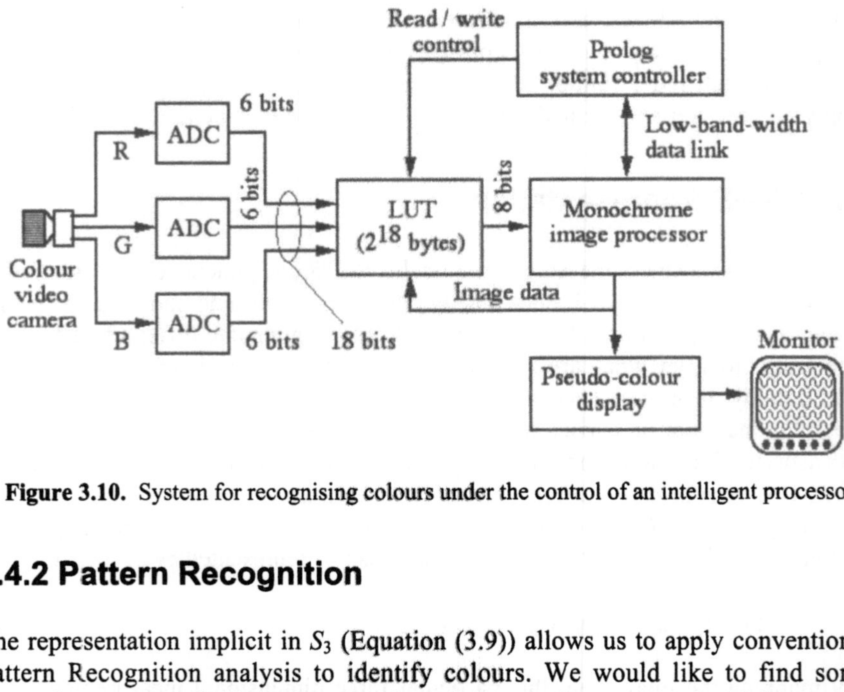

Figure 3.10. System for recognising colours under the control of an intelligent processor

3.4.2 Pattern Recognition

The representation implicit in S_3 (Equation (3.9)) allows us to apply conventional Pattern Recognition analysis to identify colours. We would like to find some convenient way to divide RGB-space into distinct regions, each of which contains all of the points associated with a given colour label, as defined by a person (p). (We might associate person p with the teacher in Figure 3.1.) We could use the Compound Classifier; in which case, the co-ordinate axes in Figure 3.4b are $[r(i,j), g(i,j), b(i,j)]$. The Nearest Neighbour Classifier, or Neural Networks, might be used instead.

3.4.3 Programmable Colour Filter

Another approach to colour recognition is explained in Figure 3.10. This derives the form of representation implicit in S_{hybrid} (Equation (3.4)). The look-up table (LUT) receives the digitised (r,g,b) inputs from the camera and generates a symbolic colour label (an integer, equivalent to *colour_name*), representing familiar names for colours. The fact that the LUT output is a number does not imply that approximately equal LUT outputs are necessarily mapped to subjectively similar colours. For example, we might arbitrarily associate names and LUT outputs as follows:

> *"turquoise"* with LUT output 136
> *"sulphur yellow"* with LUT output 137
> *"blood red"* with LUT output 138
> *"sky-blue"* with LUT output 154
> All other colours are mapped to LUT output 0, signifying *"colour not recognised"*.

Furthermore, notice that only a few of the possible output states from the LUT may actually be used. The diagram shown in Figure 3.10 describes a low-cost, high-speed colour-recognition technique that can be implemented easily in either special-purpose hardware or software. The procedure for calculating the contents of the look-up table is described in [12].

3.4.4 Colour Triangle

A graphical representation for recognising colours will now be described. This allows us to map the variability of natural products into two-dimenional space, in a convenient and meaningful way. It also allows us to classify colours in a straightforward manner. Consider Figure 3.11, which shows the RGB space as containing a cube.

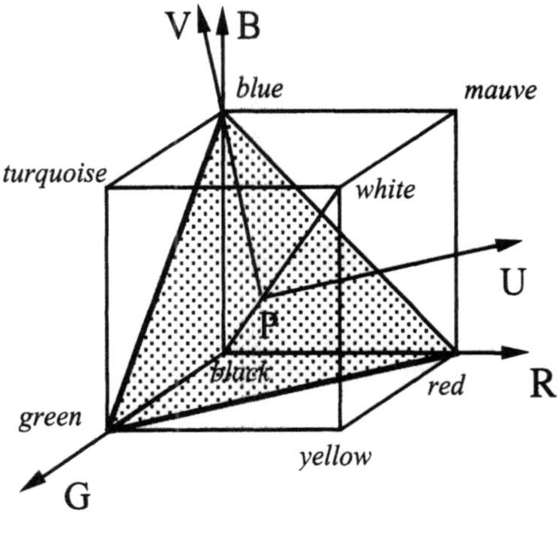

Figure 3.11. RGB space and the Colour Triangle.

The vector (r,g,b) is constrained to lie within this cube, since each channel produces an output, within a finite range: $[0,W]$. It has been found from observation that the subjective impression of colour experienced by a human being can be predicted with reasonable accuracy by noting the orientation of the vector (r,g,b) within this space. (There are many factors affecting the judgement in practice, but this is a reasonable approximation for many purposes.) Two angles are needed to define the orientation of the (r,g,b) vector. Another possibility is to note where this vector intersects that plane which goes through the points $(W,0,0)$, $(0,W,0)$ and $(0,0,W)$ in RGB space. This plane intersects the cube to form a triangle: the so-called *Colour Triangle* or *Maxwell Triangle*. Now, consider a polychrome image, such as the logo used by Apple Computer, Inc. This is a good example for our purposes, as it is composed of a number of well-separated block colours. When this is mapped into the Colour Triangle, a number of clusters are created (Figure 3.12). There are just a few points lying between these clusters and these are due to pixels that straddle the boundary between neighbouring colour bands. However, most points lie very close to the centre of one of the clusters. (Camera noise causes regions of constant colour to be mapped into clusters of finite size.) On the other hand, when a food product (e.g., a pizza, or quiche) or a natural product (e.g., apples) are mapped into the Colour Triangle, the clusters are much more diffuse and tend to merge into one another with no clear gap existing between them (Figure 3.12c – h). Blending of colours into each other is, of course, characteristic of natural products and this is manifest in the Colour Triangle by indistinct clusters.

The Compound Classifier, this time working in two-dimensional space can be used for colour recognition. This time, the inputs to the classifier (U and V) are those defining points in the Colour Triangle (Figure 3.13). This method of making decisions about the colours present in an image is unable to distinguish between bright and dark regions and, for this reason, its output is noisy in regions of low intensity. Of course, these can be eliminated by simple masking.

The heuristic techniques that have just been described do not provide a perfect solution to the task of recognising colours. If they are used intelligently and without unduly high expectations, they provide a useful analysis tool. To be most effective, they should be combined with other image descriptors, possibly in a rule-based system.

3.5 Methods and Applications

Automated Visual Inspection presents a wide variety of problems, requiring an equally diverse range of algorithmic and heuristic techniques for image processing and analysis. So far in this chapter, we have described just three of them: i. pattern classification by sub-dividing a multi-dimensional measurement space; ii. rule-based decision-making and iii. colour recognition. Of these, i and ii have never found wide application for inspecting close-tolerance engineering artefacts. It is the lack of a precise specification and the inherent variability of natural products that require their use now. Over the last 25 years, Machine Vision systems have been successfully applied in manufacturing industry and we might think that the same ideas can be applied to natural objects as well. In some cases they can, while in

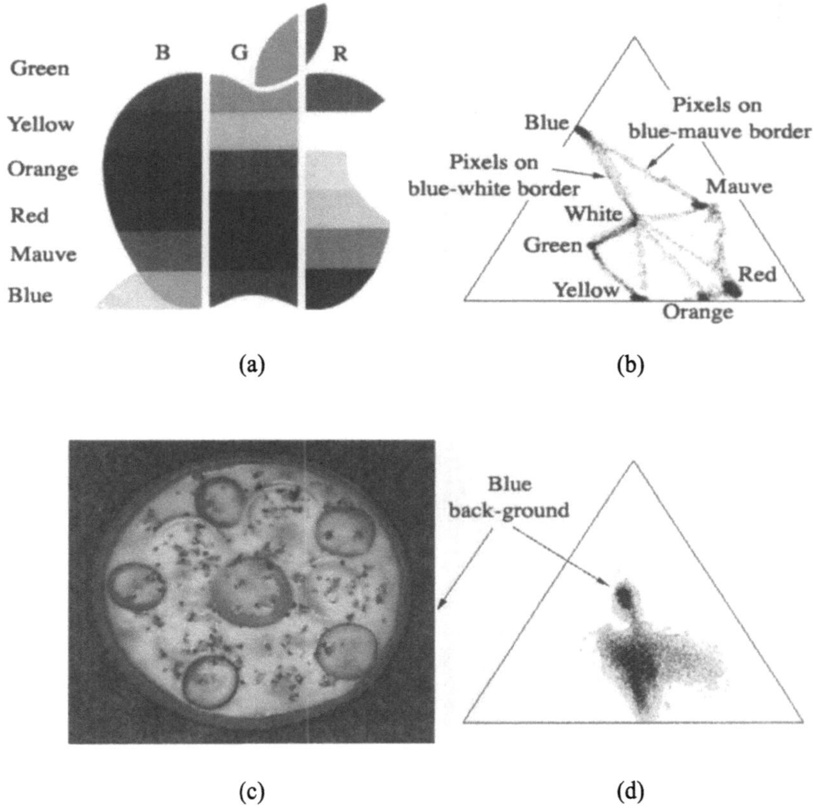

Figure 3.12. Mapping artificial and natural images into the Colour Triangle: (a) colour separations for the logo used by Apple Computer Inc.; (b) clearly identifiable clusters are generated from this logo; (c) baked quiche (this is the intensity component of the original, which is an RGB colour image); (d) mapping the colour image of the quiche into the Colour Triangle. Notice that the compact upper-most cluster is produced by the background and that no well-defined clusters are attributable to the quiche itself.

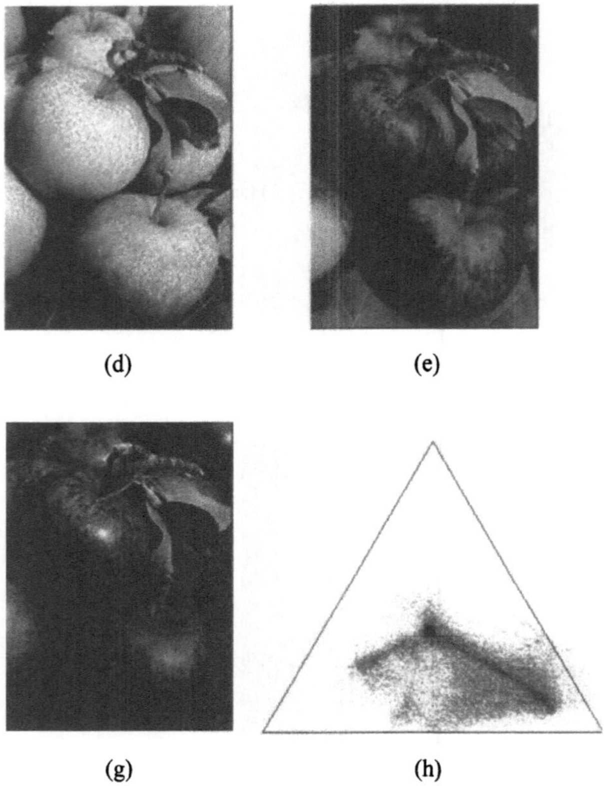

Figure 3.12 (continued) Mapping artificial and natural images into the Colour Triangle: (e) apples and leaves (R component); (f) apples and leaves (G component); (g) apples and leaves (B component); (h) Colour Triangle produced by the "Apples and leaves" image. Again, there are no well-defined clusters.

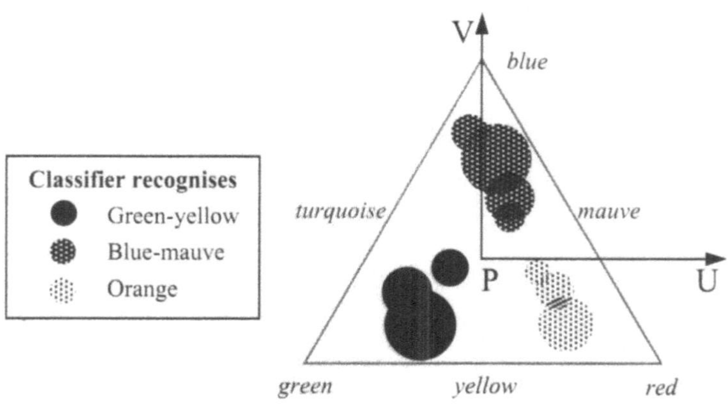

Figure 3.13. Three Compound Classifiers, working in parallel, for colour recognition in 2-dimensional space. The variables U and V are explained in Figure 3.11.

others they require some adjustment. The purpose of this section is to explain and illustrate this point, by presenting a series of brief case studies. We shall emphasise the differences that exist between inspection methods for engineering artefacts and highly variable objects, rather that the applications *per se*. The lesson is that tried and trusted procedures do not necessarily work well on highly variable objects. Some of our examples are contrived using human artefacts: a child's toy, a pair of scissors, a coil of wire. We shall use the terminology of the PIP image processing system to define image algorithms and heuristics.

3.5.1 Human Artefacts

Child's Toy

The child's toy illustrated in Figure 3.14 presents unexpected problems for a visually guided robot. Our task is to design an image processing procedure that can guide a robot, so that it can place each shape appropriately in the template. Both position and orientation have to be determined. Finding the position is easy; in each case, the centroid can be used. However, several of the established techniques for finding orientation are unable to accommodate such a wide variety of shapes. Among the popular techniques that fail to calculate a reliable value for the orientation are:

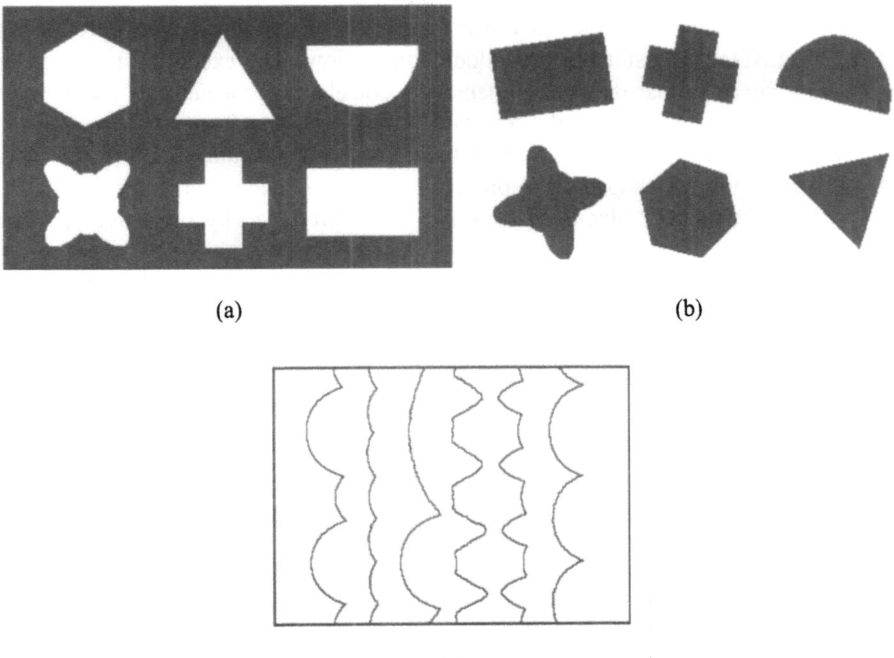

(a) (b)

(c)

Figure 3.14. Child's toy: (a) template; (b) shapes (c) plot of the distance (r(θ)) from the centroid, vesus angle (θ, vertical axis).

- Hough and Radon transforms (fails on the 4-point star).
- Principal axis (axis of minimum second moment) (fails on the cross, star, hexagon and triangle). Since mathematical moments do not measure orientation directly, some combination of measures of this kind has to be used.
- Edge-follower operator for detecting sharp corners, fails on the star.
- Convex deficiency, fails on the rectangle, semi-circle, hexagon, triangle.
- Distance of the furthest edge point from the centroid, fails on the semi-circle. (It also fails to distinguish the star, or cross, from their respective convex hulls.)

The task is made easier if the type of shape is identified first. Then, an appropriate technique for calculating orientation can be applied. Several techniques do work but with varying degrees of success:

- Fourier coefficents of the function $r(\theta)$, where r is the radius (measured from the centroid) at angle θ. (This technique will not work if the shape is more complex (e.g., crab-like) and produces a multi-value function for $r(\theta)$.)
- Fourier coefficents of the function $r(d)$, where d is the distance measured around the edge from some convenient starting point (e.g., right-most pixel). ($r(d)$ is always a single-valued function but distances around the edge cannot be measured very accurately.)
- Correlation of image arrays (three degrees of freedom) always produces a satisfactory result but the calculation is slow. The execution time can be reduced to that of a one-dimensional correlation procedure, if the position is fixed first, by using the centroid.
- Correlation of the function $r(\theta)$ (Figure 3.14c.) Again, this will not work for complex shapes. This procedure does produce satisfactory results for recognising or placing shapes like those produced by ivy leaves (Figure 3.19i)).
- Combinations of mathematical moments. It is possible to define a set of moments that are insensitive to rotation, translation and scale changes [13]. These can be used to recognise the shape first, before selecting an appropriate orientation algorithm.

Articulated Assemblies: Scissors

Of course, we expect variable objects to display significant changes of geometry but they can have a variable topology, as well. For example, a pair of scissors open by even a very small amount may have a different Euler number (a measure that is sensitive to its topology) from that of the scissors in the closed state (Figure 3.15a and b). This means that we have to be very careful when designing image analysis algorithms. For example, we might wish to build a visually guided robot to pick up the scissors, which could be closed, slightly open or wide open. It so happens that the two largest lakes in any of the silhouettes produced by the particular pair of scissors shown in Figure 3.15d correspond to the two finger holes. However, this is not necessarily the case for all types of scissors. If we decide that we can use the finger holes to pick up the scissors, it is a simple matter to calculate their centroids, as possible places to position the robot's fingers.

Trying to define a general procedure that can pick up any pair of scissors is quite difficult, as not all have closed finger holes, while some have three, or more. Closely related objects, such as shears and pliers, may have no finger holes of any sort. Here, we have two high-tolerance components, with a simple hinge connection between them. The result of this combination is an object that taxes us severely. More complex assemblies of levers are, of course, common-place in a variety of instruments, automobiles, etc. Manipulating these is even more difficult. However, the worst situation of all is found in handling objects with long pendant tubes, ropes, strings wires or cables. Plants fall in the same category. How do we cope with this huge variety? The answer is that there is no single solution that is appropriate for all situations; even our marvellous eye-brain complex finds unravelling a ball of string very difficult. In the simpler cases, a rule-based methodology is likely to suffice and, in most cases, would be our preferred solution. The trick is to sub-divide the overall task, so that each sub-problem can be tackled separately, probably using methods that are already familiar to Machine Vision practitioners. A system like CIP, or PIP, is well suited to a divide-and-conquer approach. (Each Prolog clause deals with a separate case.) However, we have to accept that some problems remain far beyond the ability of systems like this and, at the moment, have no known solution.

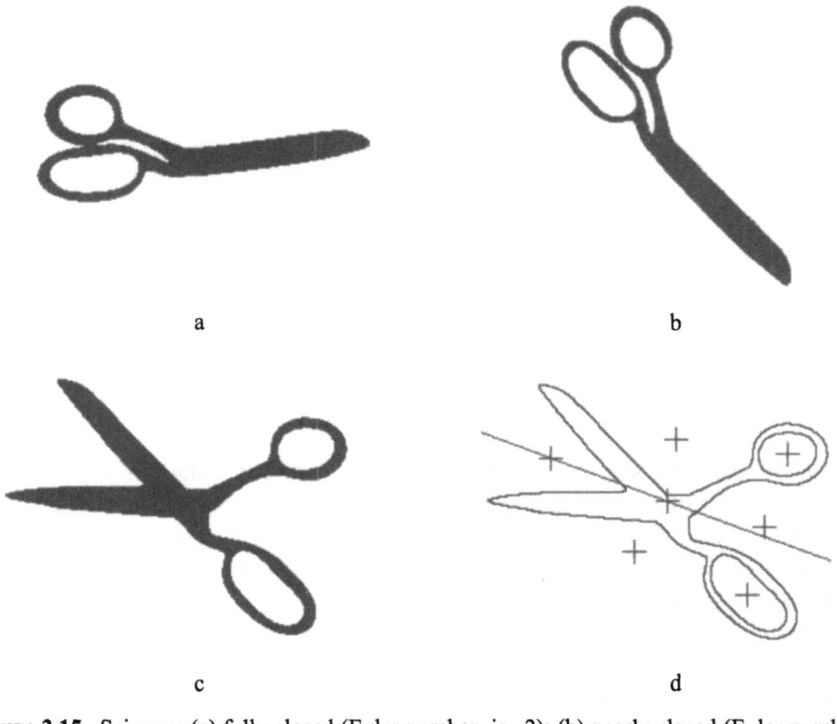

Figure 3.15. Scissors: (a) fully closed (Euler number is –2); (b) nearly closed (Euler number is –1); (c) open wide (Euler number is –2) (d) centroid and the principal axis, centroids of four bays and two lakes.

Flexible Objects

We have already pointed out the difficulties of working with flexible objects. Relatively simple situations like the coil shown in Figure 3.16 can be solved by interpreting images from several cameras. Our task is to design a vision algorithm that can locate and orientate the ends of the wires. Simple image filtering (grey-scale morphology), thesholding and skeletonisation yields a binary image that can be analysed in a straightforward way. The skeleton limb-ends are located first. Then, the ends of the skeleton limbs are eroded back a short distance, and the ends of the shortened limbs are found. This process yields two points: one at the end of each wire and another close to the end. These allow us to estimate the orientation close to the tip of the wire. However, notice that this yields only the orientation in two-dimensional space (e.g., the horizontal plane for Figure 3.16). At least one more view, this time from the side, is needed to locate and orientate the end of each wire in a vertical plane.

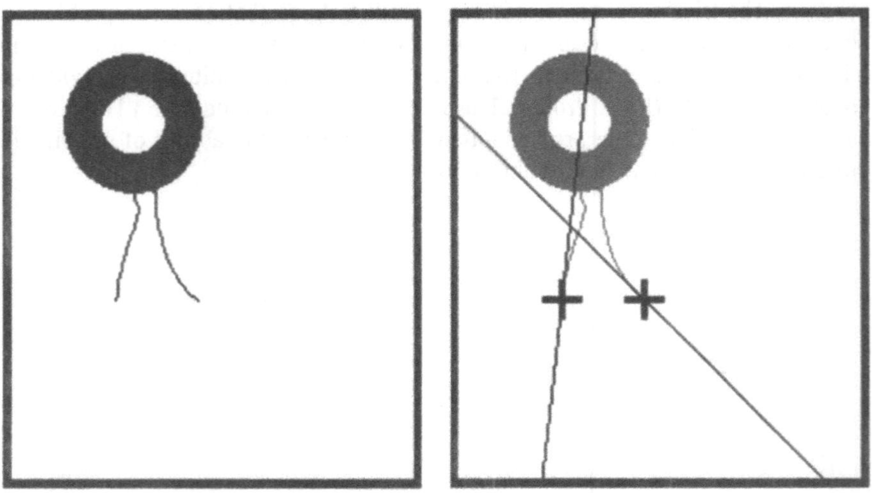

(a) (b)

Figure 3.16. Coil with two flying leads: (a) binary image (the original image was processed, by grey-scale morphological filtering and thresholding, to improve visibility of the wires); (b) ends of the wires (crosses) and tangents at the ends of the wires

More than one side-ways view of the coil is needed if there is any likelihood that the body of the coil will obscure the views of the tips of the wires. In this example, a rule-based (Prolog) system is needed to cope with all of the different situations that can occur as a result of the uncertain orientation of the coil (in the horizontal plane) and bending of the wires.

A more sophisticated approach might be needed sometimes, if for example, the wires become tangled. We can probably cope with minor entanglement, by using simple physical adjustment of the wires, prior to the process that we have just described. A robot could be used to make a slight adjustment to the flying wires, by simply inserting a finger and pulling gently. Guiding the robot as it performs this initial manipulation is another role for a vision system. An important part of this process is knowing when to give up. By now, the reader will be aware of the high level of machine "intelligence" needed to perform even simple manipulation on objects, like this coil, that have floating tubes, strings or wires.

3.5.2 Plants

There are many applications involving either whole plants, or parts of plants, that we might discuss here, for example: grading decorative house plants, harvesting fruit, trimming rhubarb, inspecting rice plants for infestation by parasitic insects, replanting seedlings, selective application of weed-killer, separating seeds from husks, "assaying" a ship-load of wheat (to ensure that it is of the correct variety), etc. We have selected just three tasks that illustrate the fact that a high level of intelligence is needed. We must not always expect established techniques, which have long been standard in industrial vision systems, to work reliably.

Leaves

To the human eye, the leaves of some plants, such as certain species of ivy, oak, and maple, appear to have very characteristic shapes. On casual observation, mature leaves of a given species often look so much alike that we not aware of the high level of variation that actually exists. This says more about our ability to perform Pattern Recognition than it does about the constancy of leaf shape. As we shall see, when we analyse leaf shape objectively, we become much more aware of the high level of variability.

Figure 3.18 shows the silhouettes of five ivy leaves, taken from the same plant. (The stems were removed manually, to make the image analysis a little easier. However, similar results could have been achieved using morphology.) In the first instance, we chose to analyse their shapes using so-called *Concavity Trees (CTs)* [14]. This method of representing shape has some attractive theoretical properties and CTs have been studied for applications such as identifying leather shoe components [5]. We might imagine that a technique that is suitable for matching shapes such as that shown in Figure 3.17 might also be appropriate for doing the same for leaves.

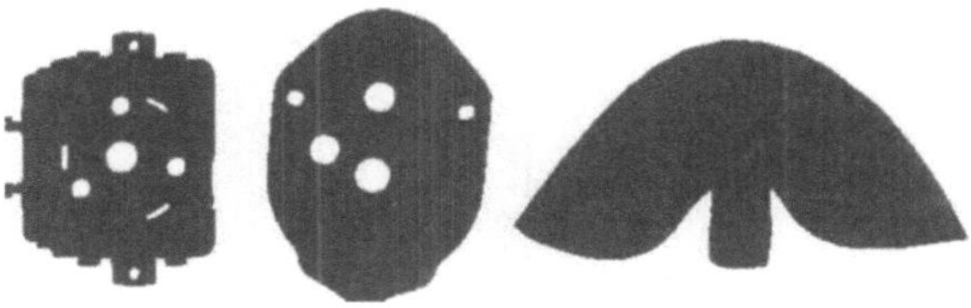

Figure 3.17. Metal castings and a shoe component are suitable for concavity trees analysis.

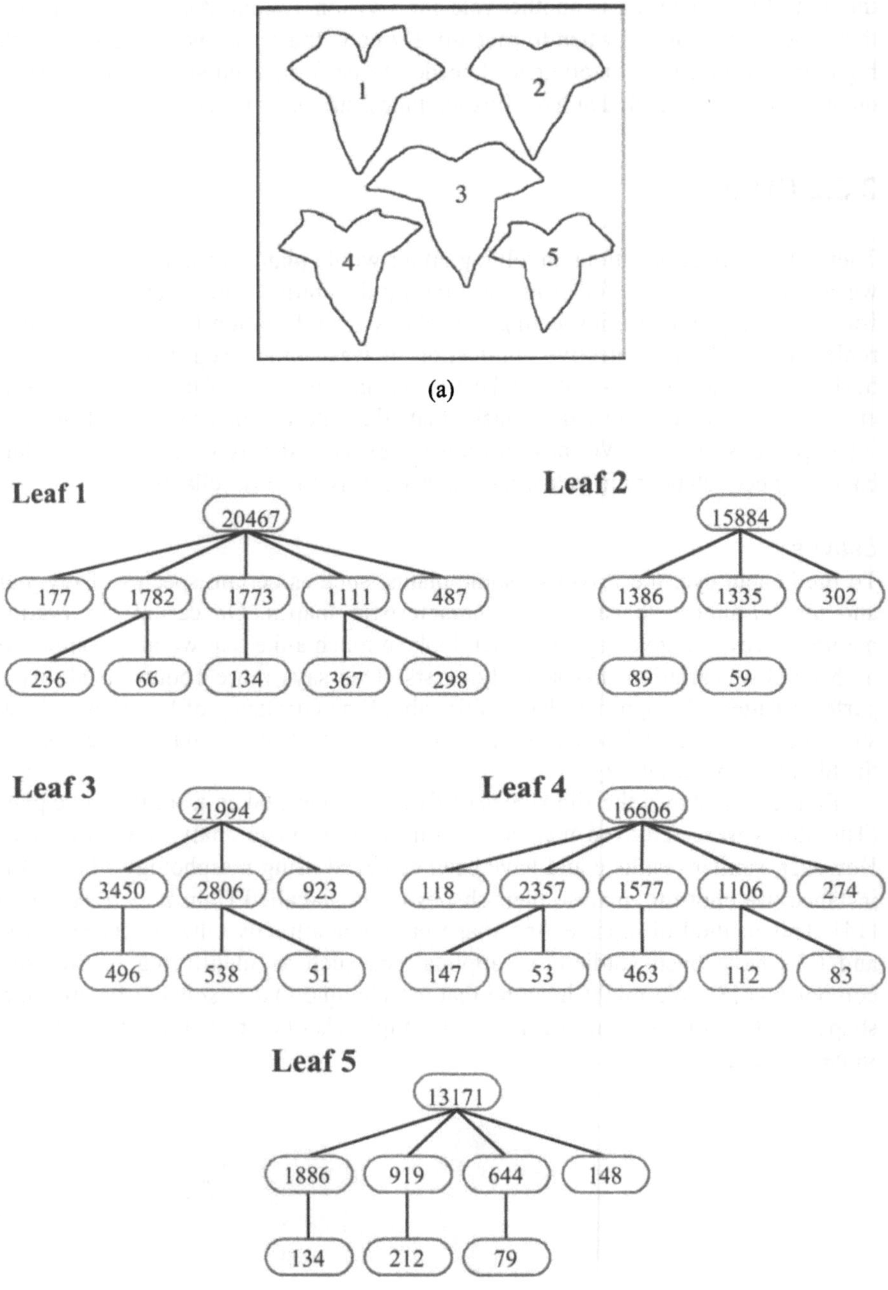

Figure 3.18. Ivy leaves analysed using concavity trees: (a) five leaf silhouettes; (b) concanvity trees. Numbers indicate the areas of the convex hulls for each shape. Minor branches (e.g., corresponding to small shapes) have been removed, for clarity.

At first sight, CTs seem to provide a very natural way to represent shapes of leaves such as ivy, since they allow us to combine global and local information in a systematic way. However, the CTs generated from ivy leaves are highly variable, reflecting differences of leaf shape that would probably elude a casual glance. In view of this, it is not possible to use CTs in a simple way, to compare the shapes of leaves. We could not, for example, compare two shapes by simply overlaying their CTs. A much more sophisticated tree-matching process is needed. It would be too much of a distraction to describe such a procedure here.

Other methods of analysis are needed, if we are to perform shape matching on highly variable objects like ivy leaves. Heuristic, rather than algorithmic, techniques for shape analysis are probably more appropriate for situations like this. Consider Figure 3.19, which shows various methods that might be used for determining the orientation and position of a leaf. Both are needed as a prelude for certain methods for shape matching. Figure 3.19(a) shows that the principal axis cannot be used to normalise the orientation reliably, while Figure 3.19(b) suggests that the centroid of the leaf silhouette and the centroid of the small bay at the top provide a better reference axis for this purpose. However, in some leaves, this bay is shallow and ill defined (Leaf 2 in Figure 3.18(a)). Another possibility is provided by the line joining the centroid to the edge point that is furthest from the centroid (Figure 3.19(c)). Which of these is more reliable needs to be evaluated by detailed study of a carefully selected sample of leaves.

Figures 3.19(d) – (f) illustrate three methods for finding reference points that might be used to control the warping of a "rubber template", as part of a shape-matching routine. It is common practice to use the centroid as a means of determining position but other methods might be more appropriate. For example, the centre of the circumcircle (Figure 3.19(g)), or the centroid of the convex hull (Figure 3.19(h)) provide possible alternatives.

Figure 3.19(i) shows a plot of radius ($r(\theta)$), measured from the centroid, against angle (θ) for three ivy leaves.[4] Simple (1-dimensional) correlation of these graphs could be used for shape matching. Since $r(\theta)$ is a periodic function of θ, it is also possible to apply Fourier analysis techniques. The Fourier coefficients could be used to provide a set of inputs to a pattern classifier of the type discussed earlier in this chapter. A set of low-order Moments could be used for this purpose instead.

The conclusion we are forced to make is that shape matching of natural objects is far from straightforward. Even though similar functions have long been standard in manufacturing of engineering products, the high level of variability makes the established methods unreliable. Moreover, greater intelligence in shape-matching procedures is needed. It is important to appreciate that the greater uncertainty about which technique is best suited to a given application makes testing and evaluating the various options more difficult and therefore more time-consuming. It is imperative, therefore, that we have a properly organised set of image analysis tools for this type of task.

---

[4] For ivy leaves, $r(\theta)$ happens to be a single-valued function of θ. If this is not the case, it may be necessary to use the function r(d) instead, where d is the distance measured around the edge. The latter can be estimated from the chain code.

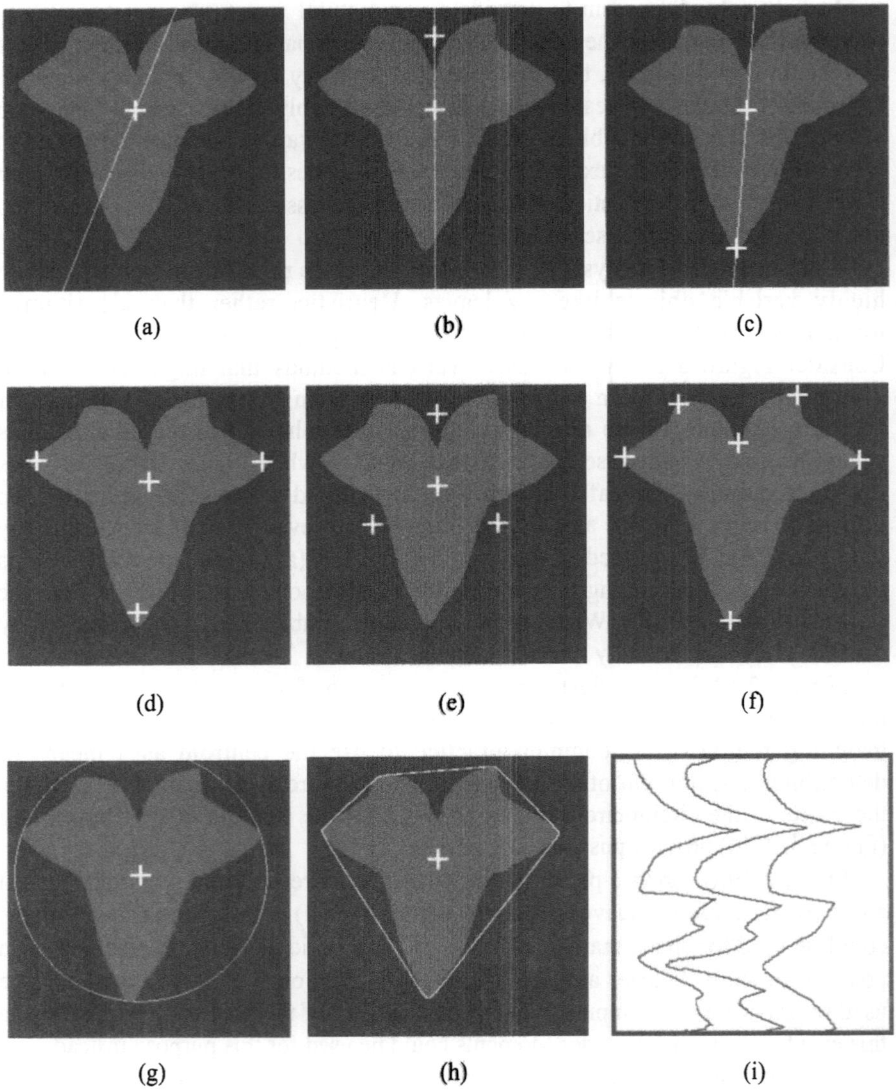

Figure 3.19. Ivy leaves: (a) principal axis; (b) line joining the object centroid and the centroid of the third largest bay; (c) line joining the object centroid to the furthest edge point; (d) tips of the protuberances; (e) centroids of the major bays; (f) points of high curvature; (g) circumcircle; (h) centroid of the convex hull; (i) polar plot ($r(\theta)$) versus θ)

Micropropagation

The technique known as *micropropagation* is used in horticulture is to "copy" a tiny living plant by vegetative reproduction, to create a large number of genetically identical plants. A plantlet is dissected, as appropriate, and the small parts are then replanted in some nutrient material, such as agar jelly. Automation of the process is of considerable economic importance, as manual handling of the plants is prone to causing physical damage and infection by micro-organisms shed by the operator.

Of course, building a visually guided robot for such a task requires considerable skills in mechanical system design, automatic control and vision systems engineering. Our sole concern here is the last of these. A plant such as that shown in Figure 3.20 has an "open" structure that is fairly easy to analyse. The specific task is locate the axial buds, located where the leaf stalks meet the main stem. There are thus two sub-tasks:

- locating the bifurcation points;
- identifying the main stem.

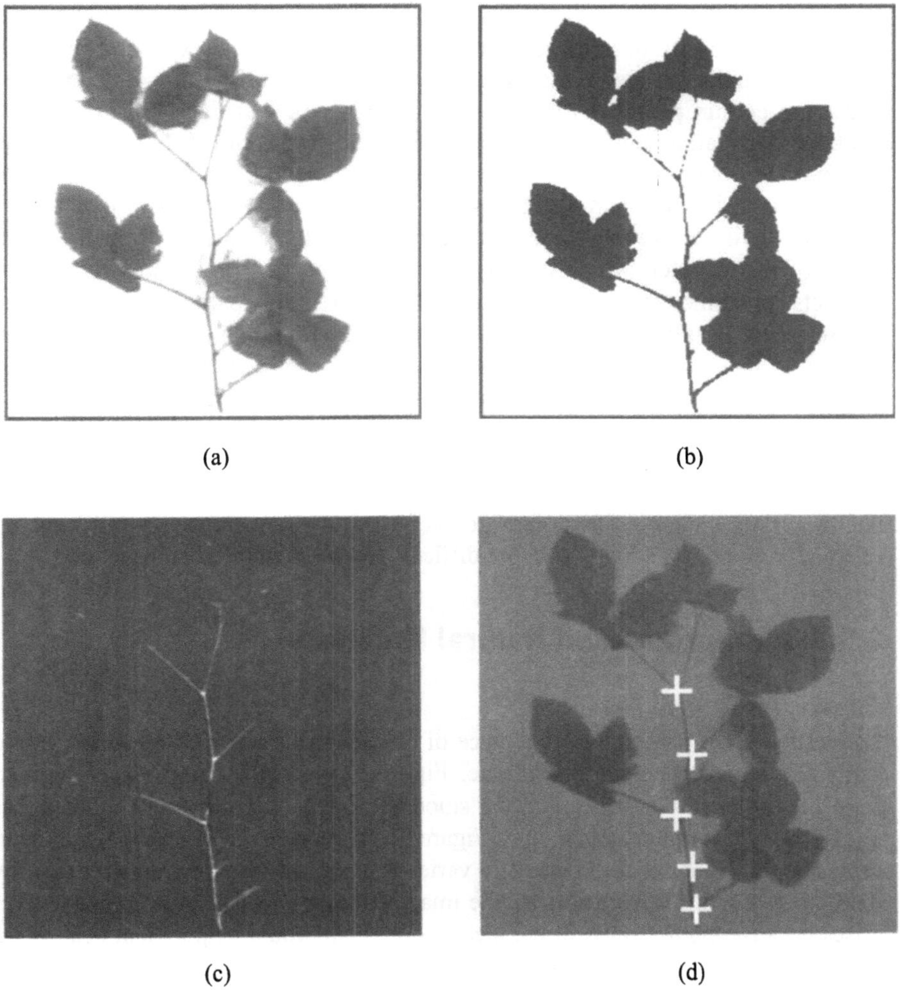

(a) (b) (c) (d)

Figure 3.20. Locating axial buds for micropropagation of a small open-structure plant; (a) original image; (b) best result obtained by thresholding (notice that the stems are "broken"); (c) crack detector applied to the original image; (d) after thresholding c and removal of noise from the binary image and then skeletonisaton, it is possible to locate the bifurcations (locations of the axial buds).

The Y-shaped junctions can be located using either (grey-scale or binary) morphology, or by skeletonising the silhouette and then locating joints on it.

A plant with slightly denser foliage will produce some situations in which the bifurcation points are obscured by leaves. In this case, it may be necessary to combine the results of processing images derived from two or more cameras. This is fairly straightforward if the leaf density is not too high. As the foliage becomes thicker, the task becomes more and more difficult. Eventually, it becomes too complex to analyse properly and a "brute force" method is needed, in which we deliberately sacrifice part of the plant in order to dissect the rest of it.

Pruning

Pruning a dense rose bush can be very difficult, since it requires careful planning beforehand and continued reassessment of the situation as the process continues. There are certain practical issues relating to this task that must be solved first: lighting, eliminating ambient light, protecting the optical sub-system, placing the camera(s) to obtain a good view of each part of the plant. Each of these is far from trivial; it is difficult obtain a good overall view and close-up views of each leaf, stem and stalk. However, these technicalities are distracting us from the main thesis of this chapter, which is that designing inspection procedures for natural products sometimes requires a fundamentally different approach from that adopted for industrial installations. The main feature here is the high level of intelligence needed. In fact, the close integration of vision and intelligent planning needed for this application is already available in PIP, which employs Prolog to implement the top-level controller. However, a very severe difficulty lies in formulating the rules needed to guide a pruning robot. Moreover, tasks such as this are likely to require the discrimination of subtle conditions in the plant (colour and structural changes), due to wilting, infection, frost damage, etc. There are numerous sub-problems, all of which make this a very much more difficult process than micropropagation.

3.5.3 Semi-processed Natural Materials

Chicken-meat Butterflies

"Butterfly" is the name given to a piece of chicken meat that is obtained by cutting symmetrically around the breastbone. Figure 3.21 shows x-ray images derived from two chicken butterflies and the smoothed intensity contours (isophotes) for one of them. Careful examination of Figures 3.21(a) and (b) will reveal one bone in each sample. Notice that the intensity varies considerably over the area covered by the butterfly and this may obscure the images of any bones (Figure 3.21(d)). This arises because the thickness of the meat changes.[5] It would help a great deal, if we could reduce this intensity variation, before applying an appropriate image filter. The *crack detector,* also called morphological *grey-scale closing,* is one possibility. Fitting a suitable model to the image of a butterfly might help us in two ways:

---

[5] This fact can be used to measure the thickness of the meat, assuming that it is homogenous.

i to compensate for those intensity changes that are predictable, knowing that chickens have a similar anatomy;

ii. to enable us to anticipate where and what type of bones might occur. (For example, wing bones are not found embedded in breast meat.)

If all chickens were identical, a fixed reference butterfly image could be produced, which could then be subtracted from each sample image. Unfortunately, chickens do not oblige us by growing to an identical size, so it is impossible to create such a reference image (Figure 3.21(e)). However, it might be possible to modify the reference image, by warping and rescaling its intensity locally, so that it fits the sample image better. To guide the model fitting, we need to define a number of "anchor points", so that we can correlate the reference and sample images. We shall therefore explore some of the possibilities for doing this.

Simply thresholding the butterfly image creates a number of "islands" (Figure 3.21(c)). It may be a good idea to smooth their edges, by applying a low-pass (blurring) filter first. We can then find the centroids of these islands. By carefully choosing which islands we analyse in this way, we can obtain a number of anchor points (Figure 3.22). The centroid of the silhouette and the centroid of its largest bay provide two more such points. In Figure 3.22(b) – (d), we see that this process works, even when the butterfly is not symmetrical. Other possibilities exist for finding anchor points. For example, it is possible to fit a standard mathematical shape, such as a cardioid, to the outer edge of the butterfly, or defined intensity contours (Figure 3.22(e)).

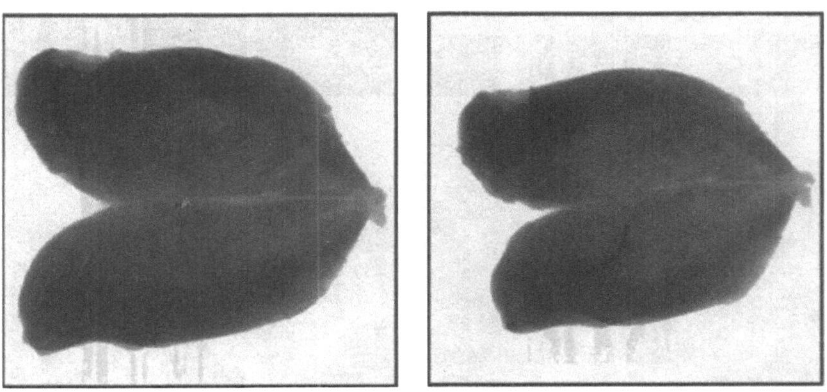

(a) (b)

Figure 3.21. Chicken-meat "butterflies": (a) unprocessed x-ray image; (b) another unprocessed image.

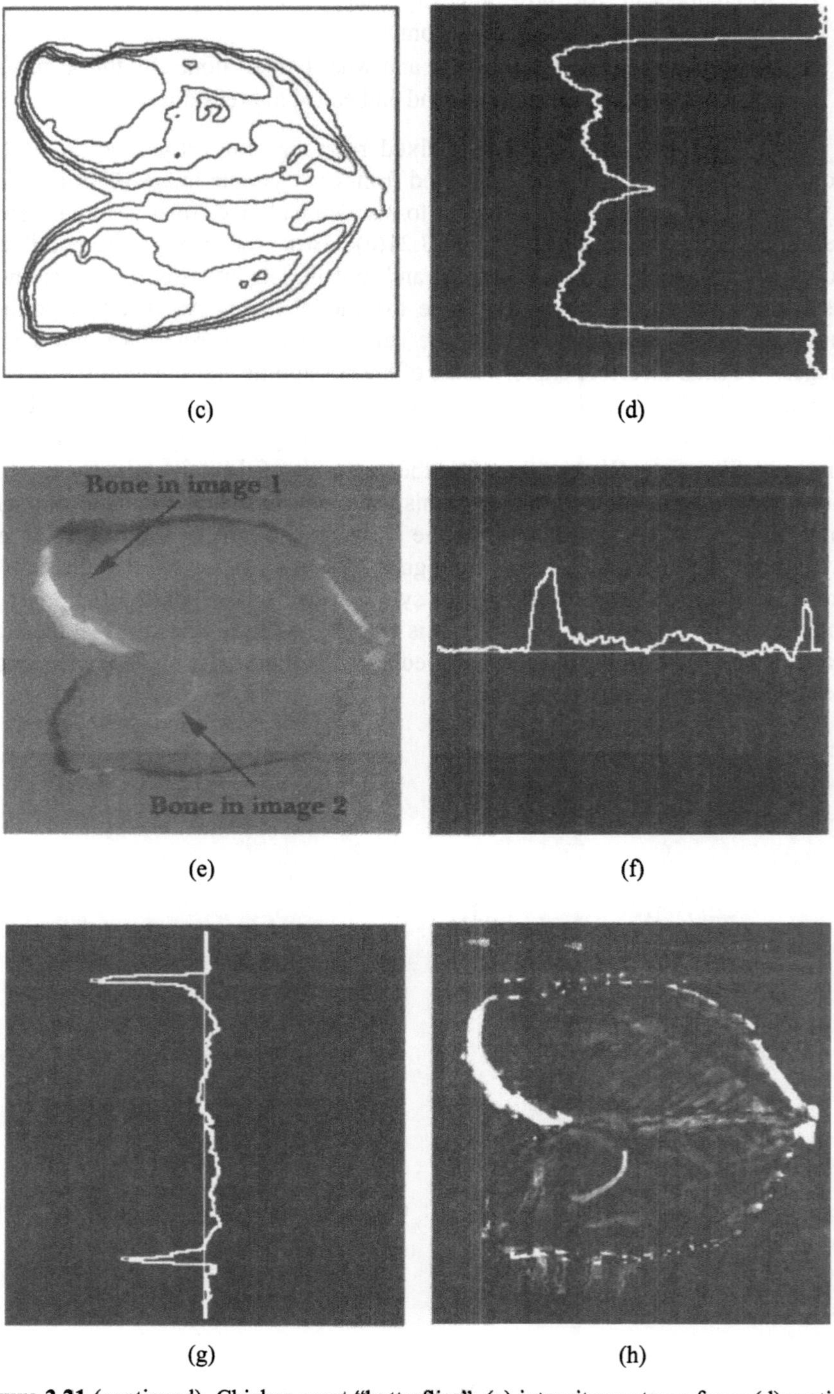

Figure 3.21 (continued) Chicken-meat "butterflies": (c) intensity contours for a; (d) vertical intensity profile for a; (e) images a and b registered (without rescaling) and subtracted; (f) horizontal intensity profile for e; (g) vertical intensity profile through the centre of e; (h) crack detector algorithm applied to e.

The parameters of the cardioid could then be used to control the model-fitting procedure. Other techniques that we have already met might also be considered, although the principal axis is unreliable, if the butterfly is not approximately symmetrical (Figure 3.22(f)).

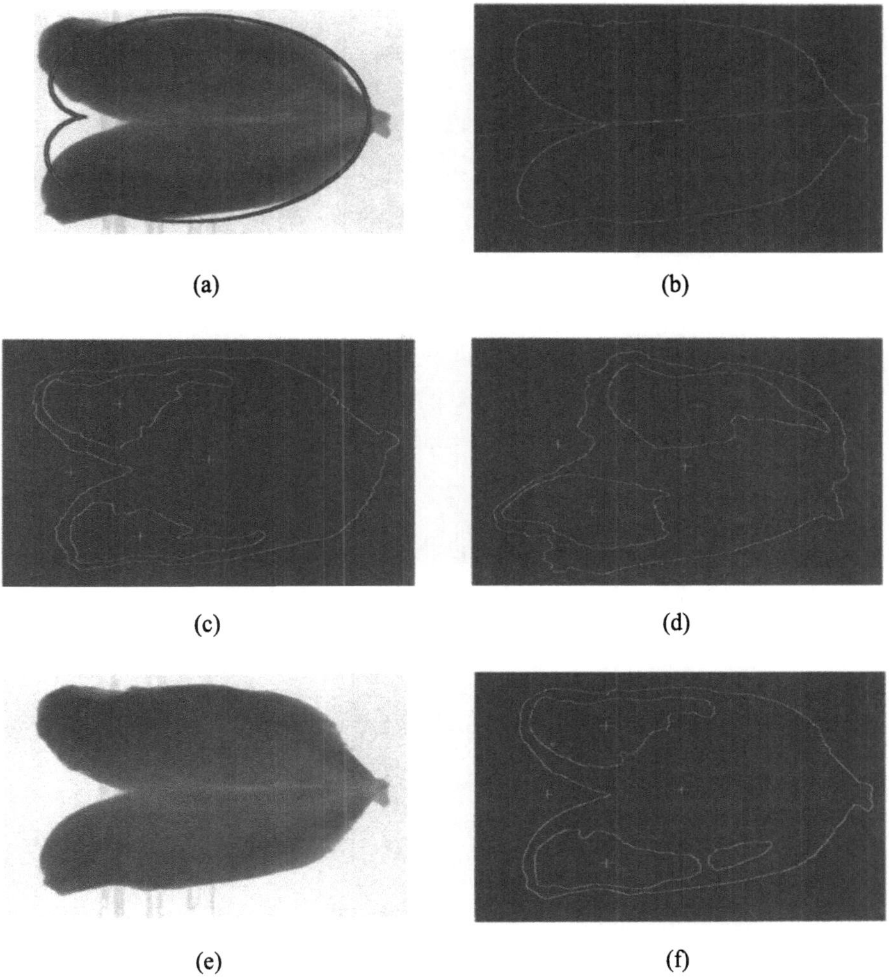

Figure 3.22. Chicken-meat "butterflies; (a) original image; (b) centroids of the outline, the largest bay and of the two largest "islands" created by thresholding at 75% of maximum density; (c) same calculations applied to another "butterfly";. (d) same calculations applied to a third "butterfly"; (e) cardioid fitted to the outline of a [Kindly provided by Dr Paul Rosin.]; f. principal axis line joining the centroid to the centroid of the largest bay.

Fish Fillets

Many of the points just described are also appropriate for processing images of fish fillets. However, the examples shown in Figure 3.23 have no axis of symmetry, or large concavity, that we were able to use on chicken butterflies. Hence some of the methods that we devised for butterflies are unsuitable for fish fillets, although new possibilities might arise The point to note is that each application involving natural products has to be considered on its merits, so that a suitable analysis procedure can be devised to take account of its characteristic features. Moreover, what may seem like similar applications might, in fact, require quite different types of solution; both the designer and the machine he/she is building must be intelligent.

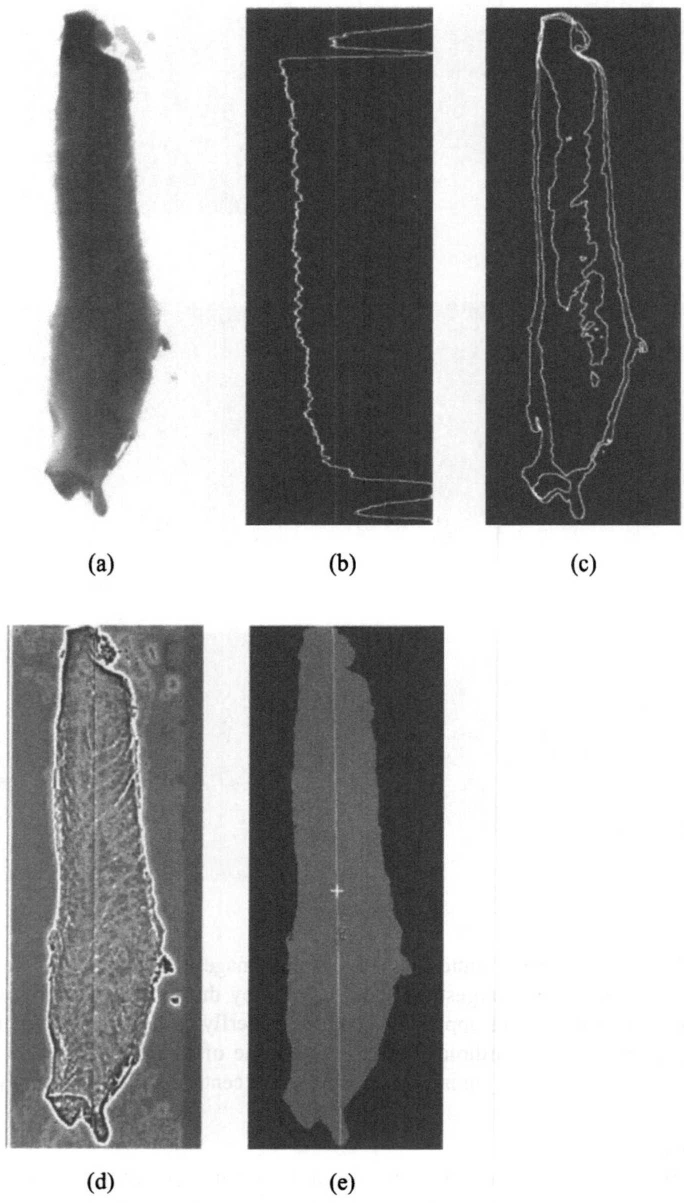

Figure 3.23. Fish fillets: (a) original x-ray image; (b) intensity profile; (c) smoothed isophotes (intensity contours); (d) high-pass filter (notice how little internal detail there is (e) principal axis.

3.5.4 Food Products

Cake Decoration Patterns

Figure 3.24 shows four decoration patterns generated by the same production machinery, making cakes in the form of a continuous strip. Patterns like these are created using one, or more, oscillating nozzles, extruding icing (frosting) onto the cake, which moves at constant speed. Among other patterns, cycloids, sinusoids and zig-zag curves can be created in this way. Although the nozzle-cake motion can be described precisely in mathematical terms, the patterns actually deposited on the cake surface are somewhat uneven. The reason is that the icing does not always flow smoothly through the nozzle. In addition, the cake surface undulates and does not provide uniform adhesion for the emerging stream of icing. The result is a pattern that follows a well-defined path but which is of uneven thickness and may actually be broken in places. Inspecting the decorated cake surface is important, because customers judge the cake, in part, by its appearance. This is an ideal application for a simple heuristic learning procedure, based on morphology and statistics such as zero-crossing frequency (Figure 3.25). Notice, however, that no single morphology operator yields sufficient information for a comprehensive test of cake decoration quality; several such filters would be needed in practice.

However, as always in Machine Vision, we need to consider the application requirements, in order to optimise the design of the inspection system. In this particular case, the nature of the manufacturing process and the types of fault it produces are significant. There are short-term glitches, when a very short section of cake deviates from the norm. These are unimportant and might actually be regarded as desirable, as a certain element of variation gives a "home made" appearance to the cakes. Small changes in the appearance of the decoration pattern should therefore be ignored. However, once the baking system starts to produce "bad" cake, it will go on doing so. Our inspection system is therefore concerned solely with long-term changes. We usually think of an inspection system as using the same algorithm, which may be quite complex, operating on every image. However, the nature of this particular application allows a different kind of inspection procedure to be used. Let us consider the continuous strip of cake as being represented by a succession of non-overlapping images, (I_1, I_2, I_3, ...). Then, we might apply a set of "mini-algorithms" [A_1, A_2, ..., A_N] to images [I_1, I_2, ..., I_N] respectively and combine their results afterwards. Typically, each mini-algorithm will use different morphology and measurement processes. The results is a set of measurements: X_1, X_2, ..., X_N, which might be used as inputs to a Pattern Recognition system such as a Compound Classifier. The output of this is a decision that the cake decoration is either "good" or 'faulty". Images I_{N+1}, I_{N+2}, ..., I_{2N} are then treated in the same way, and as are succeeding groups of N images. It is worth making several more points here:

- The process just described fits very well onto the Concurrent Processor, which is a highly cost-effective way of inspecting objects/materials on a conveyor belt [15].
- A fast implementation of the morphology operators might be accomplished using the SKIPSM technique, developed by F. M. Waltz [11].

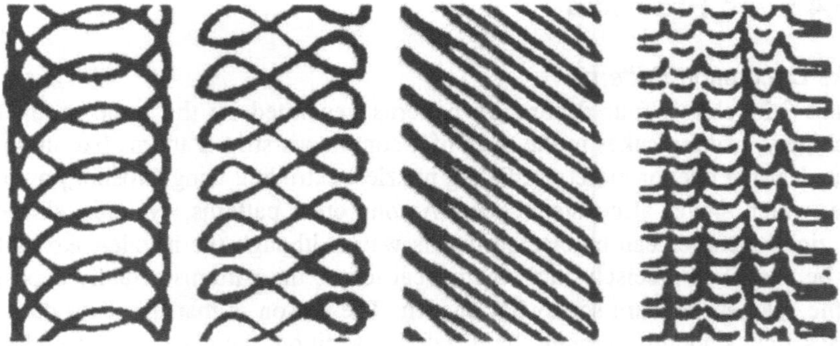

Figure 3.24. Cake decoration patterns

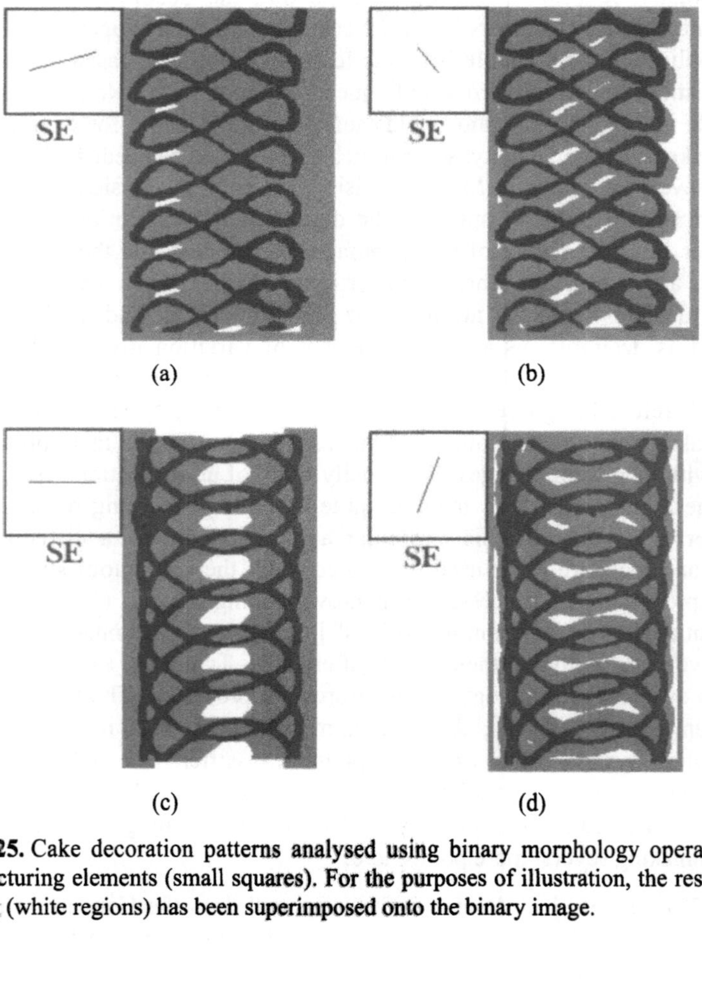

Figure 3.25. Cake decoration patterns analysed using binary morphology operators with linear structuring elements (small squares). For the purposes of illustration, the result of the processing (white regions) has been superimposed onto the binary image.

- The structuring elements used by the morphology operators might be generated. However, carefully matching the structuring element with the decoration pattern will usually result in a more efficient design.

A more complete discussion of this topic may be found in [16].

Loaf Shape

The shape of a loaf is an important criterion that affects customer choice, although nobody knows exactly what customers want. While it is extremely difficult to specify precisely what a "good" loaf is, it is possible to define some of the faults that are obviously *unacceptable.* Of course, it should not be assumed that a loaf is necessarily acceptable if we fail to show that it is unacceptable. However, we shall ignore such niceties and shall concentrate on examining loaf shape, in order to identify malformed loaves. There are two distinct approaches that we can take:

a. Analyse the shapes of individual slices and combine them to form a judgement about the whole loaf. The loaf is represented as a set of two-dimensional images, like that shown in Figure 3.26(a).
b. Analyse the shape of the whole loaf. The loaf is represented by a depth or range map [12].

Figure 3.26 shows the silhouette of a single slice of bread, taken from a loaf that is baked in a tin with no lid.

The top of such a loaf should form a nicely rounded dome. The sides should be straight and parallel, with no major indentations. The overspill (e.g., where the dough has bulged over the rim of the baking tin) should not be too small or too large (this is quantified in practice, of course). Figure 3.26 also illustrates several ways of analysing the shape of this slice. Figure 3.26(b) shows the negated Hough transform of the outer edge contour, while Figure 3.26(c) demonstrates that, if we locate the major peaks and invert the Hough transform, we can locate the sides and base of the slice. It is a straightforward matter then to test whether the sides are vertical and parallel, and have no major bulges (impossible for a tin loaf) or indentations. These same lines can be used to isolate the overspill. (Figure 3.26(d)). The top of the slice can be examined, to measure its radius of curvature. (Figure 3.26(e)).

This requires that we identify three well-spaced points on the top, to provide data for a circle-fitting routine based on simple geometry. The same principle can be used to examine the overspill, by finding the radius of curvature on both sides of the loaf (Figure 3.26(f). The radius ($r(\theta)$), measured from the centroid, as a function of angular position (θ) is plotted in Figure 3.26(g), and can be used to identify the type of slice, by cross-correlation with a standard shape. (This is the same representation of shape that we employed earlier for identifying ivy leaves.) Figures 3.26(h) and (i) show the binary bread-slice image grey-weighted in two different ways. The histograms of these two images provide useful shape descriptions that are independent of orientation and could therefore be used as part of a shape recognition procedure.

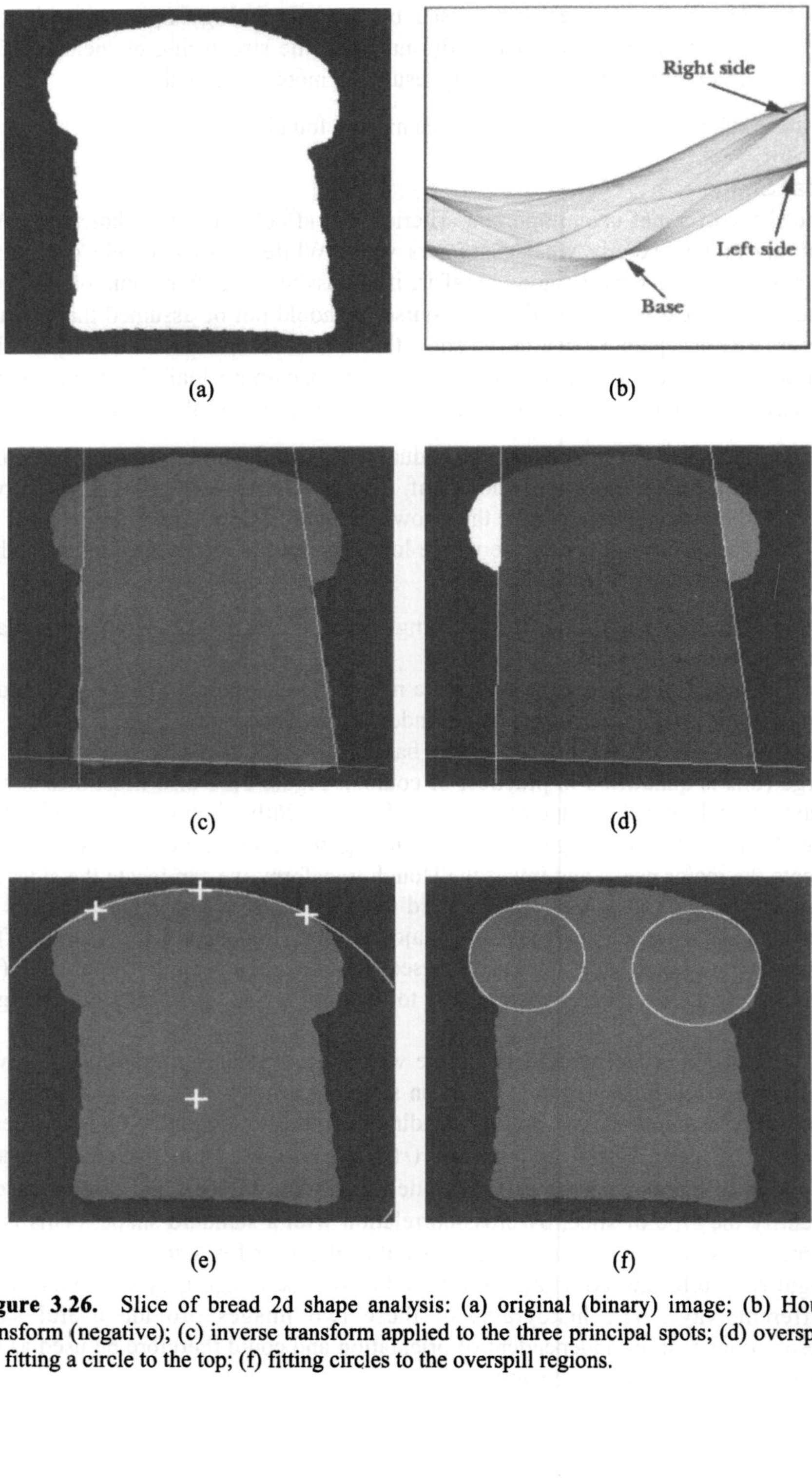

Figure 3.26. Slice of bread 2d shape analysis: (a) original (binary) image; (b) Hough transform (negative); (c) inverse transform applied to the three principal spots; (d) overspill; (e) fitting a circle to the top; (f) fitting circles to the overspill regions.

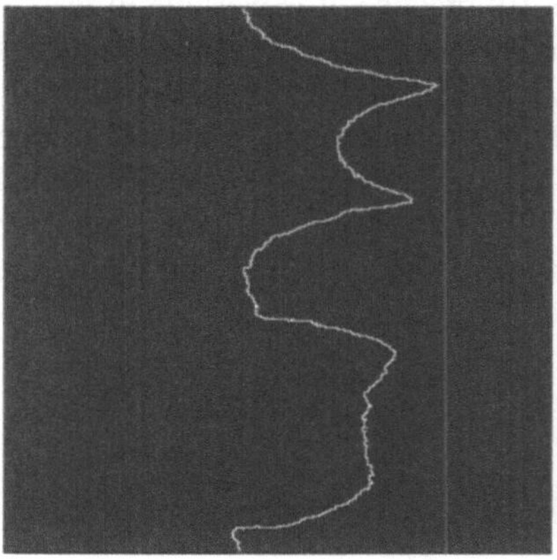

(g)

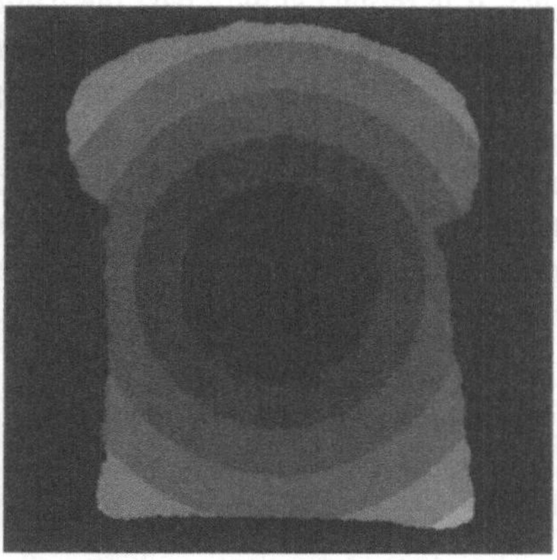

(h)

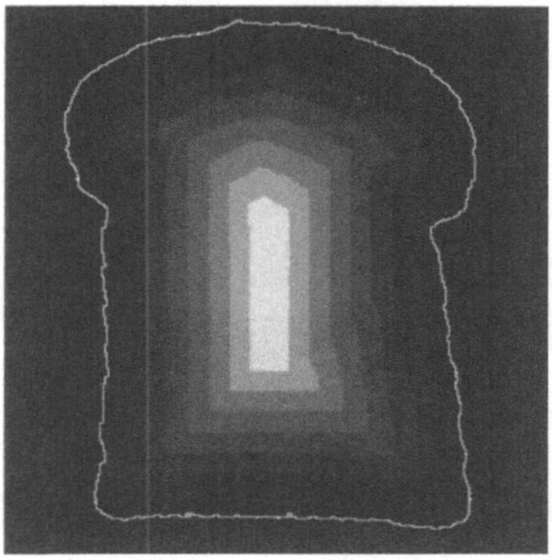

(i)

Figure 3.26 (continued) Slice of bread 2d shape analysis: (g) plot of distance ($r(\theta)$) from the centroid, versus angle (θ, vertical axis); h. grey-weighting according to distance from the centroid (the histogram of this image provides a useful way of describing shape that is independent of orientation); i. grey-weighting according to the grass-fire transform (this is another useful rotation-independent representation of shape).

The analysis described so far is based on a single slice of bread, In order to inspect the whole loaf, we could repeat measurements like those outlined above, for each slice and then relate them together. Two other ways of representing three-dimensional shape are illustrated in Figure 3.27 and require the use of specialised illumination-optical systems. The range map of a croissant shown in Figure 3.27(a) was obtained using structured lighting [16]. In this method of obtaining three-dimensional shape information, light from a diode laser is expanded by a cylindrical lens to form a fan-shaped beam. This is projected vertically downwards onto the top of the object. A video camera views the resulting light stripe from an angle of about 45°. This yields height data for just one vertical cross-section through the object, for each video frame. To build up a complete 3-D profile of an object, it is moved slowly past the camera. (In practice, it is possible to map only the top surface of a loaf with a single camera. Hence, in order to obtain profiles of the sides as well, this arrangement is triplicated.) In a range map, the height of the object surface is represented by the intensity in a monochrome image and isophotes (contours of equal brightness) correspond to height contours (Figure 3.27(b)). The pattern in Figure 3.27(c) is the result of simultaneously projecting a number of parallel light stripes onto the top surface of a loaf and applying some simple image processing, to create a binary image.

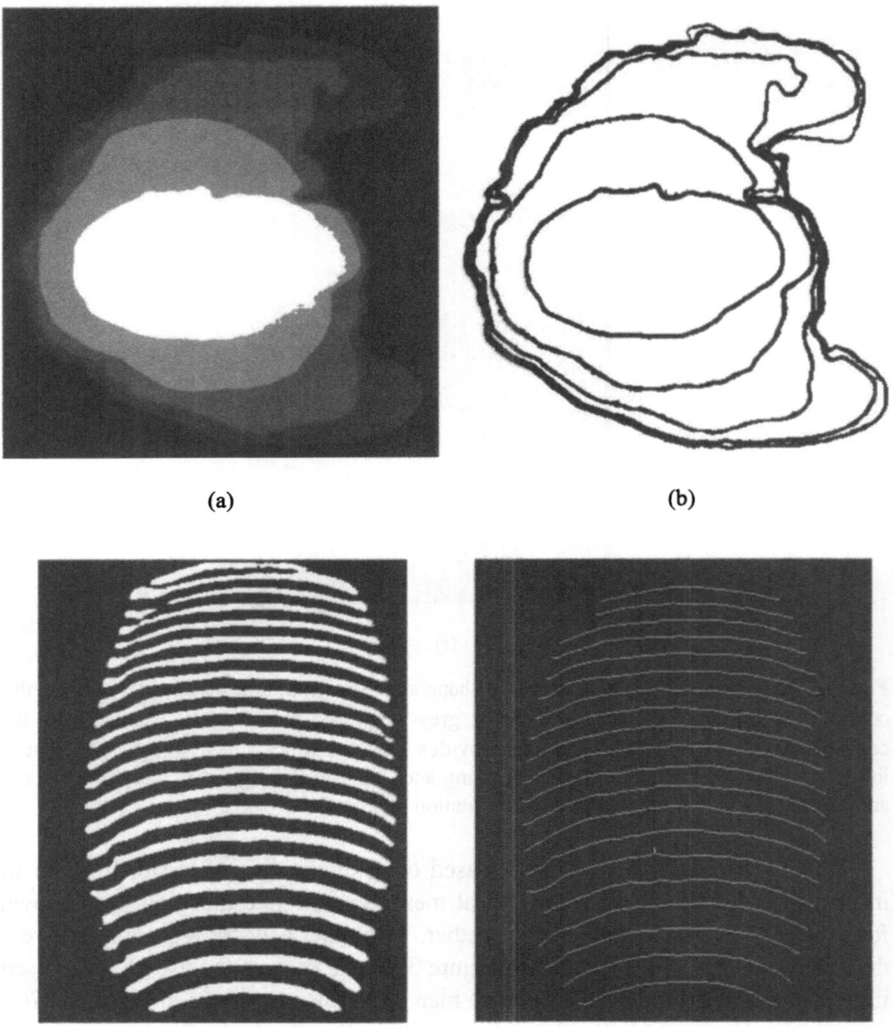

(a) (b)

(c) (d)

Figure 3.27. 3-D height profiling of a croissant and loaf.

We therefore appear to have three distinct methods for representing 3-D shape visually. In fact, these are very similar in terms of the data that they produce and, as a result, we can employ the same decision-making procedure to complete the analysis of 3-D loaf shape. In Figure 3.26, the slicing is achieved by physically cutting the loaf in a series of parallel, vertical planes. In Figure 3.27(a), horizontal "slicing" is performed computationally, by thresholding the range map. In Figure 3.27(c), vertical "slicing" is achieved optically. We therefore conclude this section with a brief discussion of methods for making decisions based on a series of binary contours, which may or may not be closed.

Let $\mathbf{X}^j = (X_{1,j}, X_{2,j}, X_{3,j}, \ldots, X_{N,j})$ denote a series of measurements derived from the jth slice of a loaf, where $j = [1, 2, 3, \ldots, M]$. The $X_{i,j}$ could be any convenient measurements, perhaps of the type that we have already discussed when referring to Figure 3.26. For our present purposes, it is the procedure that we use for combining them that is of prime importance. We begin by building a device (or writing a program segment) that makes a decision $Y^j \in$ *{good, faulty}*, based on just one slice. This might perform a series of checks of the form:

$$L_{i,j} \le X_{i,j} \le H_{i,j} \tag{3.12}$$

where $L_{i,j}$ and $H_{i,j}$ are parameters obtained by learning. They may simply be extimates of the minimum and maximum values observed for the variable $X_{i,j}$, taken over a set of training data. Alternatively, they could be limits based on statistical measures of variable spread (i.e., mean and standard deviation). We might compute the sub-decision Y^j as follows

$$Y^j = \begin{cases} good, \text{if } L_{i,j} \ ? \ X_{i,j} \ ? \ H_{i,j} \text{ for all } i \in [1,2,3\ldots,N] \\ faulty, \text{otherwise} \end{cases} \tag{3.13}$$

It now remains to combine the sub-decisions $(Y^1, Y^2, \ldots, Y^M)$. This can be done in a variety of ways, depending upon the application requirements. We might apply a strict rule demanding that all of the Y^j = *good*, or a more tolerant rule that no more than P ($P < N/2$) of the Y^j = *faulty*. Clearly, there are many variations of this basic decision-making procedure that we can apply. For example, we could make Equation (3.13) less severe, by requiring that say only Q out of N ($Q < N$) tests must be passed before we make Y^j = *good.* Which rule we choose in practice is best left to human instinct and whatever appropriate experimentation the vision engineer can devise.

The idea of making up *ad hoc* rules like this, specially to suit a given application, is an anathema to people for whom mathematical rigour is essential for their peace of mind. However, to a vision engineer working with highly variable objects, the freedom to adopt heuristic procedures is essential. The results may not be optimal but, in a situation like this, it may be impossible to do any better, or even prove that we could ever do so.

Loaf shape analysis has been reported in more detail elsewhere [17], as has the vexed subject of texture analysis of the bread matrix [18]. Much research work remains to be done.

3.6 Concluding Remarks

The important lessons of this chapter may be summarised as:

- Many of the techniques that were originally devised for inspecting engineering artefacts can be used for natural products but may well require considerable modification.
- In view of the far greater level of variation that exists in natural products, the image analysis procedures need to be selected very carefully indeed, taking into account the characteristic features of the application. The need for a good experimental image processing "tool box" is of even greater importance than hitherto [19].
- Seemingly similar applications, may, in fact, require quite different methods of solution.
- The designer of the vision system needs to apply a high level of intelligence; he/she cannot simply mindlessly "turn the handle" to design an inspection system for natural products. He/she needs to be alert to the great variation in size, shape, colour and texture that exists.
- An inspection system for natural products is likely to require a far higher level of (artificial) intelligence than industrial systems employed so far.
- The specific technologies that are of prime importance for inspecting natural products are:
    - pattern classification and self-adaptive learning;
    - rule-based decision-making;
    - colour recognition;
    - artificial Intelligence programming techniques (Prolog);
    - intelligent model fitting.
- The standard Pattern Recognition model requires some modification to take account of the fundamental difficulty of obtaining a fully representative sample of data from the "faulty" class. This may involve learning on a single class, or a hybrid combination of rule-based and traditional (hyperspace) decision-making methods.
- We may need to adopt *ad hoc* rules and may have to accept that we can do no better than achieve a *sufficient* solution. The concept of *optimal* is one that we may have to forsake at an early stage in the process of designing a vision system.
- Many applications in engineering manufacture present similar problems to those encountered with natural products. Whenever, fluid, or semi-fluid, materials are sprayed, or extruded, onto a substrate, they flow into shapes that are impossible to predict exactly. The resulting "dollops" are likely to require treatment as if they arose naturally and to require intelligent inspection procedures.

Finally, it must be made clear that all of the principles required for applying Machine Vision to engineering artefacts also apply to natural products as well. These have been expressed in the form of a set of "Proverbs" [13]. The points made above add to that list; they do not, in any way, diminish the importance of applying sound Systems Engineering principles. Machine Vision applied to the inspection of natural products must not, under any circumstances, be reduced to an abstract intellectual exercise. Machine Vision is not a scientific discipline – it is engineering.

3.7 References

[1] Batchelor, B.G. (1974) Practical Approach to Pattern Classification, Plenum, ISBN 0-306-30796-0.

[2] Haykin, S. (1994) Neural Networks a Comprehensive Foundation, Macmillan, Englewood Cliffs, NJ, ISBN 0-02-352761-7.

[3] Bartlett, S.L. *et al.* (1988) Automated solder joint inspection, *Trans. IEEE on Pattern Analysis and Machine Intelligence*, PAMI vol. 1, no. 10, pp. 31–42.

[4] Bratko, I. (2001) *Prolog Programming for Artifical Intelligence*, 3rd edition, Addison-Wesley, Harlow, UK, ISBN 0-201-40375-7.

[5] Batchelor, B.G. (1991) *Intelligent Image Processing in Prolog*, Springer-Verlag, Berlin, ISBN 0-540-19647-1.

[6] MacProlog and WinProlog, Poducts of Logic Programming Associates Ltd., Studio 4, Royal Vistoria Patriotic Building, Trinity Road, London, SW18 3SX, UK.

[7] Cyber Image processing software, Cardiff University, Cardiff, Wales, UK, URL: http://bruce.cs.cf.ac.uk/bruce/index.html

[8] CKI Prolog, S. van Otterloo, URL http://www.students.cs.uu.nl/people/smotterl/prolog/

[9] CIP Image processing software, URL http://www. bruce.cs.cf.ac.uk/bruce/index.html

[10] Clocksin, W.F. and Mellish, C.S. (1981) *Programing in Prolog*, Springer Verlag, Berlin, ISBN 3-540-11046-1.

[11] Batchelor, B.G. and Waltz, F.M. (2001) *Intelligent Machine Vision: Techniques, Implementation and Applications*, Springer-Verlag, ISBN 3-540-76224-8.

[12] Batchelor, B.G. and Whelan, P.F. (1997) *Intelligent Vision Systems for Industry*, Springer Verlag, ISBN 3-540-19969-1.

[13] M.K. Hu. (1962) *Visual pattern recognition by moment invariants, IRE Trans.*, vol. IT-8, pp. 179–187.

[14] Batchelor, B.G. (1982) *A laboratory-based approach for designing automated inspection systems*, Proc. Int. Workshop on Industrial Applications of Machine Vision, Research Triangle, NC, May 1982, IEEE Computer Society, pp. 80–86.

[15] Warren, P.W. and Batchelor, B.G. (1974) Recognising familiar patterns, *Journal of Cybernetics*, vol. 3, No2, 1974, pp. 51–5.

[16] Lighting Advisor, URL: http://bruce.cs.cf.ac.uk/bruce/index.html

[17] Batchelor, B.G. (1993) Automated inspection of bread and loaves, Proc. Conf. on Machine Vision Applications, Architectures and Systems Integration II,

Boston, MA, Sept. SPIE, Bellingham, WA, USA, vol. 2064, ISBN 0-8194-1329-1, pp. 124–134.

[18] Zayas, I.Y., Steele, J.L., Weaver, G. and Walker, D.E. (1993) Breadmaking factors assessed by digital imaging, Proc. Conf. on Machine Vision Applications, Architectures and Systems Integration II, Boston, MA, Sept. pub. SPIE, Bellingham, WA, USA, vol. 2064, ISBN 0-8194-1329-1, pp. 135–151.

[19] Batchelor, B.G. and Waltz, F.M. (1993) Interactive Image Processing, Springer Verlag, New York, ISBN 3-540-19814-8.

Chapter 4

Editorial Introduction

Despite the high level of variability in natural products, there are often certain innate characteristics that can be exploited to good effect when building the mechanical handling and image acquisition elements of an inspection system. For example, by placing potatoes, eggs or other "elliptical" products between a pair of rollers they will eventually become aligned with their long axes parallel to the axis of rotation. For centuries, people have exploited the wind and flowing water to separate materials. Similar techniques can be employed in a factory. Sieves, grids and meshes can help to separate objects of varying sizes and shapes. Vibrating tables and conveyors can help to separate granular materials into a uniform layer, so that they can be inspected easily. The general maxim is this: the more constrained the object to be inspected, the easier is the task of the vision system designer. We should always employ whatever "tricks" we can to obtain the best possible inspection conditions, while taking care to avoid damaging the product. We should use the optimal spectrum for both lighting and viewing. (These are not necessarily the same.) We should always seek and apply the relevant biological knowledge, since we are dealing with organic materials. For example, Brazil nut kernels contains a certain oil that fluoresces, while the shells do not. The identity and optical properties of this oil is known. There are many physical phenomena that can help us to differentiate between parts of an object, or between "good" and "bad" materials: polarisation, fluorescence, phosphorescence, birefringence, transmission, scattering, diffraction, refraction, specular versus Lambertian reflection, differential heating/cooling, differential absorption of water, oils, dyes, x-ray absorption, etc. The main lesson of this chapter is the need to be intelligent when designing a vision system for use in the food industry, as there can be no universal solution. Every application is different and the vision engineer should be diligent in seeking those factors that are constant and can help us to achieve the required level of discrimination. The need for high inspection speed and the ability to cope with dirt are once again evident here. We must never defer consideration of practical issues, such as the need to keep equipment clean and protect the camera and lights, as these are of fundamental importance; they are not "mere details to be considered later, when/if we have time".

Chapter 4

Using Natural Phenomena to Aid Food Produce Inspection

G. Long

"The design process is a convenient format for developing the idea of lateral thinking. The emphasis is on the different ways of doing things, the different ways of looking at things and the escape from cliché concepts, the challenging of assumptions."

Edward De Bono 1970 [1]

4.1 Introduction

It is commonly thought that inspecting natural produce by machine vision is a much harder task than inspecting manufactured items. The irregularities and variations in natural produce, together with subtle changes in texture, add complexity to the process. Manufactured items are much more predictable. Certainly the range of differences in natural produce can be enormous and mutations can really upset the logical sorting algorithms!

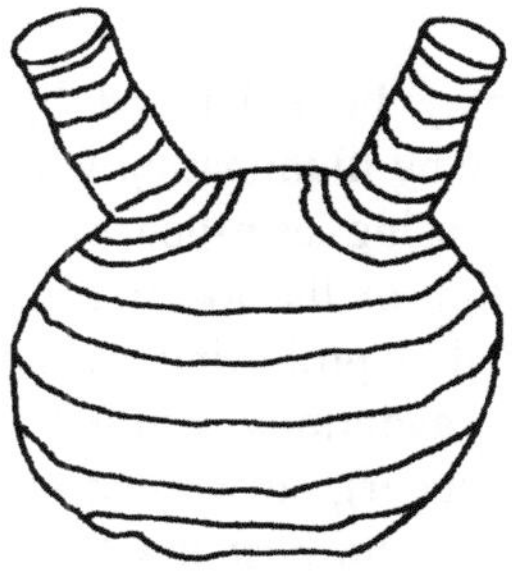

Figure 4.1. A double-stalked swede is still acceptable to the consumer.

However, by selecting inspection parameters which may depend on natural phenomena in the produce, significant advantages can be achieved which dramatically improve the classification probabilities. Other natural differences, which are not observable optically, such as density, can be used to pre-sort or post-sort during the inspection process. An example would be sorting grain from chaff by winnowing. The grain stays on the conveyor whilst the light chaff is blown away. The grain can then be projected at constant velocity to enable the heavier grains to be separated from the lighter ones which have less kinetic energy and travel less far [2]. Then, having vibrated the grains to separate them into a single layer of objects, they are ready for reliable image inspection. Likewise, stones can be removed from a potato crop either by flotation in a suitably designed tank or by x-ray transmission detection (which is really a form of machine vision).

Because it is so important for reliable inspection to ensure that the produce is presented in the most consistent manner, the natural shape features must be exploited and lighting arranged to highlight the principal features so that the contrast between what is acceptable and what is not, is maximised. The more that can be done at this stage in the process, the more efficient and reliable the machine vision and its associated processing algorithms will be. Using the natural shape of the produce – for instance oranges, peas, tomatoes and apples are nearly spherical – to pre-sort by rolling them on slightly tilted chutes allows them to be separated from leaves and stalks and to be size graded by rolling over diverging guides, prior to inspection. Advantage can be taken of the fact that potatoes, being basically ellipsoidal, will line up with their longest axes parallel to the rollers on a roller table. Another feature of some natural produce which can be exploited is that, if they can be inspected during harvesting while they are still fixed rigidly to the ground, presentation will be simplified and consistent. Surface vegetables usually grow in the same orientation and so their fruits may be checked prior to cutting.

Sorting by reflected wavelength variations (including colour) is much used for natural produce. Detection of ripeness, disease and rot or more refined classification by shades of colour use sophisticated multiple camera systems combined with carefully engineered lighting schemes. Auto-fluorescence, the natural fluorescence of materials, can be used to great effect. Mono-chromatic lighting can also be used if the colour variations to be detected vary only in the one colour axis.

Some natural features may be enhanced by structured lighting such as highlighting the surface texture by illuminating with oblique lighting or by picking out three-dimensional features using laser-generated line striping and triangulation measurement. Movement within or the location of a component within the item being inspected may be detected by projecting special light patterns.

Just as some of the best inventions resulted from observation and copying nature (e.g., cats eyes, JCB digger), so we can benefit by exploiting the natural features in produce to develop the best machine vision systems. This chapter illustrates some of these points by describing a number of practical applications, most of which provide non-contact, on-line, 100% inspection.[1]

---

[1] 100% inspection means inspecting every item rather than every feature of an item.

4.2 Techniques to Exploit Natural Phenomena

Shape is used as a primary feature for sorting. It is usually the most obvious feature and is used to control the presentation of the produce so that machine vision inspection is effective. Obviously, there is no point in the viewing camera looking at the underside of an object if it is the top of the fruit or vegetable which is the most significant. However, especially with natural produce, all round inspection is nearly always required when damage, disease or rot need to be detected. Whilst a visual inspection unit can be programmed to examine the orientation of produce and so provide instructions and co-ordinates to a manipulator, this would be a costly system and unlikely to cope with the high throughput required for natural products, especially foods. So mechanical sorting systems, using meshes, grids, chutes, rotating rollers and variable speed belts have been devised to "fit" the form of the vegetable. They are low-cost and remarkably effective but usually inflexible and can lead to damage to a crop as well as having to be designed for specific crops. Programmable sizing and sorting using electronic processing does allow instant grading size changes and complex shape classification but can rarely be justified unless included with full inspection in the same machine. The first application (Section 4.3) illustrates how shape is exploited both to improve the presentation to the inspection system and then to classify the most valuable sector of the crop.

Colour is another obvious attribute appropriate to non-contact visual inspection. Different wavelengths of light, including electromagnetic energy from UV to IR, are used in many applications for identification. The illumination or the observation can be filtered. Spectroscopic analysis of natural produce yields many interesting facts about the absorption and reflectance of healthy produce and specific information related to unsatisfactory damaged, diseased or ripening areas. This information can then be used to select specific wavelengths or colour bands in order to accentuate the contrast between the good and the unacceptable. In grain or rice analysis instrumentation, multiple IR sources at different, but narrowly selected wavelength bands, are used to determine the protein and fat concentration ratios. Analysis of the fat content in milk uses the attenuation differences of optical wavelengths. Near-surface moisture content in straw can be determined on a continuous on-line basis by measuring the absorption ratio between two closely specified IR wavelengths. Applications illustrating some of these techniques are mentioned in other chapters of this book. Fluorescence represents a special technique as the excitation wavelength (illumination) is specially chosen so that it is different from the observation wavelength. The application in Section 4.4 illustrates how this can be put to beneficial use.

Associated with colour reflectance is the translucence or opacity of an object, especially at different wavelengths. This has been used to measure the ripeness of fruit. The hypothesis that the transmission of light through fruit increases as the cell structure changes with ripening suggests that riper fruit will be more translucent then less ripe fruit. The difficulty is to eliminate the other significant differences between successive fruit to achieve a good contrast signal for the ripeness. Also a very bright light source has been needed and problems have arisen trying to avoid stray light reaching the detector when the system operates on-line [3]. Similarly in the inspection of eggs (Section 4.6), the opacity of the yolk in an intact egg is used to determine the consistency of the albumen by displacement measurement.

The texture of a surface can also influence the end-user's choice of natural produce. A shrunken, wrinkled skin on a tomato or pepper is quite unacceptable. However, it is another clue to enable an unacceptable product to be recognised. As stated previously, using oblique illumination will highlight the effect. The problem is to shine the oblique illumination in the optimum direction. Fortunately, modern lighting devices such as LEDs and laser line sources can be switched very rapidly so that successive camera image frames can be illuminated from different directions. This has been used in the packaging industry to identify wrinkles in sheet material. Texture classification, using FFTs (Fast Fourier Transforms) for spatial frequency analysis on images has demonstrated the potential in this area. It is very appropriate for wood grain observation.

Another form of oblique lighting can be used to make the height of objects apparent to the 2D camera. This is done by using structured lighting. A laser-generated line source can be conveniently used to project a line at an oblique angle onto the field observed by an area camera [4],[5]. The horizontal deformation of the line enables the machine vision system to estimate the height of the object. If the object is passing under the camera on a conveyor, then each subsequent image frame will enable a volumetric picture (data set) to be built up as the object passes through. It can be seen from Figure 4.2 how the orientation of the scampi can be determined, its width measured, its approximate volume estimated and its location on the conveyor for subsequent treatment, quantified. It is not necessary to use a diode laser source – any project light source would do but the diode laser produces a very neat line without heat and it can be modulated or synchronised with the camera easily. If the disadvantage of "speckle" has to be overcome, this can be done with a rotating diffusion filter, but usually natural produce inspection does not require such fine measurement resolution.

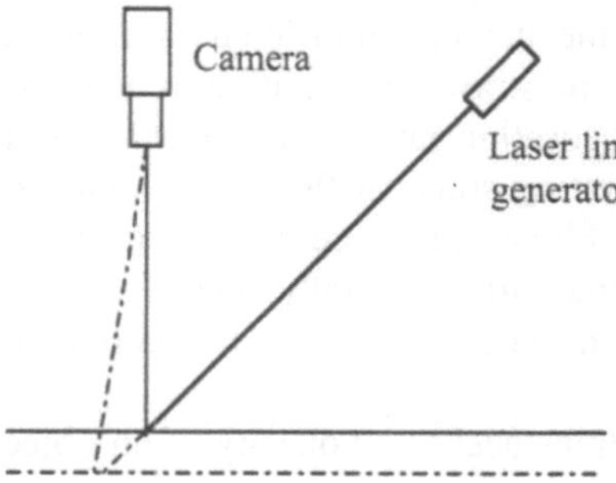

a

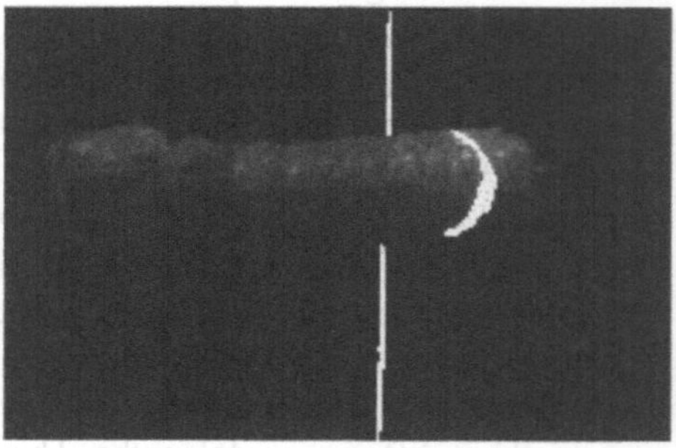

b

Figure 4.2. (a) line striping technique; (b) line striping on scampi.

4.3 Potato Sizing and Inspection

Conventional sizing methods for potatoes use a riddle, sieve or screen to separate the smaller from the larger potatoes. These screens can be made from a mesh of steel wires on a grid pattern and, to enable the crop to be separated into different size grades, a number of screens with different mesh sizes can be installed in a grading line, one above the other. Such screens can handle large quantities of potatoes per hour and are normally reasonably efficient. But there are disadvantages too. For example, during a growing season in which there have been alternating periods of wet and dry weather, secondary growth may cause slightly misshapen tubers so that one end of the tuber may be slightly larger than the other. When this type of potato is jogged along on top of the screen, the smaller end may well lodge in the mesh, like an egg in an eggcup. Subsequently, another potato knocks it further into the mesh and then the grid becomes blocked or the potato is knocked repeatedly until it passes through with flat sides and becomes unattractive to a potential consumer.

Another disadvantage is that marketing requirements relating to sizing vary. The crop is commonly separated into at least three sizes; the largest potatoes are preferred by the chip makers, the smallest are used as seed potatoes for the next season and the majority, in the middle section, are sold for general consumption. Sometimes this category may be divided again into three or more bands by a further set of screens. Depending on market requirements, the farmer may wish to adjust the grading sizes from time to time. There are many occasions when it would be convenient to be able to adjust the screen size from day to day to accommodate natural variations in the crop being processed. Many grading lines are run by farmers' co-operatives and therefore the crop being processed may not be just from different fields but also from a variety of farms with different soils and farming methods. If mesh screens are used it can take up to four hours to strip down the grader and replace the screen with an alternative size. This laborious task is hardly an incentive to the farmer to 'tune' the machine for the best output as much worthwhile production time would be lost during the screen replacement.

There is a third disadvantage to the mesh screening system. Because of the nature of the process, the mesh sizes the produce according to the two minimum dimensions of a potato and no measurement of the longest dimension is made. Therefore, in terms of volume or weight, the sizing is very approximate. Except for the largest potatoes, which will be peeled and sliced into chips before cooking, the majority of the crop will be cooked whole. As the cooking time for a potato depends on its volume, particularly for baked potatoes, it would be advantageous if all the potatoes in one sack were of the same shape and size. As the potato tries to minimise its surface area to a spherical form while growing, selection by weight achieves a much better approximation to equal shape than mesh sizing. However, it is still possible, because of environmental changes during growth, for a long thin tuber to grow. Such a long thin tuber and a spherical tuber can have equal weights. Weighing every potato at a total processing speed of several tonnes per hour would require very ingenious and rapid handling mechanisms. The ideal would be to measure the three prominent dimensions of each potato and grade them according to shape and size. This can be achieved by using a roller table to present the potatoes and machine vision to measure each row of tubers.

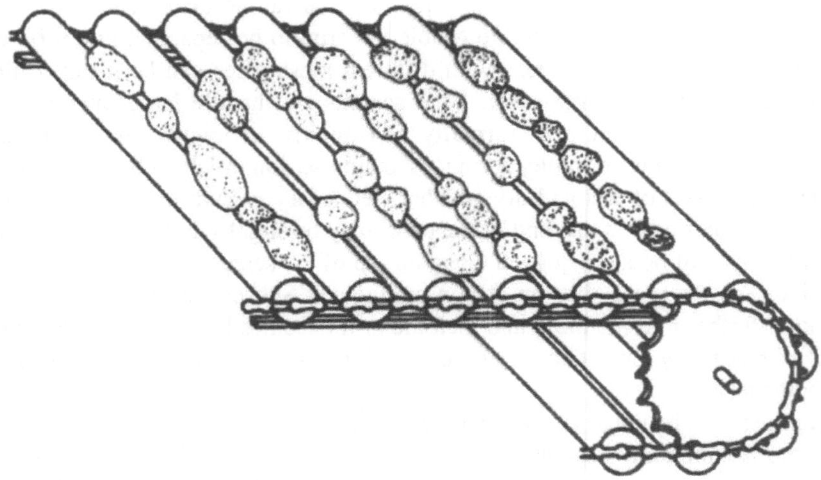

Figure 4.3. Sketch of the roller table.

The roller table has a capacity for separating out the crop into single advancing lines. Because of the continuous rotation of the rollers and the approximately elliptical form of the potatoes, any extra potato is soon accommodated into an adjacent line by a shuffling process and, in each line; the rotation causes the potatoes to orientate themselves with their longest axes in line with the parallel rollers. Both orientation and singularisation are achieved within a small distance.

To measure the dimensions of the potatoes, a linescan camera can be positioned above the end of the roller table, just before the potatoes fall off. This camera scans transversely along each row of advancing potatoes. Because the roller table is moving forward continuously at a constant rate, the linescan camera can be used to acquire a whole picture of the potatoes by recording successive scans. Obtaining a high contrast image, from which measurements could be taken, of a row of unwashed potatoes on a dirty and dusty background, is done by highlighting the potatoes against an unlit background using illumination from an oblique angle. Figure 4.4 illustrates the lighting arrangement. Because the rollers are not illuminated within the camera observation area, the colour shade of the rollers or of a coating of mud is immaterial.

A microcomputer is used to sample the camera's signal at regular intervals so that, as the potatoes advance, successive scans enable a 'picture' or data set to be built up and stored in electronic memory. Using this 'picture' the system can determine the maximum length of each potato and, by assuming symmetrical proportions about the longest axis, it can measure the approximate shape and volume. If two potatoes touch end to end, the software can determine that there are two potatoes rather than one by checking the V formation in the 'picture' formed between two potatoes. The "picture" acquisition is synchronised with the roller table movement using a shaft encoder so that each potato's volume can be predicted.

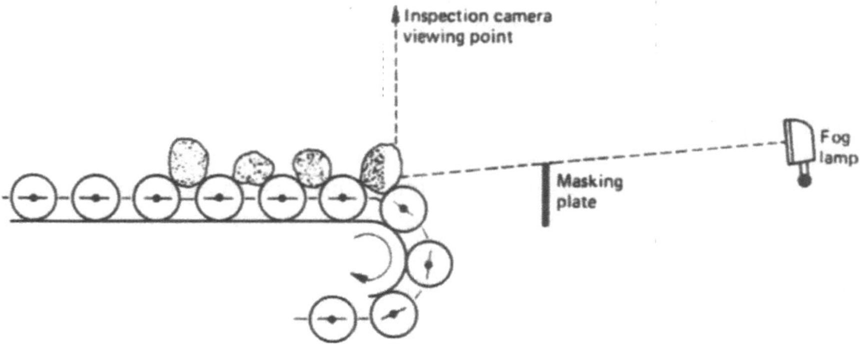

Figure 4.4. The last row of potatoes is illuminated without the roller table being lit in the camera observation area.

By using this machine vision sizing system, the potatoes are not subjected to vigorous mechanical shaking and forcing through sieves or meshes. The grading sizes are numbers set in the computer and therefore can be altered quickly and easily. The potatoes are deflected to the appropriate receptacle by pneumatically operated fingers, controlled by the computer, as they leave the end of the roller table. By being able to measure the maximum length of each potato, the seed potato-planting problem of the tuber sticking in the mechanical planting cup during replanting can be avoided by not allowing any potato exceeding a certain length to be classified as 'seed'. Those potatoes with ideal dimensions for baking, can be easily selected from the main crop so that they can be sold at a premium. None of these features can be accomplished with conventional mesh screens.

The application also lends itself to further automated inspection, using other cameras mounted upstream above the roller table. As well as using monochromatic cameras for inspecting peeled potatoes for black spots prior to chipping, cameras have also been used for colour selection and detecting diseased, damage and rot. The absorbency of different wavelengths, including both the visible and IR bands, can be used to detect diseases which cause water loss, stress in the plant tissue or the presence of chlorophyll caused by greening. A multi-spectral camera is used and as the whole of every tuber has to be inspected, a fast camera and processor combination is required, with high-speed programmable logic (FPGAs) to handle the discrimination algorithms and matrix arithmetic [6].

4.4 Stone Detection in Soft Fruit Using Auto-fluorescence

Cherry halves were used by a confectionery manufacturer for one of the sweets in a popular box of chocolates. Whole cherries were shipped in barrels from Italy to

England. The first process, on arrival at the factory in England, involved cutting the fruit in half and poking out the cherry stone or pit. The de-stoning and slicing machine was very efficient, in fact so efficient that occasionally it sliced the pit in half too, leaving small broken pieces of pit with the freshly cut cherry halves. It was necessary to develop an inspection system in order to prevent these small pieces of stone travelling on with the cherry halves, so that no one would damage their teeth when chewing the end product. The maximum particle size which was allowed to pass through undetected was a cube 1.5mm across.

The x-ray contrast ratio between the pit and the flesh had been proved to be insufficient for adequate discrimination. Image analysis, based on object shape, was also considered on the basis that broken pieces of cherry stone would have well-defined straight edges compared with the softer edges of the fruit flesh and enable the sharp, pointed objects to be detected. However, cost and the difficulty of achieving the throughput required, ruled this approach out. Taking advantage of multi-disciplinary skills available at the University of York, a microbiologist quickly identified a fluorescence technique that would offer a good contrast ratio. The technique used auto-fluorescence, the native fluorescence owing to excitation of the cells or tissue constituents, which is commonly used to highlight features under a microscope [7]. Light in one wavelength band (blue) excites the specimen and the emitted fluorescent light is observed through a barrier filter (orange) which prevents the incident light being seen. An initial study revealed that the cherry pit fluoresced far more than the cherry flesh, contrasting the pieces of pit against the background (see Figure 4.6). This is primarily due to the concentration of oils in the stone compared with the fruit's flesh. Later quantitative measurements established a ratio of 4:1 between the pit fluorescence amplitude and the light emanating from other parts of the cherry.

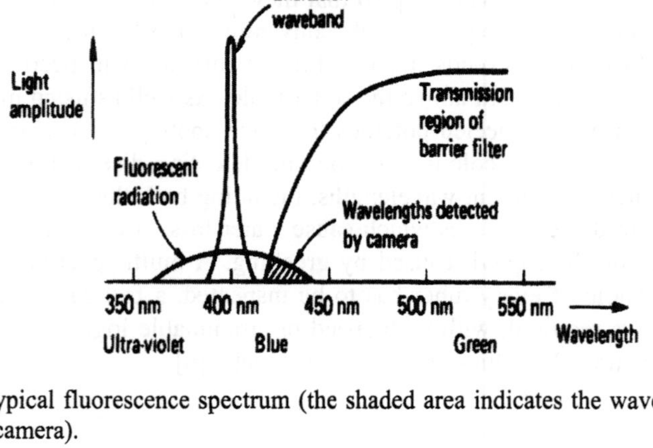

Figure 4.5. Typical fluorescence spectrum (the shaded area indicates the waveband of light falling on the camera).

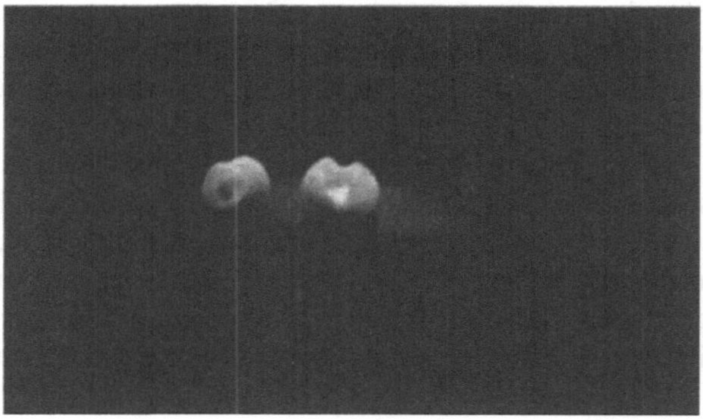

Figure 4.6. Two cherries, illuminated with blue light, viewed through an orange barrier filter, the one on the right showing the contrast between the triangular piece of bright pit and the fruit's flesh.

The fluorescent light was very dim so a sensitive linescan camera had to be developed to discriminate between the very weak fluorescence signal and the general background noise. Linescan cameras were positioned above a flat conveyor belt on which the upturned cherry halves passed. A straightforward threshold comparator was used to detect the presence of pieces of pit. To achieve the process requirement's throughput, it was necessary to scan a belt width of approximately 600mm, which was divided into fifteen parallel channels. This meant that multiple contiguous cameras were used, scanning transversely across the conveyor. The lines viewed by the cameras overlapped slightly so that no piece of pit could be missed (see Figure 4.7).

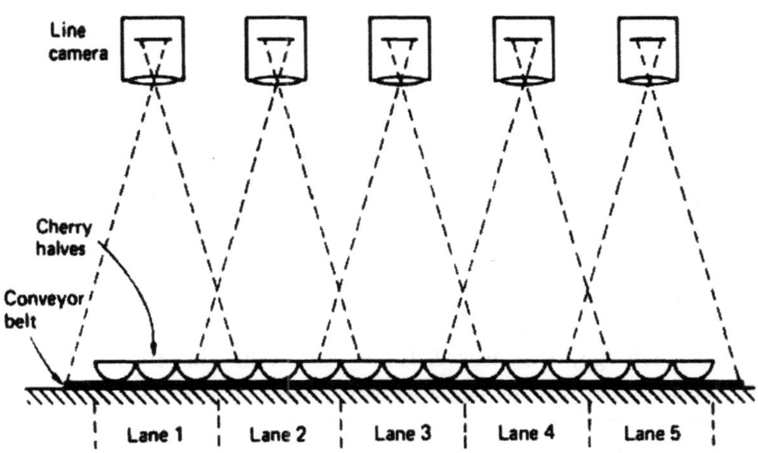

Figure 4.7. Transverse arrangement of the five line cameras above the conveyor belt carrying cherries.

Repetitive scans at 0.7mm intervals along the belt were needed to detect cubes of pit as small as 1.5 mm. With a maximum belt speed of 10 metres per minute, the cameras had to scan in about 4.5 milliseconds. The machine was capable of inspecting 3600 cherry halves per minute. The information recorded about the faulty items was propagated electronically in parallel with the conveyor belt, so that air jets could deflect them from the falling stream when they reached the end of the conveyor belt.

To create an even, brightly illuminated line of blue light (excitation) across the belt in the inspection area was achieved by using four line-filament halogen lamps, heat resistant and blue glass filters and focusing the resulting light, using Fresnel cylindrical lenses, onto the belt at angles to avoid a Gaussian distribution along the line. Although forced-air lamp cooling was required, at least this arrangement produced a flicker-free, consistent uniform band of light across the conveyor.

4.5 Brazil Nut Inspection

In the separation of Brazil nut kernels from their shells, it is important to ensure that no shell remains adhering to the kernel. The nut has three main parts: the hard, dark brown shell, a thin, brown skin, which lies between the shell and the kernel and the soft, cream coloured kernel. Often, during shell separation, this skin remains attached to the kernel but as the skin is edible, these nuts are acceptable. Therefore an inspection system cannot use colour discrimination at normal visible wavelengths.

As the kernel contains a considerable amount of vegetable oil, auto-fluorescence is an appropriate technique to use for this application. The skin is also saturated with the oil so under suitable excitation conditions, both the kernel and the skin fluoresce but the shell does not. A similar fluorescence inspection system to that used for the soft fruit can therefore be used, provided all the surfaces of the nuts can be viewed.

4.6 Intact Egg Inspection

Making sure that eggs are of the highest quality and kept in prime condition on their way to the consumer is most important for the poultry industry. To ensure this, samples of eggs are taken regularly from an egg packing house to a quality control laboratory. There the shell colour and weight are recorded and then the eggs are broken out onto a flat plate and the height of the egg white (albumen) near the yolk for each egg, is measured. It has been found that freshness of an egg is directly related to the way the egg albumen spreads on a flat surface. The further it spreads, the less fresh the egg is. As it is often the appearance of an egg, when broken out in a frying pan, by which the consumer will judge its quality, this is very relevant. The post-laying storage temperature and the age of the bird can also affect the breakdown in the firmness of the albumen.

An egg quality laboratory measurement system may work very well but it still means that samples of eggs have to be taken from the packing line and destroyed during testing. Only a minute fraction of all the eggs is tested and although the sample is statistically large enough to be valid, this still means that none of the eggs that actually go to the consumer, has been tested for freshness. To be able to measure the albumen consistency of all the intact eggs on the packing line would be highly desirable.

A number of methods were tried to measure the consistency of the albumen in intact eggs, such as measuring the egg's rotational damping factor when twisted about its vertical axis and trying to measure the natural resonance of the yolk when subjected to an external sinusoidal vibration over a range of frequencies. But the most promising one, and one more closely correlating with the accepted static albumen measurement (Haugh Unit[2] [8]), turned out to be a method which measured the position of the yolk in the intact egg when in various orientations and from this determined the consistency of the albumen. The internal structure of an egg is shown in Figure 4.8 [9].

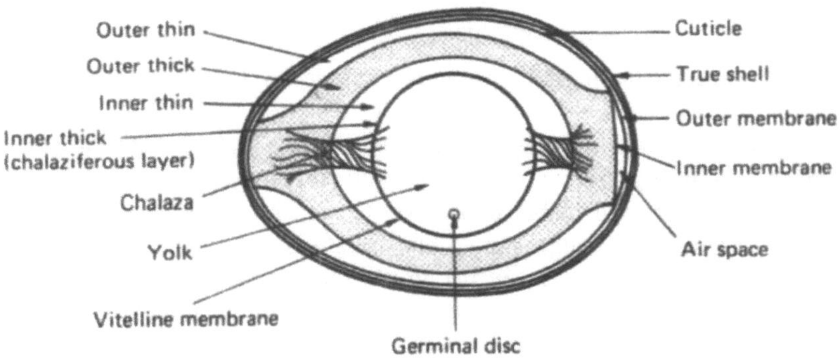

Figure 4.8. Internal structure of an egg.

When the egg is newly laid, the outer thick white and the chalazae are in a firm condition. The air sack, shown in Figure 4.8 at the right-hand end, forms as the egg cools because the contents shrink and the sack gradually increases over the following days as moisture is lost through the porous shell. Over a period of time the thick white and the chalazae disintegrate, allowing the yolk to move around the egg more freely. Although the air sack grows with the age of the egg and therefore could be used to estimate the age of the egg if the storage humidity and temperature conditions were known, there is no simple relationship between albumen consistency and air sack size. As was mentioned earlier, the age of the bird and hereditary factors also play a part.

To determine the albumen firmness it was decided to measure the position of the yolk from the same end of the egg, but firstly with it point uppermost and secondly point downwards. The yolk is less dense than the albumen because it contains a significant fat content so it floats upwards in the egg, pulling at the

[2] The Haugh unit was devised by an American, R.R. Haugh in 1937 and is the accepted international measurement unit for egg freshness. It is a logarithmic scale based on an egg's weight and albumen height when broken out on a flat plate.

lower chalaza and squashing the upper chalaza and thick white. Figure 4.9 shows the measurements (x and y) that were needed to determine the yolk displacement.

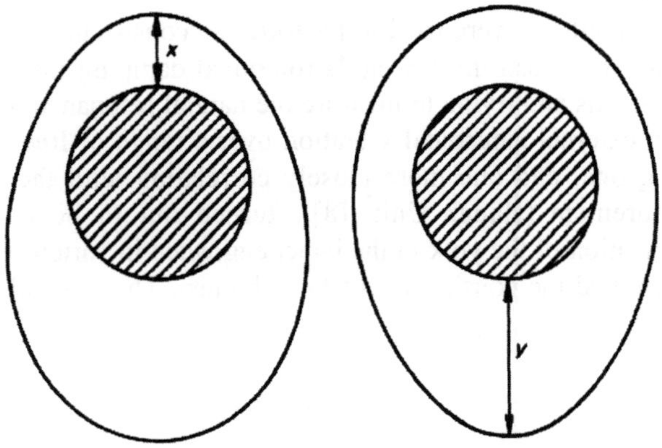

Figure 4.9. Diagram showing the displacement of the yolk with the egg upright and inverted.

To locate the yolk in a white shelled egg is relatively straightforward. Rear illumination of the egg with a bright light, preferably most intense towards the blue end of the spectrum, will enable an observer to see a blurred shadow of the yolk on the shell. The reason for suggesting a bluish light is that Rayleigh scattering from the particles in the yolk causes the transmitted light through the yolk to be attenuated more as the wavelength decreases, therefore creating a darker shadow. Unfortunately, blue light is the worst colour to use for the much more common, brown-shelled eggs because brown shells are almost completely opaque to blue light. To see a yolk shadow in brown eggs is therefore very difficult. Bright white light was therefore used in a structured lighting arrangement.

Instead of illuminating the rear of the egg over its whole surface, a spot of very bright light was projected onto the rear surface. Inside the egg, light shone onto the yolk from just one point on the shell, so casting a clean shadow on the opposite wall. On moving the spot of light along the surface of the egg, the shadow cast travelled in the opposite direction. When the light spot and the shadow were opposite each other, the edge of the yolk would be on the same line. Therefore the position of the upper and lower edges of the yolk could be pin-pointed accurately (Figure 4.10).

A machine vision system could have been implemented to observe this effect but in this case just two photosensors were used to measure the yolk position automatically. They were placed close to the egg on the opposite side from the spot of light. One of these was positioned diametrically opposite the spot of light and the other a short distance further along the path of travel.

In this particular case the egg was moved and the sensors remained still, but identical results would have been obtained if the egg had been static and the sensors and light source had moved. The difference signal derived from the two sensors could be used to indicate when the observed light changed to shadow and back again (Figure 4.11).

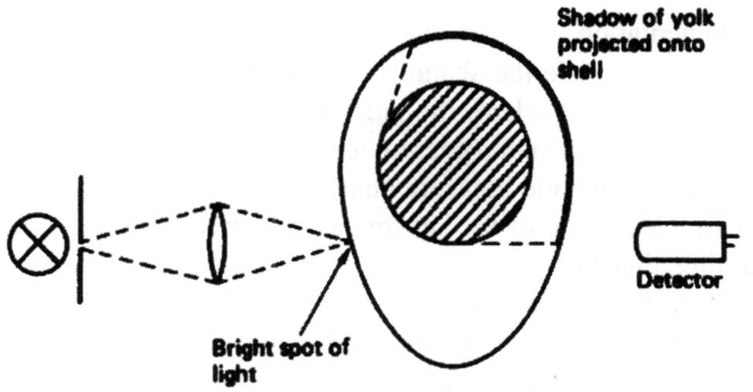

Figure 4.10. How the internal shadow is generated.

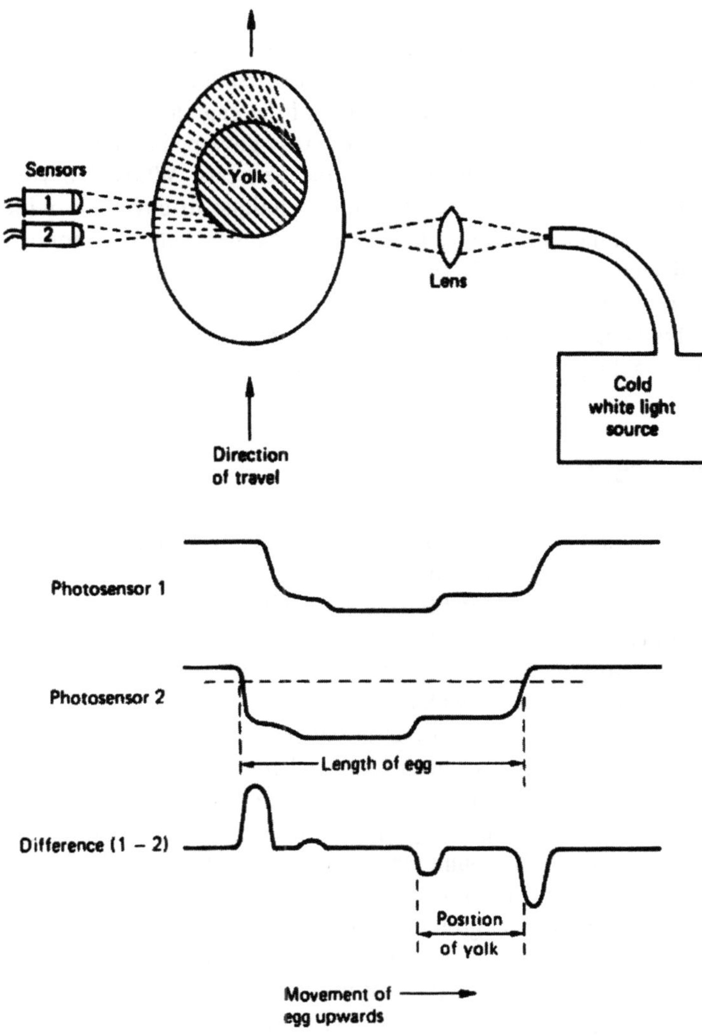

Figure 4.11. Sketch of yolk position measuring system with timing diagrams.

In this implementation only the change from shadow to light was used as the egg passed upwards because the shadow was much more distinct at the centre of the egg than near the point where the sides sloped in rapidly. The egg was then inverted and on the return path, passed identical sensors, but connected in the reverse order, so that a second measurement of yolk shadow could be made in the centre of the egg. Using one of the sensors also to measure the total length of the egg and assuming that yolk size variations from egg to egg were minimal, an answer relating to yolk displacement was found.

When these measurements were compared with the broken-out albumen height recordings for the same eggs, the correlation was not good enough to enable this technique to replace the laboratory system. However, the clustering of the comparative results did show it was possible to use the on-line system as a "go/no-go" system to make sure that no egg had an albumen firmness less than the normally accepted level.

As this chapter is about using natural phenomena to aid classification, it is perhaps appropriate to mention here egg quality laboratory instruments, which were developed for measuring the brownness of the egg shell and the yolk colour from broken eggs. Eggshell brownness is primarily governed by genetic factors but yolk colour is primarily affected by what the hen eats. Whilst the yolk colour measurement required a tri-colour stimulus technique to cover the wide colour range from red orange to pale yellow, due to the hen's diet, a monochromatic source and detector at a specific waveband was found to be possible for the egg shell measurement by observing the whole range of brown eggs commonly available. This simplified monochromatic instrument achieved the egg industry's requirements enabling a relatively low-cost, portable instrument to be produced.

4.7 Wafer Sizing

Chocolate is a natural product and is refined by grinding cocoa to a precise particle size. In subsequent liquid form it behaves as a non-Newtonian fluid. Making chocolate bars requires considerable skill to maintain the correct "temper" and to control the thickness of the enrobed chocolate layer poured over a biscuit or "centre" or to control the amount deposited in a mould. Flour is a major constituent of the batter used to produce the wafers.

The automated inspection of flour is covered elsewhere in this book but a machine vision system has also been produced for checking the size, shape and brownness of biscuit wafer sheets before they are cut up into "fingers". The inspection system uses structured lighting to accentuate the contrast between the biscuit surface and the background which may gradually become as brown as the biscuit due to dust from the process. To the camera, the biscuit appears bright compared with the unlit background. In the diagram, Figure 4.12, the biscuit is conveyed under a linescan camera, scanning transversely, positioned over a narrow gap between two belt conveyors such that the camera "looks" through the gap. The biscuit is long enough to travel over this gap without deviating horizontally. Both the lights and the camera are focused on the biscuit position so only objects at that height are illuminated and in focus. The background is not illuminated nor in focus so that no matter how many biscuit pieces collect underneath, they will not be "seen" by the camera.

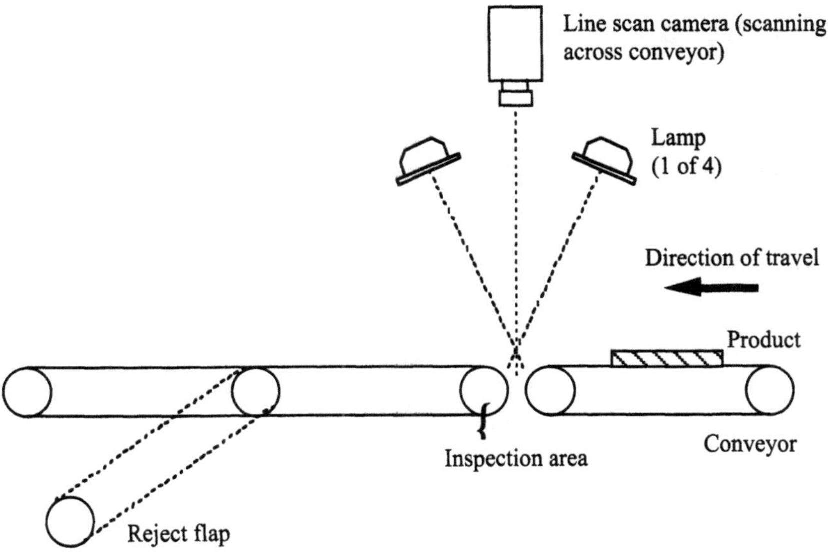

Figure 4.12. Diagram of wafer inspection machine showing structured lighting.

Figure 4.13. Photograph of installed equipment on line.

The linescan camera checks the width of the wafer, the length (by using a conveyor shaft encoder), whether the wafer is skew on the conveyor, for broken corners and even whether any layers have slid with respect to others. Using separate monochromatic photosensors, surface brownness is also measured to monitor the consistency of the automated baking process. A reject flap is operated if any wafer is substandard and measurement data is recorded to show trends and also transmitted to a remote computer. As many parameters as possible are measured in the one inspection station to keep the costs low.

4.8 Enrobed Chocolates

Triangulation, using a technique called line striping, in which the camera views the scene from a different angle than the direction of illumination, can be used on-line to determine the volume of a wet enrobed chocolate bar. It could also have been used to measure the height of the wafer in the previous application, had that been required. By trigonometry, the vertical displacement of a surface can be calculated by knowing the angle of the incident beam and the camera viewing angle. It is important to remember, especially for natural products, that the camera will be viewing diffuse reflected radiation from the surface so if the surface is highly reflective, little radiation will be diffusely reflected. Fortunately most natural products are not shiny but careful consideration should be given if applying this technique to products that have been freshly washed and are still wet. Fortunately, chocolate's particulate nature ensures that the incident light is absorbed into the upper liquid surface and retransmitted as diffuse radiation. The disadvantage is that the resulting observed line is broadened.

Figure 4.14 shows a typical enrobed chocolate bar illuminated with a line stripe. It is necessary to set up the optical arrangement to view the bar from an angle so that the area camera observes both the sides and ends. A double system is needed to ensure all the required surfaces are seen. Multiple images are taken as rapidly as possible as the bar progresses through the inspection area to include the whole bar. Bringing the two data sets together to complete the volumetric calculation is therefore very complex.

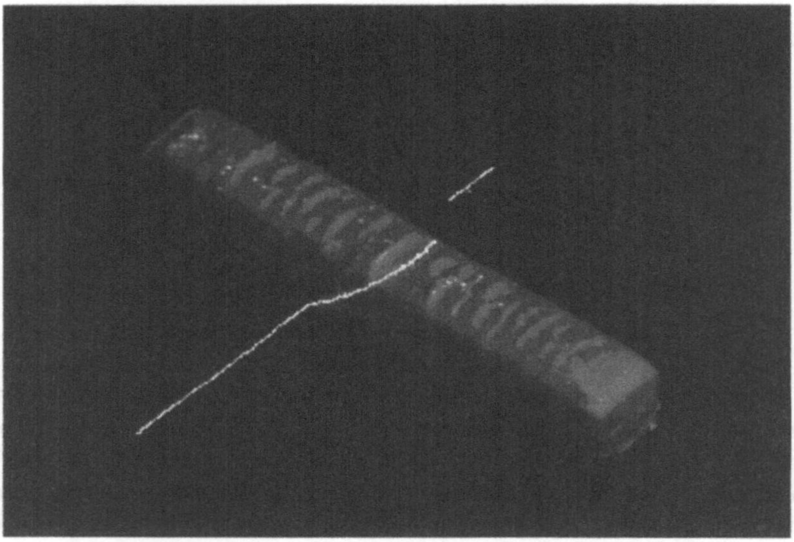

Figure 4.14. View of line striping on a chocolate bar.

Assuming that the internal biscuit or wafer is produced to a consistent size or using a similar inspection system to measure the "centre's" volume before the enrobing process, it is possible to calculate and therefore control the volume of chocolate applied on a continuous basis.

There are many other applications for line striping in the food industry. It can be used to estimate the volume of natural products for grading purposes, for instance as an alternative technique in the potato grading example given in Section 4.3. Like the dark field method, it has the advantage of being a high-contrast technique unaffected by the colour of the background and the accuracy of the result is not directly dependent on the amplitude of the signal observed. This is illustrated by the picture of sweets on which a line is projected (Figure 4.15). It is useful to be able to measure the height of the sweets as they are subsequently stacked and packed in a tube. If the pile height for a nominal number exceeds the packaging machine's capacity, there will be a jam or if the height is too small, the sweets will rattle around in the tube and the customer could be disgruntled. The picture (unfortunately in black and white so the bright red laser line is not as visible as in a colour version) deliberately shows how the laser line can be clearly distinguished crossing the sweets on both a black and a white conveyor background but it is only the deviation in the line that provides the sweet height and location information. Using a laser line generator means that a very narrow waveband interference filter can be fitted to the camera lens to minimise the intrusion of ambient light [4]. For this picture such a filter has not been utilised.

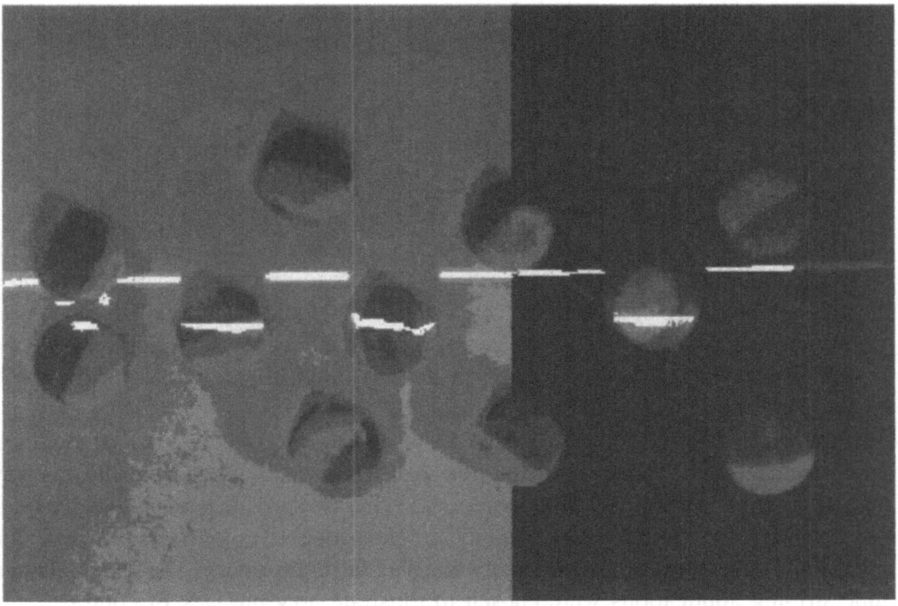

Figure 4.15. Illustration of line striping on a white and a black background.

Figure 4.16 shows line striping applied to control the deposition of whipped cream in a sundae container. It is necessary to ensure that the foam height is not too high as it would interfere with the lid when fitted and so the cream topping would appear squashed and unattractive to the consumer. Such a non-contact technique is ideal for monitoring foams in the food industry.

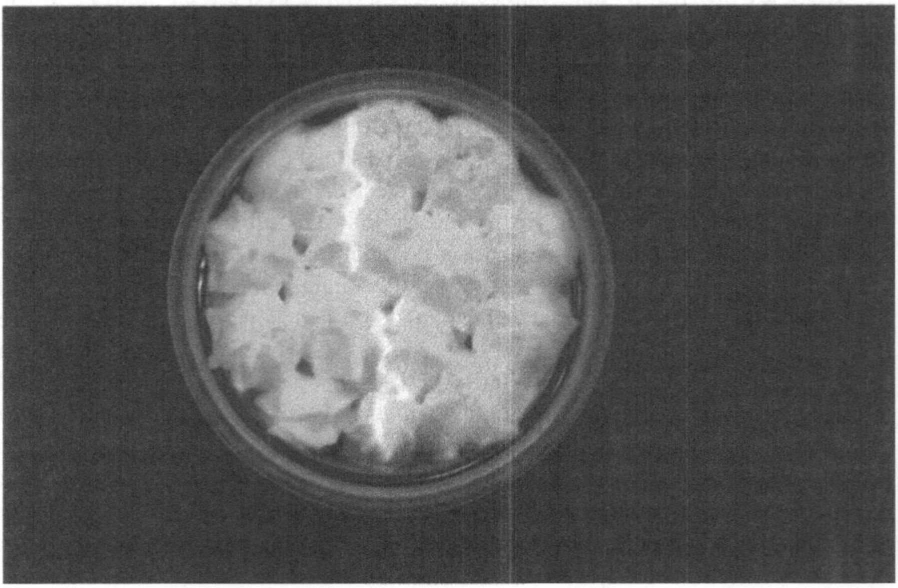

Figure 4.16. Height control of whipped cream in a sundae container by line striping.

4.9 Conclusion

This chapter has described just a few attributes, such as spectral reflection, auto-fluorescence and shape that can be exploited to enhance the performance of machine vision inspection for natural produce. There are others, such as detecting volatile organic compounds produced from diseased or rotting tissue (electronic nose) [10], moisture content detected by electromagnetic absorption (microwaves) or weight (check-weighers), which can also be used but are not within the machine vision remit. However, it is hoped that mentioning all these possibilities will encourage lateral thinking in future designers.

The presentation of the objects to be inspected, the effective use of lighting, good camera lenses, adequate resolution and firm mounting arrangements are key to a successful optical inspection system. The first two are the hardest to achieve. It is essential to engineer the system to maximise the contrast between the identifiable attributes of unsatisfactory and satisfactory produce. The dark field and line striping applications were chosen to illustrate how the lack of contrast from dusty or similar coloured backgrounds, so common in agricultural environments, can be overcome.

There are still many challenges for inspection by machine vision of natural produce. The latest request is to determine the size of the flower (curd) in a cauliflower while the plant is still growing in the field, with the curd enclosed in leaves!

4.10 References

[1] De Bono E. (1970) *Lateral Thinking*, Penguin Books.
[2] Drury R. (1989) *Overall Winner, Toshiba Year of Invention Awards*, Design Council Publication.
[3] Greensill C.V.,Walsh K.B. (2000) A remote acceptance probe and illumination configuration for spectral assessment of internal attributes of intact fruit, *Meas. Sci. Technology* 11 pp. 1674–1684.
[4] Long P.G., Price T.P.W. (1997) Scanning Height Gauge: On-line Inspection of Packaging, IEE Colloquium Digest for "Industrial Inspection, Ref. No. 1997/041.
[5] Long G. (1995) How to grow a systems integrator, *Image Processing*, Vol 7, Issue 1, pp. 48–49.
[6] Muir A.Y., Ross D.W., Dewar C.J., Kennedy D. (1999) Defect and disease detection in potato tubers, Scottish Agricultural College Crop Science Research Report, pp. 5–9.
[7] Ploem J.S., Tanke H.J. (1987) *Introduction to Fluorescence Microscopy*, Oxford University Press, Royal Microscopical Society.
[8] Haugh R.R. (1937) The Haugh unit for measuring egg quality, *U.S. Egg Poultry Magazine*, No. 43: pp. 552–555 & 572–573.
[9] Overfield N.D. (1982) *Quality testing of Egg*, Her Majesty's Stationery Office, London, Ref Book 428.
[10] De Lacy Costello B.P.J., Ewen R.J., Gunson H.E., Ratcliffe N.M., Spencer-Phillips P.T.N. (2000) The development of a sensor system for the early detection of soft rot in stored potato tubers, *Meas. Sci. Technology* 11, pp. 1685–1691.
[11] Schumacher E.F. (1973) *Small is Beautiful*, Blond & Briggs Ltd.
[12] Long G. (1989) *Real Applications of Electronic Sensors*, Macmillan Education.

Chapter 5

Editorial Introduction

Many bulk food materials consist of large quantities of small particles, such as seeds of rice, wheat, sesame, mustard, etc. Inspecting these requires a completely different approach from that needed for larger items (e.g., apples, potatoes etc.), which typically are inspected at much slower rates. In applications of bulk material sorting, the number of individual items to be inspected is usually very large indeed and, as a result, the range of algorithms that can be performed on each one in the limited time available is rather limited. In addition the processing is usually carried out in dedicated electronic hardware rather than PC-based software solutions. Multi-lane inspection machines already exist that can examine individual peas for shape and colour defects at a rate of 16 tonnes per hour. Single-camera machines that can examine seeds of a given kind at a rate of 4.10^4 per second are in use.

There are several reasons why we need to inspect such small items individually. Firstly, food safety depends on the identification of dangerous bacterial and fungal infections (e.g., ergot in rye). Secondly, for commercial reasons, we need to verify that a shipment of a material such as rice is of the variety purchased. (Unscrupulous dealers may be tempted to substitute a cheaper variety that does not possess some desired qualities.) Thirdly, even small amounts of certain contaminants can seriously alter the physical/chemical characteristics of a batch of material that is to be used for processes such as baking (e.g., bran in flour).

Feeding large quantities of small particles past the camera at high speed requires customised handling techniques. Moreover, specialised reject mechanisms are needed for different kinds of materials. The latter include multi-jet pneumatic "blowers" for small seeds, "suckers" for slurries, piston-operated "pushers" and deflector paddles for larger items. There is a limited range of options for equipment that can operate with the required level of reliability (over 10^9 cycles for some reject mechanisms). It must also be borne in mind that even dry organic materials present a hostile working environment for mechanical, optical and electronic equipment.

Obtaining an all-round view of small particles is problematical and specialised lighting techniques are required. At various times, visible, infrared and ultra-violet imaging is used, either individually or in combination. Narrow-band and multi-band optical filters can often be used to good effect to enhance image contrast, by selecting wavelengths indicative of certain materials or conditions (e.g., rotting).

Multi-spectral images are also valuable on occasions. Which wavelengths should be used and how they should be combined for a given application is not always obvious. In this type of situation, learning techniques may be employed. There are many other factors to be considered when designing inspection systems for bulk food materials. There are many varied applications of this general type and most require a customised system, in view of the difficulties involved in designing fast, effective and reliable equipment.

Chapter 5

Colour Sorting in the Food Industry

S.C. Bee and M.J. Honeywood

5.1 Introduction

Good food can usually be distinguished from bad food by colour. This may appear to be an obvious statement, but the implications for the food industry are significant. Human perception of colour has proved very effective in determining food quality. Sorting of food products using the human eye and hand is still widely practised in regions where labour rates remain low. However, where the cost of labour has increased, so automated techniques have been introduced.

As a consequence of increasing consumer awareness of food hygiene, it has now become a basic prerequisite for all optical sorting machines to identify and remove all gross contaminants (glass, stones, insects, rotten product, extraneous vegetable matter etc.). In addition, optical sorting provides a cosmetic enhancement to the product by removal of blemished, discoloured and misshapen product. Contemporary consumers are also demanding increased quality and in conjunction with this, a litigation culture has developed. Especially in recent years, tighter EU and American Food and Drug Administration requirements on food quality have been implemented. Food processors benefit from using automated systems for food sorting, since a machine can maintain greater levels of consistency than hand sorting and frequently offers reduced labour costs [1]. Food processors are able to provide a premium quality product at increased margins, allowing their competitive positioning to be strengthened.

5.2 The Optical Sorting Machine

Colour sorters generally consist of four principal systems:

i. feed system;
ii. optics;
iii. ejection process;
iv. image processing algorithms.

Figure 5.1 shows a typical layout for an optical sorting machine.

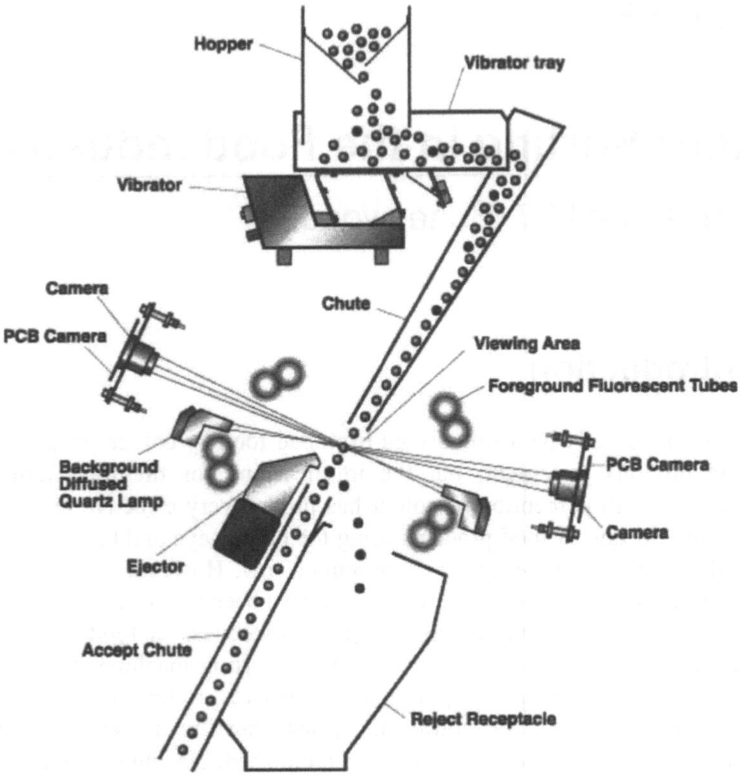

Figure 5.1. Schematic layout of a typical optical sorting machine

5.2.1 The Feed System

In a bulk sorting system, dry products (rice, coffee, nuts) are fed from a vibrating hopper onto a flat, or channelled, gravity chute. To prevent excessive clumping, fresh or frozen products are fed from a vibrating hopper onto an accelerating belt. Both methods separate the product into a uniform "curtain", or monolayer. This ensures the product is then presented to the optical system at constant velocity.

5.2.2 The Optical System

The optical inspection system measures the reflectivity of each item. The inspection components are housed within an optical box and the objects under inspection travel either through, or past the optical box. Objects should not come into direct contact with any part of the optical box and are separated from it by toughened glass windows. The optical box contains one or more lens and detector units, depending on the number of directions from which the product is viewed. Early optical-sorting machines viewed the product from only one side, which meant that they could only detect surface defects facing the optical system.

Nowadays, two or three cameras are used to view the product from different angles as it leaves the end of the chute. Obviously, this increases the efficiency at which the system can identify defects. Lamp units, designed to provide even and consistent illumination of particles, are also usually contained within the optical box.

5.2.3 The Ejection System

The ejection system must be capable of physically removing unwanted product items from the main accept stream. The ejection process typically takes place while the product is in free fall; accept particles are allowed to continue along their normal trajectory, and rejects are deflected into a receptacle. Deflection is usually achieved by emitting short bursts of compressed air through nozzles aimed directly at the rejects, although large or heavy objects (e.g., whole potatoes) may require some sort of piston-operated device to mechanically deflect the rejects.

5.2.4 The Image Processing Algorithms

The image processing system classifies particles as either 'accept' or 'reject' on the basis of colour, or both colour and shape.

5.3 Assessment of Objects for Colour Sorting

The size, cost and complexity of sorting machines varies, depending on the size range of particles to be handled, the throughput requirement and the complexity of optical measurement.

Machines are employed in sorting particles as small as mustard seeds; however, rice grains are among the smallest particles to be sorted on a large commercial scale. At the other end of the size range are fresh and frozen vegetables (peas, green beans, cauliflower florets, etc.) and fruit such as apples or potatoes.

Seeds are usually sorted on a single- or double-chute machine at a throughput of 60 to 600 kg/hour. A higher throughput can be achieved on a multi-chute or a conveyor belt machine; rice at 16 tonnes/hour (320,000 objects/second) and peas at up to 16 tonnes/hour are typical examples.

The products that can be handled by today's automatic sorting machines include seeds, coffee, rice, breakfast cereals, nuts and pulses; fresh, frozen and dehydrated vegetables; cherries (with and without stalks); olives; tomatoes; prawns; biscuits and confectionery. Foreign material such as stones, sticks and organic matter can be removed, as well as objects with defects such as discoloration and damaged skin. Figures 5.2 and 5.3 show two typical sorting machines.

Figure 5.2. Sortex 90000 machine for rice, coffee, nuts and grains

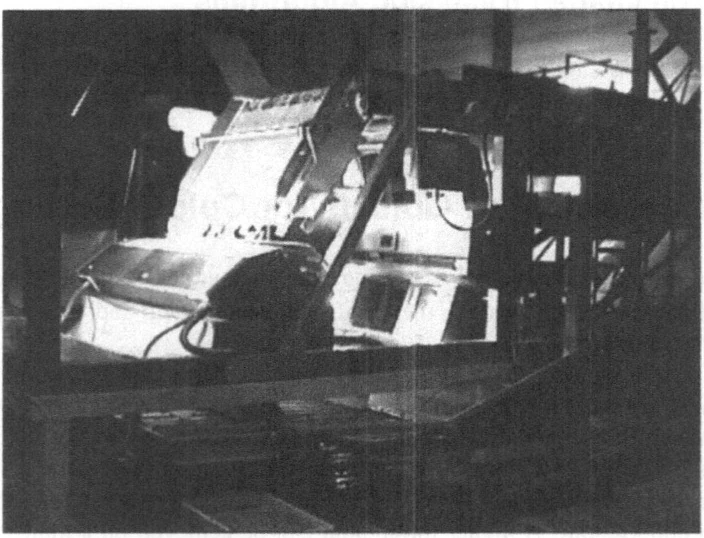

Figure 5.3. Sortex Niagara machine sorting frozen peas

5.3.1 Spectrophotometry

To determine whether a particulate food product is suitable for colour sorting, and which type of sorting machine and optical configuration is most suitable, samples of both acceptable and unacceptable produce must be measured and assessed in the laboratory.

The term "colour sorting" arises from the effect on the overall product appearance as a result of optical sorting. Unfortunately, the term is misleading. The criterion the sorting machine measures when it inspects the product is spectral reflectivity at particular wavelengths, rather than the colour as a whole. Figure 5.4 illustrates typical spectral curves obtained from white rice and white rice grains with yellow colour defects. The relative reflectance signal varies from black (zero and therefore no reflectance) to white (100% reflectance). The wavelengths cover the visible spectrum (400 to 700 nm) and extend into the near infrared (700 to 1100 nm). Optical sorting exploits the region of the spectrum where the reflectance values for all acceptable products are either higher or lower than values for all unacceptable material. If this feature is present, then with the aid of band-pass optical filters, this part of the spectrum can be used as a basis for optical sorting.

Conventional spectrophotometry involves the measurement of carefully prepared surfaces under controlled optical conditions and illumination. However, practical industrial, bulk-sorting machines must deal with naturally occurring surfaces, viewed under non-ideal illumination conditions.

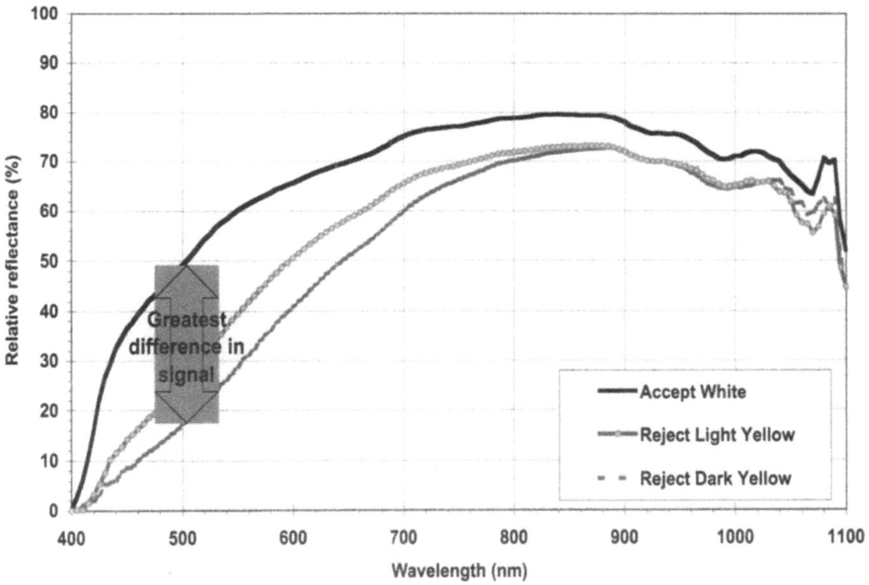

Figure 5.4. Visible reflectance spectra for white rice. In this example, a blue band-pass filter would be used for monochromatic sorting.

Computer controlled reflection spectrophotometers are now widely available and enable measurement of the appropriate optical properties of naturally occurring surfaces. Diffuse spherical broadband lighting is used to uniformly illuminate the item under test. The reflected light is then passed through a computer controlled scanning monochromator, which splits the light into its constituent wavelengths.

The output is measured using a suitable detector and sent to the computer. When the equipment is appropriately calibrated, the results can be plotted showing the variation of reflectance (or transmission) with wavelength, for both acceptable and defective product.

5.3.2 Monochromatic Sorting

Monochromatic sorting is based on the measurement of reflectance at a single isolated band of wavelengths. For optical sorting to be effective, there must be a distinct difference in reflectance within the selected waveband, between all the acceptable particles and all the reject particles (Figure 5.4).

Removal of dark, rotten items from product like peanuts, or dried peas and removal of black peck from rice are typical applications of monochromatic sorting.

5.3.3 Bichromatic Sorting

Sometimes it is not possible to find a single section of the reflectance spectrum where the intensity levels of accept and reject material are clearly separated. Therefore, it becomes necessary to compare simultaneous measurements at two different wavelength bands; a technique called bichromatic sorting.

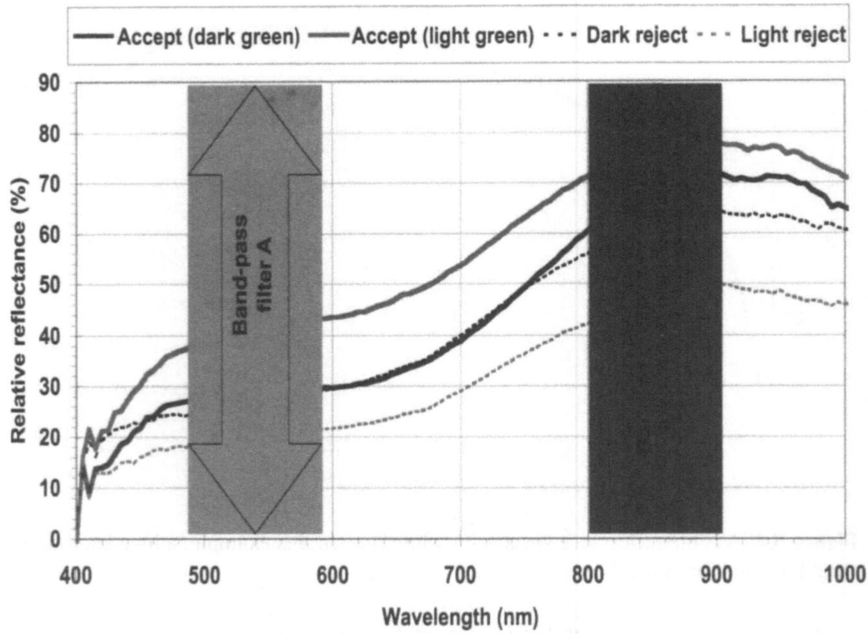

Figure 5.5. Spectral curves for green Arabica coffee. Bichromatic sorting is necessary, since there is no one region of the spectrum where reject material can be successfully separated from the accept material.

Figure 5.5 shows two sets of spectral reflectance curves obtained from green arabica coffee. One set of data (solid lines) represents the lightest and darkest of acceptable beans, the other (dotted lines) represents the lightest and darkest of discoloured beans. In this case, no region of the spectrum allows successful separation of the two sets of curves. However, between 500 nm and 600 nm, the difference in the gradients of the two sets of curves is at its greatest. This is repeated between 800 nm and 900 nm. If measurement *A* is taken at 540 nm and measurement *B* is taken at 850 nm, the ratio *A*:*B* can be calculated. This ratiometric approach will yield a distinct signal difference between the "accept" and "reject" reflectance spectra and allow effective optical sorting. (In principal, measurement *A* could be taken at 510 nm, but this would give lower signal intensity.)

By measuring at two, rather than one region of band-pass wavelengths, bichromatic sorting involves twice as many optical components. At each optical inspection point, besides simply duplicating many of the optical components (e.g., filters, lenses), additional light-splitting devices and more complex signal processing are also required. Consequently, bichromatic sorting is used only when a simple monochromatic measurement is not adequate for effective optical sorting.

5.3.4 Dual Monochromatic Sorting

Dual monochromatic sorting is similar to bichromatic, in that two wavebands are measured but instead of a ratiometric approach, the dual system sorts monochromatically in each of two separate wavebands. This type of sorting is used when it is necessary to reject two distinct types of defect or, defects and foreign material, each of which exhibit different spectral characteristics.

Dual monochromatic sorting is employed with white beans. Maize is rejected by detecting blue reflectance, and white stones are rejected using near infrared.

Some optical sorting applications require both monochromatic and bichromatic decisions to achieve a successful sort. Therefore, bulk optical sorting machines are available which are capable of making both types of measurement simultaneously.

5.3.5 Trichromatic Sorting

Bichromatic sorting techniques can obviously be extended to trichromatic applications. The information gained from the third band-pass filter is often used for detection of gross defects such as the presence of foreign material such as: glass; stones; thistle heads; caterpillars; insects and mice! Trichromatic sorting almost always uses either green, red and infrared band-pass wavelengths. It is unusual for the food industry to use the traditional machine vision community choice of red, green and blue.

Trichromatic sorting allows objects to be sorted according to their size or shape, by suitable modifications to the sorting algorithm. In this way, objects of the same colour, but different shapes can be separated. For example, pea pod can be distinguished from peas and green stalks, or green caterpillars from green beans. Under or over sized objects along with mis-shaped objects, with holes or cracks, can also be detected and effectively removed. The Sortex Niagara machine is capable of simultaneously sorting for both colour and shape at a rate of 40,000 objects/second.

5.3.6 Fluorescence Techniques

Of course, not all bad food is a different colour from good food. It has been found that certain non-visible defects (e.g., bacteria) fluoresce when irradiated with long-wave ultra-violet light (350 nm), and this property may be used as a basis for sorting.

This technique was originally developed for removing 'stinkers' from green arabica coffee beans, but has found applications in sorting peanuts, almonds and cranberries. However, the fluorescence effects can be short-lived and may also depend on the circumstances and time elapsed since the product was harvested.

5.3.7 Infrared Techniques

Over the last decade, the wavelength range used by sorting machines has been extended from the visible, further into the infrared region. Here, both water absorption and other chemical effects play an important part in determining the reflectivity characteristics of food particles.

Bichromatic infrared machines are proving particularly effective in removing shell fragments from a variety of tree nuts.

5.3.8 Optical Sorting with Lasers

Incorporating the use of lasers into bulk food sorting is a technique that is still in its relative infancy. A laser beam is used to illuminate the product and the reflected light is affected by the amount of laser light that is either scattered from the surface, or diffused within an object. Since the laser produces narrow beams of coherent light at a single wavelength, there is no need to use optical band-pass filters. However, a disadvantage with this technique is in maintaining the high capacity demanded by the bulk sorting industry, in conjunction with the necessary resolution to accurately detect defects. The linear scan rate of the laser across the width of the view and the velocity of the product determine the vertical resolution. To date, laser scanning is limited to approximately 2000 scans/second. Therefore, for product travelling at 4 m/s, the vertical resolution is of the order of 2 mm. By comparison, linescan CCD (Charged Coupled Device) technology offers around 5000 scans/second and a resolution of the order of 0.3mm.

Laser scanning also suffers from problems associated with the drop in illumination intensity and therefore, signal to noise levels that result by fanning out the laser beam across a line of sight. It can be quite a design challenge to reliably mechanically scan a laser, by say a rotating polygon mirror, in the hostile temperature and debris ridden environments that are often encountered in food processing plants.

To a limited extent, some successful sub-surface and texture inspection can be carried out on certain soft fruits and berries with laser light. In fact, the technique has already been commercially deployed for some product areas in the food industry, notably for dried fruits like raisons, or certain vegetables, nuts and tobacco. There is certainly scope for further study and possible wider exploitation of the technique.

5.4 The Optical Inspection System

The range of wavelengths measured by an optical sorting machine is defined by the choice of light source, the properties of the optical filters (if used) and the properties of the detector itself. Similarly, at any particular wavelength, the characteristics of the electrical signal from the detector will also depend on these components.

Once the optical characteristics of a product on which the basis for optical sorting have been identified and selected, the relevant wavelength bands must be isolated by selection of appropriate filters and illumination. A primary objective of selecting filters and lighting is to obtain the maximum possible signal to noise ratio from the detector at the required wavelengths, and the minimum possible signal at all other wavelengths

5.4.1 Illumination

When dealing with irregularly shaped objects, uniform, diffuse illumination is required to minimise highlights and shadows, since these would obviously detract from the measurement of true surface reflectivity. To eliminate shadows and highlights at the point of measurement, the particle should be surrounded by a spherical surface of uniform brightness. However, in practice this is just not possible due to the following constraints:

i. To allow a path for particles through the optical inspection chamber, there must be entry and exit points.
ii. The position of the optical components will result in areas of different brightness, compared with the main chamber wall.
iii. The use of light sources of finite size leads to non-uniform illumination.

Specular reflection is almost always a problem, even with a perfect diffuse illumination sphere. If a particle with a diffuse reflective surface is placed in such a sphere, then its true colour will be observed. However, if the particle surface is not diffuse, specular reflection will occur, giving highlights which do not exhibit the true colour of the surface. Clearly the highlights can adversely affect the optical system and consequently result in the incorrect classification of a particle.

The most cost-effective form of illumination is with fluorescent tubes and/or incandescent filament bulbs. A number of lamps are arranged to provide as uniform a distribution of light as possible. With discrete lamps, diffusing windows are used in front of the bulb to diffuse the high-intensity point source of light emitted by the filament. To overcome some of the inefficient heat loss issues associated with incandescent lamps, arrangements using glass rods in conjunction with reflecting ellipses can be implemented.

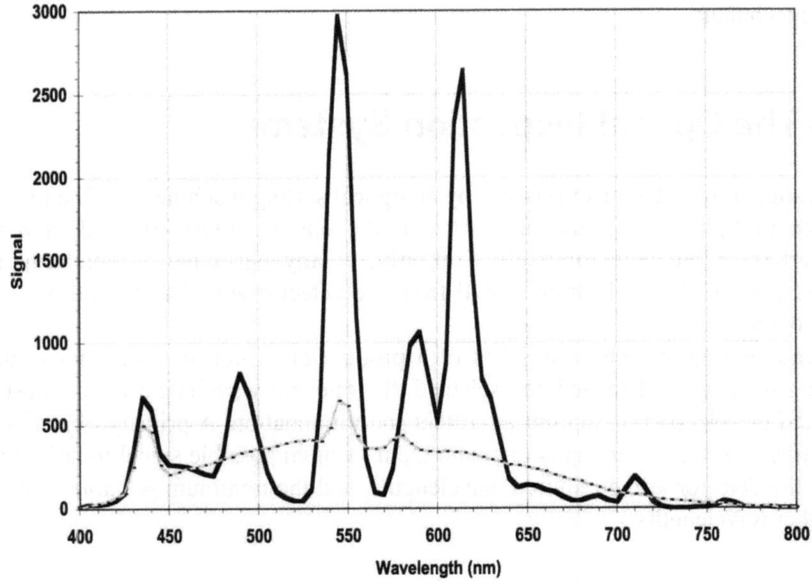

Figure 5.6. Emission spectra for two broadband fluorescent tubes. Both are attempting to simulate pure sunlight, or natural daylight, where all wavelengths of the visible spectrum are present.

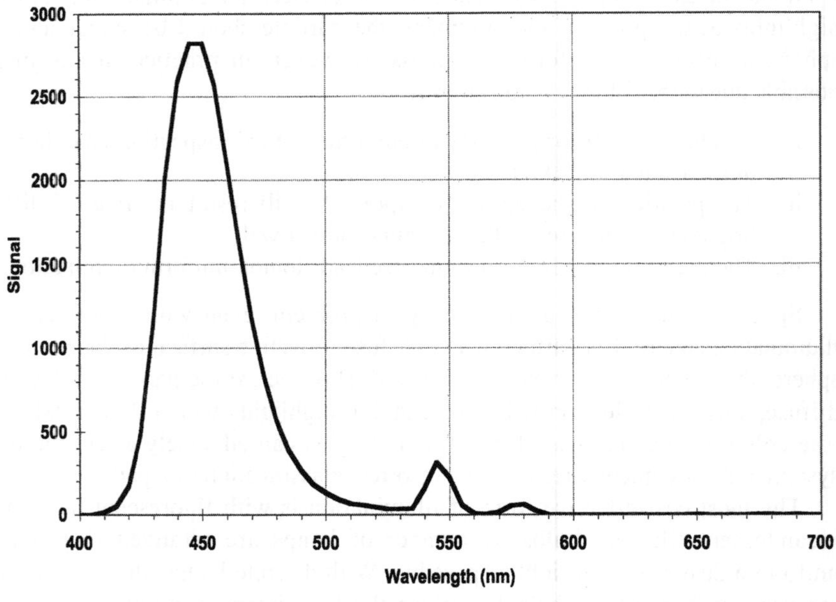

Figure 5.7. The emission spectrum from a fluorescent tube where special phosphors have been used to enhance emission in the blue region (450 to 500 nm) of the spectrum and suppress emission in the red regions (650 to 700 nm).

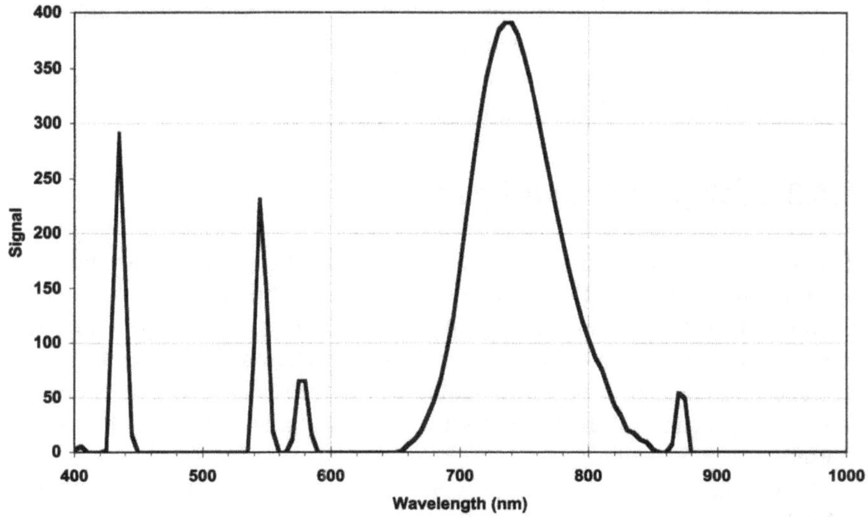

Figure 5.8. Different phosphors can also be used to enhance the emission in the red region of the visible spectrum

Fluorescent tubes can be manufactured with different spectral characteristics depending on the phosphors used. The spectral range of fluorescent tubes spans the ultra-violet to the far visible red (Figure 5.6, Figure 5.7 and Figure 5.8). The advantages of the fluorescent tube are its long life, diffuse light, low cost and relatively cool operation. Its disadvantages are that it is limited to the visible and UV wavelengths and requires a special power supply to prevent flicker.

The advantages of incandescent lamps are their inherent broad spectral range, from blue to the near infrared and their DC operation. However, they suffer from being point sources that dissipate large amounts of heat. A typical emission spectrum is shown in Figure 5.9.

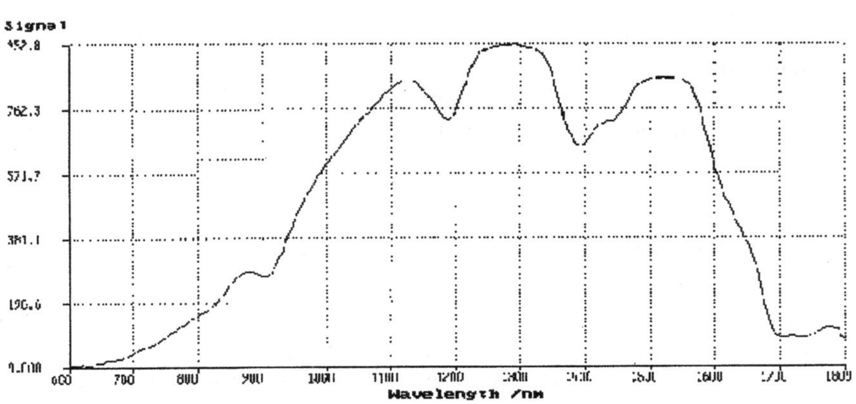

Figure 5.9. A typical broad-band emission spectrum from an incandescent bulb.

In general, for optical sorting applications, fluorescent tubes are the most widely used, except in cases where near infrared or infrared measurements are required. The wider spectral range required for bichromatic machines necessitates the use of incandescent lamps, often in combination with fluorescent lamps.

5.4.2 Background and Aperture

The simplest form of inspection system views the particles through a small aperture and against an illuminated background. The brightness of the background is adjusted so that the optical system measures the same average value, with or without product. This is known as a "matched" background because it matches the average brightness of product, including any defects. For effective shape sorting, where the boundary of each object must be apparent, an "unmatched" background is used.

Matching the background offers an advantage in that measurement of reflectance is independent of object size. For example, consider the case of a stream of particles containing reject items that are darker than accept items. With a matched background, whenever a dark defect passes the aperture there will be a decrease in signal amplitude and with a light particle there will be an increase in amplitude. Hence an unequivocal decision can be made by the electronics. However, if the background were lighter than the average of the product, then all product items would give a decrease in signal. In particular, small dark defects would give signals identical to those of large light particles and the two could not easily be distinguished. The intensity of the light reflected from a particle via the aperture is the product of the size of the particle and its reflectivity, including any area of discoloration.

The background usually consists of an array of suitably located lamps (or LEDs – Light Emitting Diodes) behind an optical diffusing material. In some cases a white, diffuse reflecting plate is used to reflect light from rear-mounted lamps or LEDs towards the detection optics.

The aperture is usually a rectangular slit. The width of the slit must be sufficient to allow for scatter in the trajectories of the particles and for the range of anticipated object sizes. The height of the slit is maintained at a minimum, although allowing for sufficient signal to provide maximum signal to noise for detection, in conjunction with resolution and accurate timing of the delay between detection of a defect and rejection of a particle.

5.4.3 Optical Filters

An optical filter is essentially a piece of coloured glass. Extensive ranges of optical filters are readily available as off-the-shelf components. Alternatively, custom filters can be made at a higher cost. Four basic types of filter are used for optical sorting applications:

i. low-pass: transmitting only below a certain wavelength (Figure 5.10);
ii. high-pass: transmitting only above a certain wavelength (Figure 5.11);

iii. band-pass: transmitting only within a band of wavelengths (Figure 5.12);
iv. combinations of the above in a single filter, e.g., a double band-pass filter (Figure 5.13).

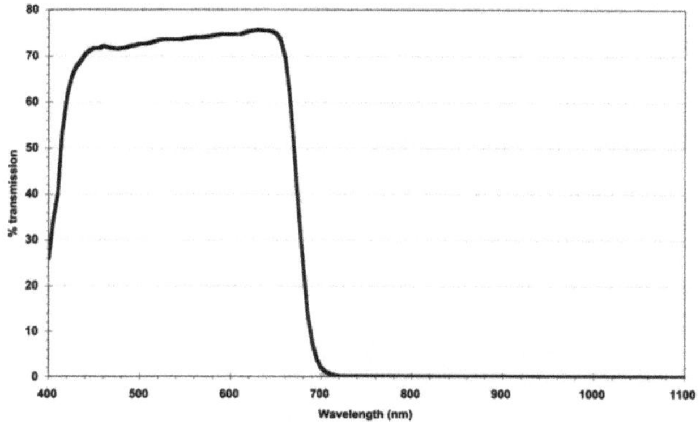

Figure 5.10. The transmission spectrum for a low-pass optical filter.

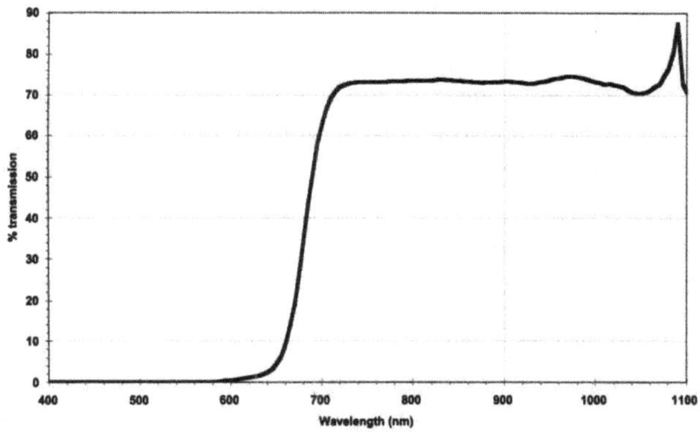

Figure 5.11. The transmission spectrum for a high-pass optical filter.

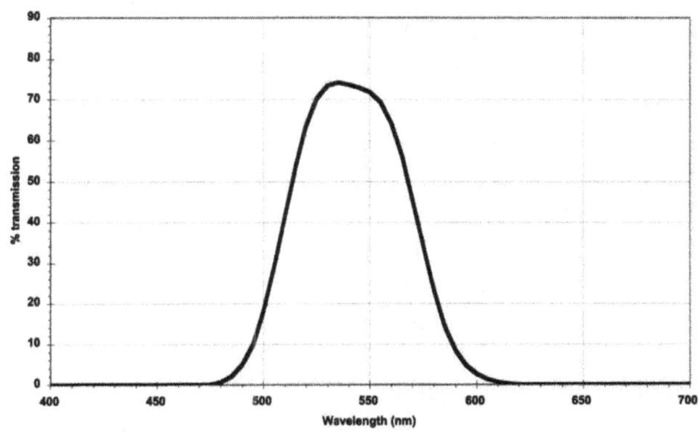

Figure 5.12. The transmission spectrum for a band-pass optical filter.

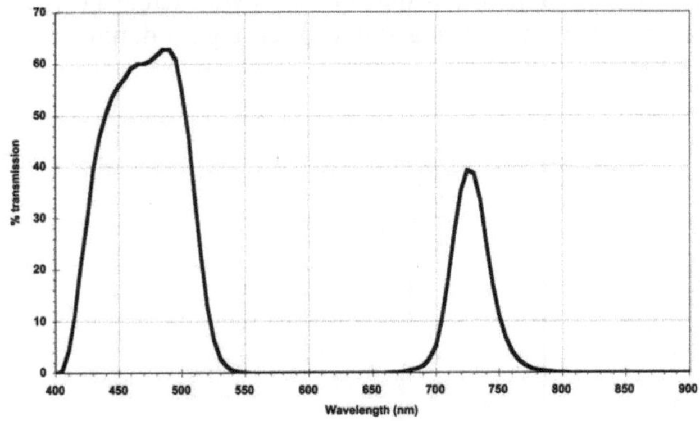

Figure 5.13. A double band-pass filter.

5.4.4 Detectors

Prior to the advent of solid-state detectors, the photomultiplier tube was the best detector of visible radiation. The photomultiplier has a good signal to noise ratio that allows detection at low light levels, and a satisfactory response in the violet–blue (400 to 500 nm) region of the electromagnetic spectrum. However, due to their fragile mechanical construction, photomultipliers are not very robust. They also suffer from limited life, high operating voltage and poor deep red and near infrared response (650+ nm).

Following photomultiplier tube technology, solid-state technology now dominates the optical sorting industry. Initially the photodiode was used due to its comparative cheapness, mechanical robustness and almost indefinite life. However, compared to photomultipliers, photodiodes have a poor blue response.

Contemporary optical sorting machines now employ high-speed linescan CCD (Charge Coupled Device) technology. Silicon CCD technology offers the advantages of high sensitivity, good broadband response (400 to 1000 nm), high spatial resolution and good quantum efficiency. Although CCDs are analogue sensors, their output is easily converted to digital form. Consequently, state of the art, low-noise, high-speed digital processors can be used for subsequent signal processing. Unfortunately, CCDs continue to suffer from relatively poor response in the blue region.

In order to extend the detection range into the infrared domain, other detector materials besides silicon are used. Having a good infrared response (up to approximately 1700nm), germanium detectors are readily exploited by the food sorting industry for detection of foreign matter. Historically single photodiodes were used. However, Germanium linear arrays are now commonplace and generally used in combination with silicon CCD technology. In fact much of the infrared technology developed for the telecommunications industry is now readily exploited for optical sorting applications.

5.5 The Sorting System

5.5.1 Feed

The product feeding system in a sorting machine should provide three basic functions:

i. metering: to ensure that the optimum numbers of objects per unit time are fed through the optical inspection area;
ii. acceleration to a constant velocity. The time taken for objects to travel from the optical inspection point to the ejection point must be constant so that activation of the ejector can be accurately coincided with the position of the object. Typically the velocity of the product is of the order of 4 m/s. The delay between detection and ejection is between 0.5 and 100 ms;
iii. alignment: to ensure a controlled trajectory through the inspection and ejection points.

In reality, metering of product is achieved by a vibrating feeder tray, situated just below the output of a hopper. The following feed systems are commonly employed:

- an inclined gravity chute;
- a flat belt;
- an inclined belt (unique to Sortex);
- a "C" shaped belt;
- contra-rotating rollers;
- a narrow grooved belt.

To reliably detect small blemishes on fruit and vegetables, it is necessary to inspect the product from two sides. The traditional architecture of an optical sorting machine is to feed the product along a horizontal conveyor and then observe the product from top and bottom as the product flies off the end of the conveyor. The drawback with this approach is that the bottom camera is soon covered in product. Sortex' Niagara machine overcomes this problem with the PowerSlide™ feed system (Figure 5.14). The belt conveyor on the Niagara is inclined at 60° to the horizontal, such that the cameras can view the product from either side, and remain clean.

The flat belt or gravity chute approach presents the product in a single layer, restricting the view to two sides, but allowing a much higher throughput of product to be achieved. In contrast, some feed methods channel and separate the particles into a single stream, each object dropping down after the other in "single file." This feed technique allows an all-round view of each object's surface, since three cameras can be positioned around the foot of the chute. Obviously three views allow a very high quality sort with an excellent yield. However, one disadvantage is the relatively low throughput of product compared to wide flat belt or chute techniques (a few hundreds of kg/hour compared to several tonnes/hour). As a consequence, single channel feeding is usually only employed for high value products (Blue Mountain coffee, nuts like almonds or macadamia, selected beans and pulses, etc.). Throughput can be increased in single channel feed systems by adding two or more channels to a machine.

5.5.2 Ejection

The usual method for removing unwanted items from the main product stream is with a blast of compressed air from a high-speed solenoid or piezoelectric valve, connected to a strategically positioned nozzle.

Pneumatic ejector valves must have rapid action, reliability, long lifetime (a minimum of one billion cycles) and mechanical strength. The fastest (a Sortex patented piezoelectric design) operates at a frequency of 1 kHz, firing a pulse of air for 1 to 3 ms. Ejectors operate at input pressures between 200 to 550kPa (30 to 80 psi), depending on the size of the object to be removed. Typically, the ejection point is located outside the optical inspection area, because the action of the air blast on a rejected object could cause dust particles and skin fragments to be blown around that could create false rejections. However, at the same time, it is advantageous to eject objects as soon as possible after the optical inspection point, due to unavoidable variations in the trajectory of each individual item.

Figure 5.14. The Niagara PowerSlide™

The appropriate time delay between the inspection and ejection point is generated by electronic circuits. The accurate timing required, to coincide the ejector air blast with that of the object to be ejected, relies on the objects having constant velocity as they fall in front of the ejector nozzle. In practice the tolerable variation in product velocity is about 5%. The trajectory of each particle also becomes harder to predict between the viewing point and the ejection point. It can become a major design challenge to position the chute, optics and ejection system as close together as possible.

The operational lifetime of the ejectors must be in the region of at least a billion or more cycles. Food processing is usually a 24 hour-a-day, all-year-round operation. Operators cannot afford to regularly shut down a machine for even a few minutes to replace faulty ejectors. Under these circumstances, machine reliability and stability of operation are critical.

For certain large or heavy objects a solenoid valve may be used to control a pneumatically operated flap or plunger to deflect rejected items. Specialized ejectors have been developed for pulps or slurries to remove rejects by suction and are mounted above a flat belt, downstream from the inspection unit.

Smart ejection Systems

To accommodate long, or irregularly shaped objects, Sortex has developed the SmartEject™ system. With most optical sorting machines, the air blast is fired from one or more ejectors in an array, spanning the width of the belt or chute, just after the optical inspection area. The air blast is aimed solely at the centre of the defect (also known as "centroid ejection"). If the defect is a small blemish on a large object, then the blast may not be sufficient to remove the item. To combat this problem, Sortex's Niagara machine computes the location of the object encompassing the defect and fires the appropriate number of ejectors so as to fire at the entire object – this improved ejection system is known as SmartEject™ (Figure 5.15). SmartEject™ fires one or more of 160 high-speed ejectors positioned across the line of view at the profile of an object, rather than at a defect, which improves both accept quality and yield.

Three-way Separation by Two-way Ejection

Sortex has also pioneered the ability to perform a three-way sort by adding a second bank of ejectors to the Niagara vegetable sorting machine. To date it has been the convention in optical food sorting to have only two-way separation, into accept and reject categories. The three-way separation by two-way ejection allows an additional product classification. For example, three-way separation for green beans now enables the following three categories:

i. accept;
ii. reject (rots, blemishes, foreign material, stalks, etc.);
iii. accept with stalks.

The advantage of the new third category is to allow recovery of otherwise "good" product that would normally be rejected. Accept green beans with uncut stalks can be returned to the "snibbers" (stalk cutters). In this way significant savings on recovered volumes can be made. Approximately 2% of green beans, typically running at 5 to 10 tonnes/hour are rejected on uncut stalks alone.

5.5.3 Cleaning and Dust Extraction

The successful application of optics in an industrial environment pervaded by dust, oil, starch, food debris or water poses major design issues for optical engineers. Considerable expertise is necessary to design an optical sorting machine capable of successful commercial operation under in-plant conditions. The operating temperature range encountered in a food processing plant varies between –5 and +40°C, making optical, mechanical and electrical tolerances critical to the effective operation of the machine. If the cameras of an optical sorting machine become obscured by debris, then the performance of the machine rapidly deteriorates.

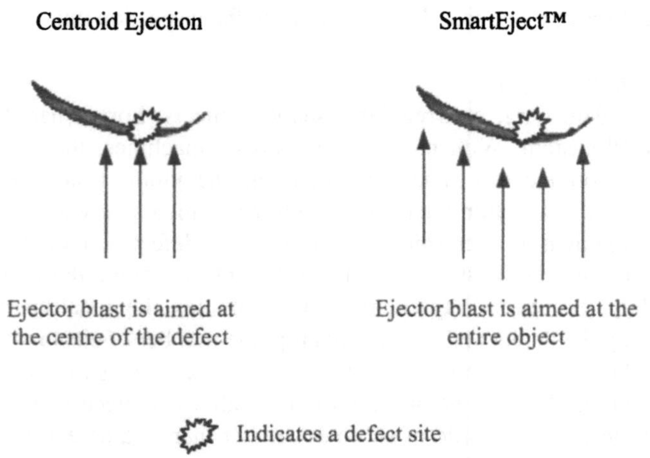

Figure 5.15. The SmartEject™ system for precise removal of larger objects with small defect sites that may not be located at the centre of the object.

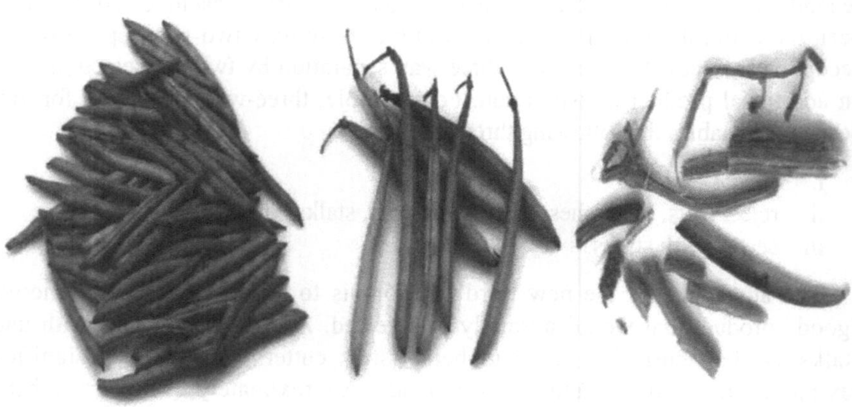

Figure 5.16. Three-way separation by two-way ejection for green beans. From left to right: accept; accept with stalks; reject.

To protect the optical components from dirt or moisture, they are contained in an "optical box" with a glass window. The position of this window in the optical path should be such that any small particles, which may settle on the surface, are out of focus and therefore create minimum noise in the optical signal. However, it is essential that this window is kept as clean as possible and a number of facilities to achieve this may be provided on the machine.

Firstly, the product being fed to the machine should be as dry and dust free as possible. However, the action of storing it in a hopper and feeding it on a vibrating tray will usually create some dust. Hence, a dust extraction nozzle is often fitted at the end of the vibrating tray. In the case of a chute feed, the top of the chute may be perforated so that air can pervade the product stream to remove dust particles.

In addition to dust extraction, the optical box window can be cleaned by means of compressed air jets. These "air knives" as they are also sometimes called, provide a continuous curtain of air to prevent particles settling on the surface of the glass. If necessary, they can also provide a periodic high-pressure blast which removes any particles that may have settled on the window. Pneumatically driven blades or brushes can also be used to periodically wipe the window. In some machines this may be combined with an air "blow-down" facility.

As a final precaution, any dirt created by the action of the ejector blast on the particles, may be drawn away from the window area by a dust extraction nozzle, positioned just below the optical box.

Similarly, for wet or frozen product applications water jets and wiper blades can be substituted for air-based systems. Machines for sorting wet product are periodically hosed down with water, so must be water and dust proof to IP65 standards [2]. Similarly, dry product machines are also manually cleaned with an air hose. For hygiene reasons, all potential "bug traps" must be designed out of all sorting machinery.

5.5.4 The Electronic Processing System

The electronic systems in sorting machines have progressed from the simple analogue circuits of the early machines to the advanced digital microprocessor based circuits found in the present generation of machines. In contrast to many "machine vision" applications, it is common for the optical data processing system of a bulk optical sorting machine to be hardware, rather than PC based. At the present time it is simply not practical to process 40,000 objects/s for colour and shape followed by effective control of the ejection process with a PC based system.

Most of the setting up of the sorting parameters can be done by the machine itself, including in some cases the ability of the machine to "learn" the differences between good and bad product. However, the operator is always given the opportunity to fine-tune the final result.

A sophisticated optical sorting machine will track the average colour of the product so that, even though the average product colour may change with time, the machine will continue to remove only the predefined abnormal particles. Optical sorting machines are often provided with a white calibration plate which is either manually, or automatically placed in the optical view at user defined intervals. The machine is then able to correct for any measurement drift that has occurred.

Once a machine has been set up for a particular product, all the machine settings can be stored in memory. This can be repeated for a number of different products and then, at a later time, the machine can be made ready to sort any of these products simply by recalling the appropriate settings from the memory. Alternatively, the settings can be used as a coarse starting point from which to fine tune a machine towards an optimum setting for a particular set of circumstances. Most food plants sort one particular product type. For example a rice mill may sort different varieties of rice, but would not for example suddenly switch to coffee. It would be unusual for a food processor to be sorting many different and diverse types of product.

Advanced sorting machines have a memory capability that can be exploited to provide information about the product for the operator. For example, the number of rejects that have occurred in a certain time, or information about any drifts in colour in a certain batch of product. Information about how the machine itself is operating can also be provided to assist with preventative maintenance.

Optical Detection and Differentiation by Shape

There are many applications in food sorting where the defects are similar in colour to the good product. For example: insect larvae in amongst blueberries take on the same colour as the berries; the stems on green beans are the same colour as the bean (Figure 5.17); similarly pea and pea pod; or green caterpillars among green beans or peas. In order to be able to solve these types of applications, Sortex has pioneered the ability to sort objects on the basis of size, roundness, area, length and therefore, shape. In addition to colour and shape, the minimum size of the discoloration necessary for a particle to be rejected can also be defined.

Figure 5.17. Stems and stalks found in green beans.

In the above examples, the larvae are elliptical in shape whereas the berries are round, and the stems on green beans are much thinner than the beans. A major innovation for the food sorting industry has been the development of new vision algorithms for computing the size and shape of objects. This has required the optical sorting industry to implement these algorithms in specialised electronic hardware. The Niagara machine's ability to sort objects on the basis of shape as well as colour, at high speed, is the basis of one of the major innovations of the machine. Up to 40,000 objects per second can be simultaneously sorted for shape and colour, across a 1100mm wide line of view.

User interfaces

A typical machine will have either a keypad and a display unit, or more commonly in contemporary machines, a touch-screen user interface. A good user interface should allow the operator to set up and control the machine by means of an easy to follow series of menus. In addition, the display unit will provide the operator with information regarding the settings of the machine while it is sorting, together with details of any faults that may occur.

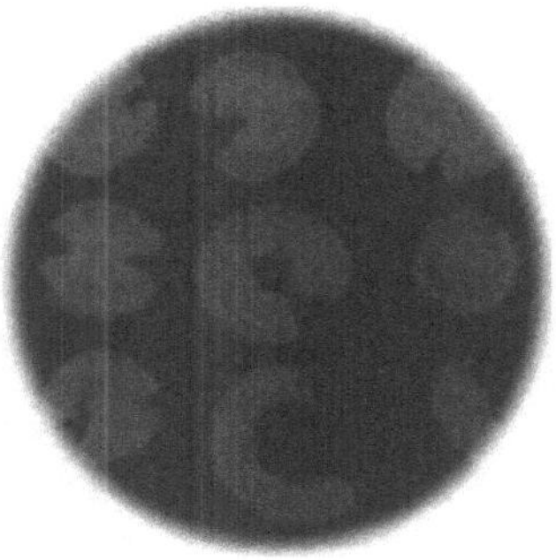

Figure 5.18. Advanced shape processing allows removal of many of the problems encountered in sliced carrots such as "polos", ellipses (oblique slices), cracks and those with tangential misshapes. This is simultaneous with colour defect removal.

Mapping Techniques

A bichromatic sorting machine using two band-pass filters, say green and red, makes a decision based on the ratio of the two signals in conjunction with the intensity of the individual signals. The situation can be represented as a two-dimensional "colour" map by plotting the reflectivity of colour 1 versus that of colour 2 (Figure 5.19).

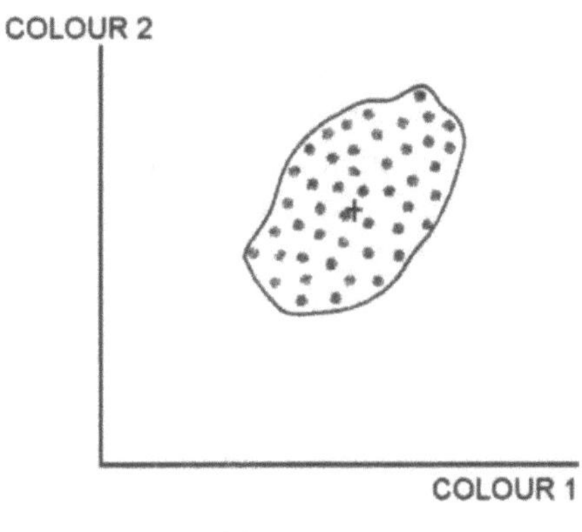

Figure 5.19. A bichromatic colour map representing the two channel colour distribution.

The bottom left-hand corner of this map represents the reflectivity from a black particle (0% reflectivity) and the top right-hand corner represents the reflectivity from a white particle (Figure 5.20) (100% reflectivity). The boundary curve in Figure 5.20 is the reflectivity map contour, outlining the acceptable product, as seen by the sorting machine, for a typical product. The contour line represents the chosen accept/reject threshold. The "+" within the map contour is the background "balance point", which represents the average colour of the product.

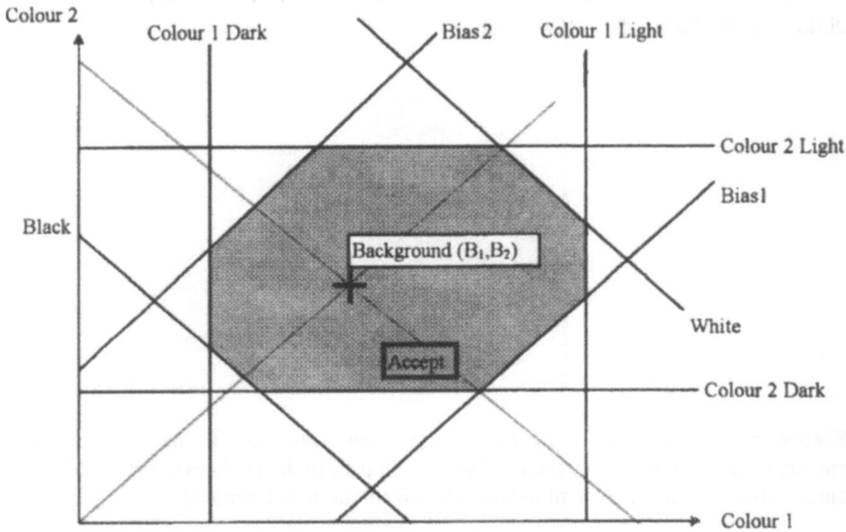

Figure 5.20. Bichromatic sensitivity thresholds.

A major part of setting up an optical sorting machine is to achieve the best overall accept/reject ratio for the product being sorted. The operator can do this by using the user interface to adjust the shape and size of the map contour, to match as accurately as possible the map contour of the product batch (Figure 5.21). The sorting sensitivity increases as the machine map contour is decreased in area, as it approaches the area of the map contour of the product batch. Product within the area bounded by the threshold levels is accepted and product outside is rejected.

These techniques allow an optical sorting machine to remove a far greater range of defects, with greater accuracy and without the penalty of removing large amounts of accept product. Obviously, these techniques can be extended into three dimensions for trichromatic colour sorting.

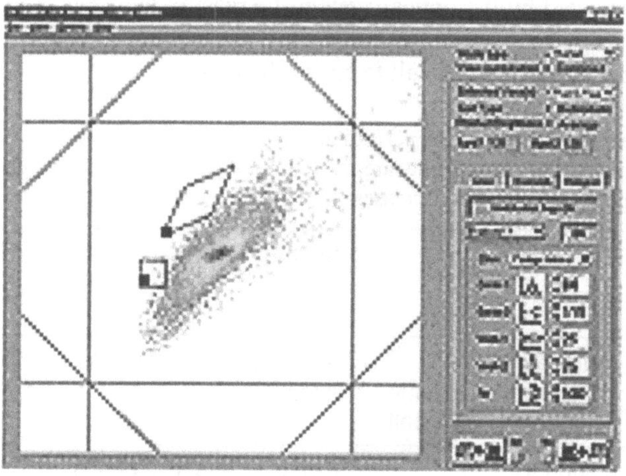

Figure 5.21. A bichromatic colour map, as displayed on the user interface.

5.6 The Limitations of Colour Sorting

There is often a misunderstanding that a colour sorter can remove 100% of the defects from a given batch of product. In practice this is impossible. A colour sorter will reduce the concentration of defective product, but it can never be 100% effective.

All colour sorters are bound to remove some acceptable objects and fail to remove some of the defective objects. There are several reasons for this. Sometimes, the physical size or the colour difference of the defect from the product may be too small for accurate detection. Occasionally the machine may detect a defect and remove the object, but the object re-enters the accept stream after it has been ejected as a consequence of a random collision. Ejector performance and minimal positional pitch of the ejectors in the array below the optical system can also become a limitation for accurate ejection. At present, the smallest ejectors have a 3mm pitch. This limits the ejector "resolution," especially for small products like rice or sesame seeds.

Machines can be adjusted by operators to optimise their performance. Sensitivity is one of the principal parameters that the operator can change. Increasing the sensitivity will result in the machine rejecting more defective material. However, a greater proportion of good product will also be rejected as the sensitivity threshold approaches the average product colour. There is normally a compromise point between achieving a high sorting efficiency and optimum yield (the ratio of good to bad material that is rejected). This compromise point is primarily achieved as a result of operator experience and training.

There are physical limits to the product throughput that a sorting machine can successfully achieve. If the product flow is increased above the upper limit, the product sheet will no longer be a monolayer. Objects will overlap and sorting performance will deteriorate, since many defects will be obscured and therefore, will not be detected by the optical system. Increasing the flow of product through the machine will also result in increased good product being lost, since overlapping and colliding products are difficult to eject efficiently.

Table 5.1 Illustrates some typical performance figures for a variety of products sorted on different machines. The throughputs are quoted in ranges, since the throughput increases as the level of input contamination decreases.

Product	**Machine**	**Sorting Criteria**	**Throughput (Tonnes/Hour) per Machine**
Whole green beans	Niagara - Trichromatic colour sorter, 1m wide belt	Remove attached stems and blemish	8
Green coffee	90003Bi - 48 channel, gravity chute bichromatic sorter	Remove defective beans and foreign material	3 to 6
Parboiled rice	90004 mono - 128 channel, 4 gravity chute, monochromatic sorting machine	Remove spotted, discoloured rice and foreign material (stones, glass, paddy etc.)	5 to 10
Frozen Peas	Niagara - Trichromatic colour sorter, 1m wide belt	Remove foreign material, pea pod, sticks etc. by colour and shape	10 to 16

5.7 Future Trends

Computer vision systems are increasingly being used in general manufacturing, for example in pick and place applications such as printed circuit board (PCB) population and manufacture. However, the demands of the food industry are generally far greater. At present, there is only a limited range of computer vision equipment available for use in the food industry. However, in the future this is likely to change.

Two factors are limiting the rate at which computer vision systems are being introduced to the food industry:

i. The data processing rates required in a sorting machine for the bulk food processing industry are very much higher than those in a similar inspection machine for manufactured objects.
ii. The development of improved materials handling and separation systems is not keeping pace with the dramatic advances being made in computer hardware technology.

A computer vision system potentially offers many benefits over a conventional colour sorter. The ability to sort objects simultaneously on the basis of several different criteria would be a primary advantage. However, for the immediate future, the most likely application of advances in electronic hardware is the gradual improvement in performance of the present generation of sorting machines.

The optical sorting industry readily exploits advances in components, manufacturing processes and designs. At the present time, defect detection is mostly carried out in the visible and near infrared wavelengths, mainly because of the added cost of infrared detector technology. However, other wavelengths are already used in other areas of the food-processing industry. X-ray techniques are often employed as a final check for foreign material in packed or processed foods, or to detect hollow potatoes, for instance. Ultra-violet light can be used in some nut sorting applications, especially to detect fungal-infection sites that fluoresce when exposed to UV light. The natural progression of monochromatic to bichromatic sorting will inevitably lead towards wider use of trichromatic technology. As the cost of lasers continues to decrease, so laser technology may become commonplace for texture or sub-surface inspection.

Meanwhile, advances in detector resolution, valve technology, ejector-duct materials and design, will all help to optimise the ejection process. In the future, unwanted objects may be removed with rapier-like precision. Improvements to the operational stability of the sorting machine are likely to have a big impact, increasing the product throughout and ensuring that the machine optics need to be calibrated less frequently.

A consequence of the increased pace of technological advances will be a reduction in the working lifetime of sorting machines. New machines will have to be developed and manufactured under faster cycle times to keep pace with the market. Some components, particularly electronic chips, can quickly become obsolete. Similarly, the falling price of high technology is already allowing new competitors to enter the marketplace. Any optical sorting company that ignores these factors can only expect reduced profit margins. In some ways, the real future challenge will be to provide integrated solutions that fulfil the demands of the food processing industry, at a price that can be justified.

5.8 References

[1] Anon. (1987) Electronic sorting reduces labour costs, *Food Technology in New Zealand* 47.

[2] British Standards Institute (1991) Degrees of protection to EN 60 529/IEC 529 (1991).

Chapter 6

Editorial Introduction

Inspecting decorative ceramic tiles for the bathroom and kitchen presents certain difficulties, which can only be overcome by considering all aspects of the design of the vision system. This includes the illumination and optics, as well as the image processing algorithms and their implementation. Since the human eye can detect very small defects, the inspection system must also be capable of doing so. This requires that the image resolution be very high indeed: far greater than that achievable using a single CCIR/RS170 camera. Moreover, many tiles have a printed pattern, such as a stylised flower or similar motif. Tiles may have mirror-like or heavily embossed surfaces. The time available to inspect each tile is also short, as the throughput rate is determined by modern high-speed manufacturing techniques. Although ceramic tiles are not natural products, the surface condition is only loosely controlled. This is true of many coating/spraying processes. As so often happens, the inspection task cannot be defined in precise objective terms; designing the vision system requires compromise based on engineering judgement. Provided the illumination is chosen appropriately, it is possible to detect very bright surface features that are much smaller than a single pixel. Since this requires illuminating and viewing at an oblique angle, the geometry of the optical sub-system must be adjusted to obtain an image that is sharply focused everywhere. Although this produces geometric distortion of the image, it can be corrected by image processing, based on a suitable calibration target. It is this type of interplay between the different parts of a system that makes the design process so fascinating for the vision engineer. It is this feature above all others that makes the process of designing a vision system almost impossible to automate and very difficult to teach to other people.

Chapter 6

Surface Defect Detection on Ceramics

A.K. Forrest

6.1 The Problem

Ceramic tiles are manufactured in a huge variety of types ranging from surfaces almost as rough as a house brick to a polished mirror finish. The tile surfaces must be blemish free, but the definition of blemish and the possible types of blemish vary with tile type. There are a very large number of types of blemish and the smallest blemish visible to a human can be surprisingly small, for example less than 10μm deep, which may be on the surface of a heavily embossed tile. The detection of blemishes is therefore both a poorly defined and poorly constrained task. Although these are not strictly natural materials the techniques necessary are very similar to those for true natural materials because of the free-form nature of the problem.

The first approach one might take is to determine the smallest blemish and from this calculate what resolution is required to obtain two pixels across this blemish (e.g., the Nyquist criterion). The minimum size of the image in pixels can then be calculated and image processing algorithms devised. Unfortunately the tiles may be 50cm across and the smallest blemish may be 100μm or less implying an image size of $10{,}000 \times 10{,}0000$. Tile production lines run at up to 1 m/s, this implies a pixel rate of 2×10^8 pixels a second. This is an extremely high rate to even capture the data. Most image grabbing electronics do not work faster than 50MHz and even if the image can be acquired the processing power necessary to separate blemishes will require many operations per pixel, if indeed it is possible at all. Obviously this straightforward approach will not work for this problem.

In many cases surface blemishes are not visible perpendicular to the tile. Changes in surface texture and small scratches are two examples that occur very frequently in practice. The reason that these apparently invisible blemishes are important at all is that in use the tiles will be viewed obliquely as well as perpendicularly and will be lit in a large variety of ways. This is, of course, an obvious clue as to how to solve this problem. By arranging the lighting and viewing angles we can make the blemishes much more obvious, which will allow much lower performance image processing to work effectively.

6.2 Oblique Imaging

6.2.1 Oblique Lighting

Lighting and illuminating a surface at an oblique angle makes the surface appear as if it is more reflective, i.e., more specular, than it is at perpendicular incidence. This effect has a long history and is described by the Rayleigh criterion.

$$h < \lambda/8\cos\theta$$

where h is the height of the surface roughness variations and θ is the angle of incidence of the light and λ is its wavelength. The criterion is that if the above equation is true the surface acts as a specular surface i.e., a mirror. If rays reflected from the surface combine randomly the effect is that of a scattering surface. If they combine constructively the surface acts like a mirror. The nearer the light is to grazing incidence the smaller the phase shift between rays hitting the surface, hence the above equation. Making the surface appear more specular is useful because light is reflected from the surface but scattered by the blemishes. If an image is made in scattered light the blemish appears as a bright mark against a dark background. This has the added advantage that even a blemish that is less than a pixel in size, will still be detectable if it raises the brightness of the pixel above the noise on the dark background. In practice this occurs very often.

A practical oblique lighting system will almost certainly need to use collimated light. Because the angle to the surface is small the range of angles from the light source must also be small, implying collimation. Conventional light sources are not good at producing a collimated beam. This is because through any optical system the product of area and solid angle is constant (or increases). To obtain a small solid angle we therefore need a large area, which implies lowering the intensity and also, of course, implies big optics. Lasers have a small area and small solid angle output at high intensity (e.g., they have a high luminosity). They are therefore very well suited to this task.

6.2.2 Oblique Viewing

The viewing camera should be placed so that it is not illuminated by the reflected specular light, but it should be as close to this reflected beam as possible so that light which is scattered only a small amount will be collected by the camera (Figure 6.1). The camera will have to image the surface at a very large angle to the optical axis and on first sight this would appear to give a very small depth of field. By tilting the image plane of the camera, however, it is possible to get the whole field of view in focus. The condition for this to occur is called the Scheimflug condition. It is fulfilled if lines through the object plane, image plane and centre of the lens meet at a point as shown in Figure 6.2.

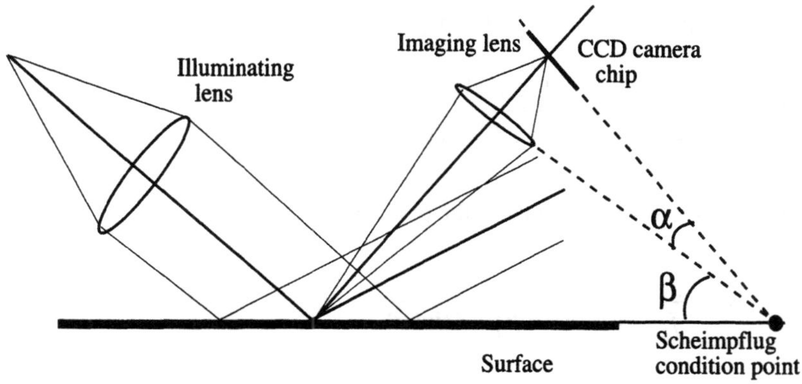

Figure 6.1. Oblique imaging set-up.

It might appear that special lenses would be needed in this situation. In fact it can be shown [1] that any lens that images with a flat field in the normal imaging configuration will work just as well in the oblique imaging configuration.

6.2.3 Image Rectification

The image obtained obliquely will have different object scales at different parts of the image and the image intensity will also vary across the image. To obtain images that have constant scale and intensity the images must be processed. The resulting images can then be processed by conventional methods.

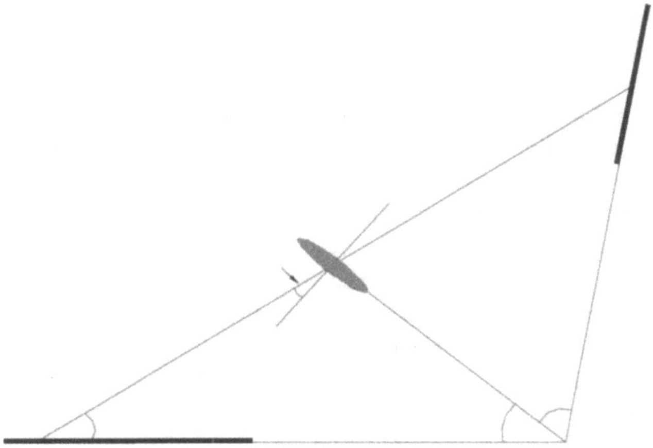

Figure 6.2. Geometry of oblique imaging,

The relationship between image and object positions [1] in Figure 6.2 is

$$x_1 = \frac{d x_2 \sin(\beta)}{x_2 \sin(\alpha+\beta) - d\sin(\alpha)} \tag{6.1}$$

The magnification in the plane of the paper is then

$$M_t = \frac{\delta x_1}{\delta x_2} = \frac{-d^2 \sin(\alpha)\sin(\beta)}{\left[x_1 \sin(\alpha+\beta) - d\sin(\alpha)\right]^2} \tag{6.2}$$

The magnification perpendicular to the plane is

$$M_n = \frac{f_1}{f_2} = \frac{d\sin(\alpha)}{x_2 \sin(\alpha+\beta) - d\sin(\alpha)} \tag{6.3}$$

The area magnification which can be used to normalise the image intensity, which is inversely proportional to the magnification, is

$$M = M_t M_n = \frac{-d^3 \sin^2(\alpha)\sin(\beta)}{\left[d\sin(\alpha) - x_2 \sin(\alpha+\beta)\right]^3} \tag{6.4}$$

which can be rewritten as

$$M = \left[\frac{x_1}{x_2}\right]^3 \left[\frac{\sin(\alpha)}{\sin(\beta)}\right]^2 \tag{6.5}$$

If a two-dimensional CCD camera is used the image rectification can be quite onerous, however, because in this application the surface was already moving, a one-dimensional linescan camera was used leaving the rectification to be calculated in one dimension only [2].

Real lenses do not usually have coincident input and output principal planes, as is assumed in Figure 6.2. This does not cause any fundamental problems but for a practical system this must be taken into account. Measuring the required angles and distances accurately so that image rectification can be accomplished can be difficult. For this reason an algorithm was developed that, when given an image of a known grid, the parameters of the system would be returned automatically. This system failed in practice due to the badly behaved nature of the equations involved, that is the parameters interact in a very complex way which, when combined with even very small amounts of noise, leads to a system with many local minima. The correct minimum could therefore not be found. The problem was solved by applying a polynomial fit of image magnification using the known grid image. A seventh-order polynomial gave good convergence of the algorithm as well as the correct results in the presence of noise. Image brightness across the field of view can vary very greatly if very oblique angles are used. The 8 bit dynamic range of the camera can be insufficient to cover this acceptably well. For this reason the illumination was varied in intensity across the field of view. Computation of the theoretical brightness was therefore not necessary in the practical system. The reflectance/scattering was assumed to be equal across the field of view and was simply normalised.

Figure 6.3. Normal view of "flower" tile.

Figure 6.3 shows a normal CCD picture of a 20 × 20cm glossy ceramic tile. The aim of the optical arrangement was to image tiles like these for scratches and surface blemishes using the minimum amount of image processing. Figure 6.4 shows the image obtained with the tile illuminated and viewed obliquely. The very bright tile edge is due to light being scattered into the camera directly from the edge. All the linescan images have been inverted to allow better printing of subtle detail; therefore the bright edge is black.

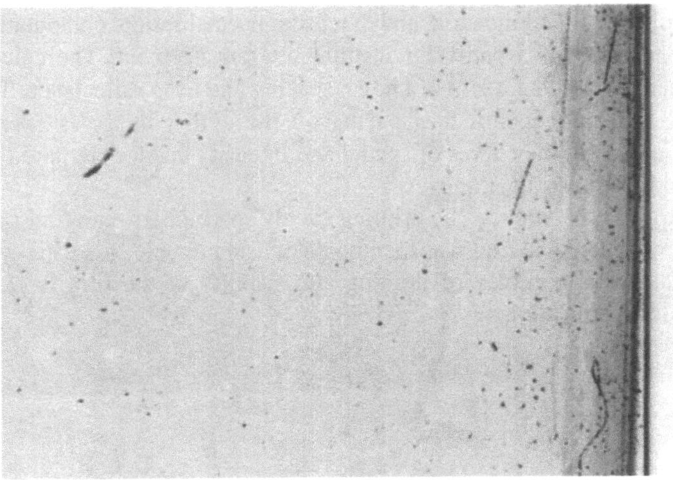

Figure 6.4. Linescan image of flower tile taken from oblique angle.

Even without image rectification or intensity normalisation it is clear that surface blemishes are much brighter than the pattern on the tile. Most of the blemishes in this case are dust with a surface abrasion top left. This image has been taken at quite an extreme angle so dust is very prominent. Less extreme angles give a reduced effect but there appears to be no way of distinguishing against dust. There is a possibility that polarisation might be used to differentiate dust and blemishes.

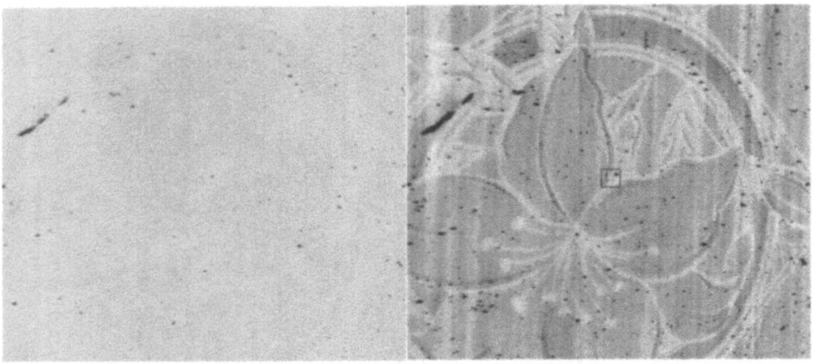

Figure 6.5. Rectified images of the flower tile with different contrast and brightness settings. Blemishes can be thresholded from the intensity-normalised image.

When the images have been rectified and normalised it is easy to threshold out blemish information. The contrast and brightness have to be specially adjusted to show the tile pattern as in the right section of Figure 6.5. The rectified image using the whole 8 bit dynamic range is the image on the left. The small-boxed part of image 5 is shown in Figure 6.6. This shows the large amount of image information available. The images shown are for a 20cm tile. This occupies about a quarter of the available 5000 pixel image. An image fully utilising the camera would contain 16 times as many pixels as shown in Figure 6.5. Note that somewhat counter-intuitively the parts of the surface farther from the camera are brighter.

The geometrical rectification and intensity normalisation computations need only be done once for a particular instrumental arrangement. The calculation is accomplished by imaging a grid of known spacing and even reflectance. The image of the grid is distorted until the spacing on the output image is even and the intensity is uniform. The curves of spatial and intensity changes required to do this are then used for subsequent images.

The test grid was produced by printing evenly spaced bars onto transparent film using a laser printer. The film was then bonded, using double sided sticky film to a sheet of glass. The accuracy of printing and stability of the film is easily good enough for this application.

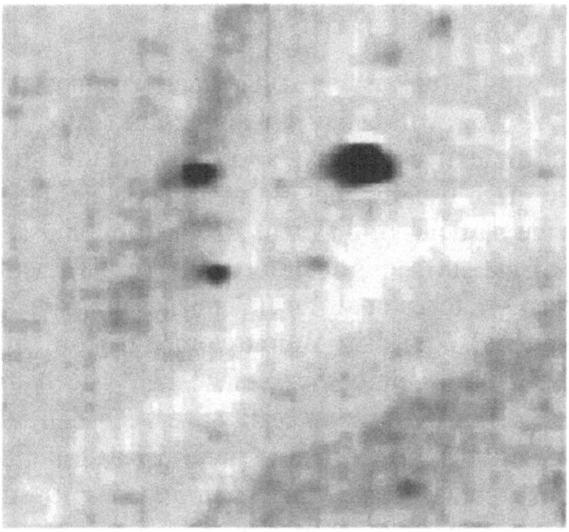

Figure 6.6. Enlarged portion of Figure 6.5 showing the detail resolved, which in some cases comes from blemishes less than a pixel across

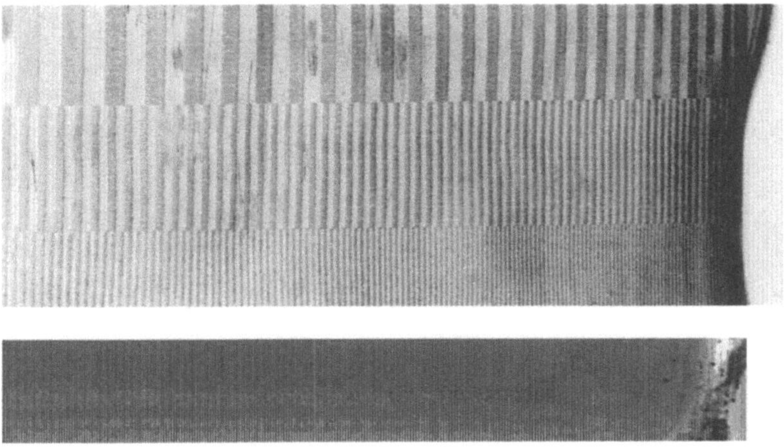

Figure 6.7. Test grid imaged with linescan camera and rectified fine grid after processing

Figure 6.8. Enlarged portion of Figure 6.7 showing recovered grid pattern

6.2.4 Properties of Tile Surface

Up until this point in the discussion it has been assumed that the surface of the, typically glazed, tile is a well-behaved Lambertian scatterer. This is far from the truth. This is a dielectric surface and will scatter different polarisations differentially. The glaze usually consists of a glass-like dielectric film containing metal or metal oxide particles of different size distributions at different depths. The optical response of the surface is therefore complex. The situation becomes even more complex when scratches or surface blemishes are added to the picture. Figure 6.9 shows the scattered light distribution from a narrow laser beam shone at an angle of about 15 degrees onto the surface [3]. The distributions are clearly affected by the scratches. Follow-up tests have been done on these distributions with a scatterometer and these confirm the unusual distributions in the presence of scratches. In most actual cases the undamaged tile surface is quite well approximated by a Lambertian component added to a specular component. The relative sizes of the two parts are determined by the roughness of the tile surface.

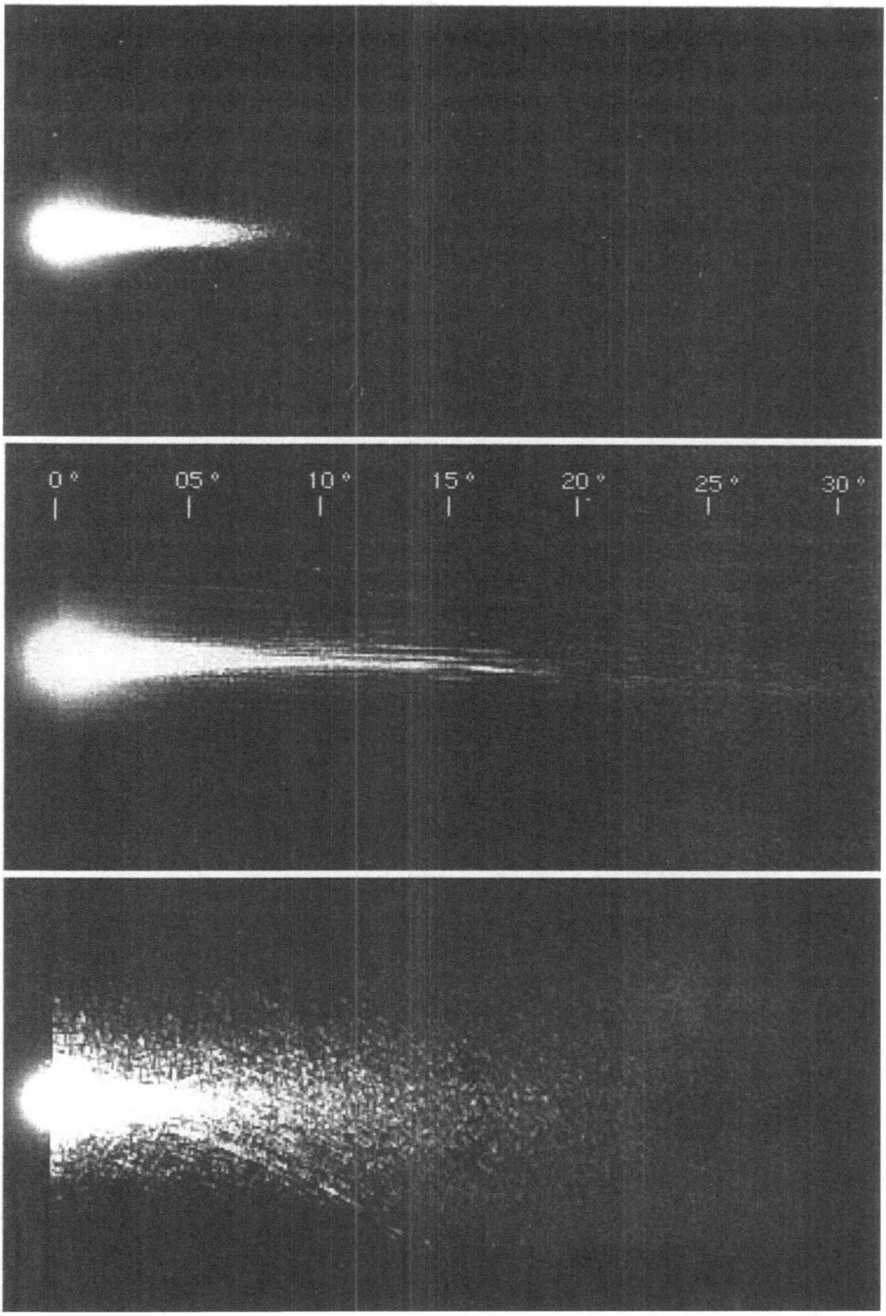

Figure 6.9. Scattering from a glossy tile [3] with no scratches (top), a scratch perpendicular to the plane of incidence (middle) and parallel to the plane of incidence (bottom).

6.2.5 A Practical System

The ideas above were used to produce a prototype production instrument [4], which was used to produce most of the oblique images in the figures. The design of the prototype is shown in Figure 6.10. Because this is a prototype the illumination and viewing angles were made adjustable by means of stepper motors. The instrument is cased to prevent illumination from outside and to conform to safety standards for lasers. Two cameras are used in the prototype to research possible techniques to distinguish between types of blemish by the ratio or other combination of the two images. For this to be successful image rectification needs to be accomplished to pixel accuracy.

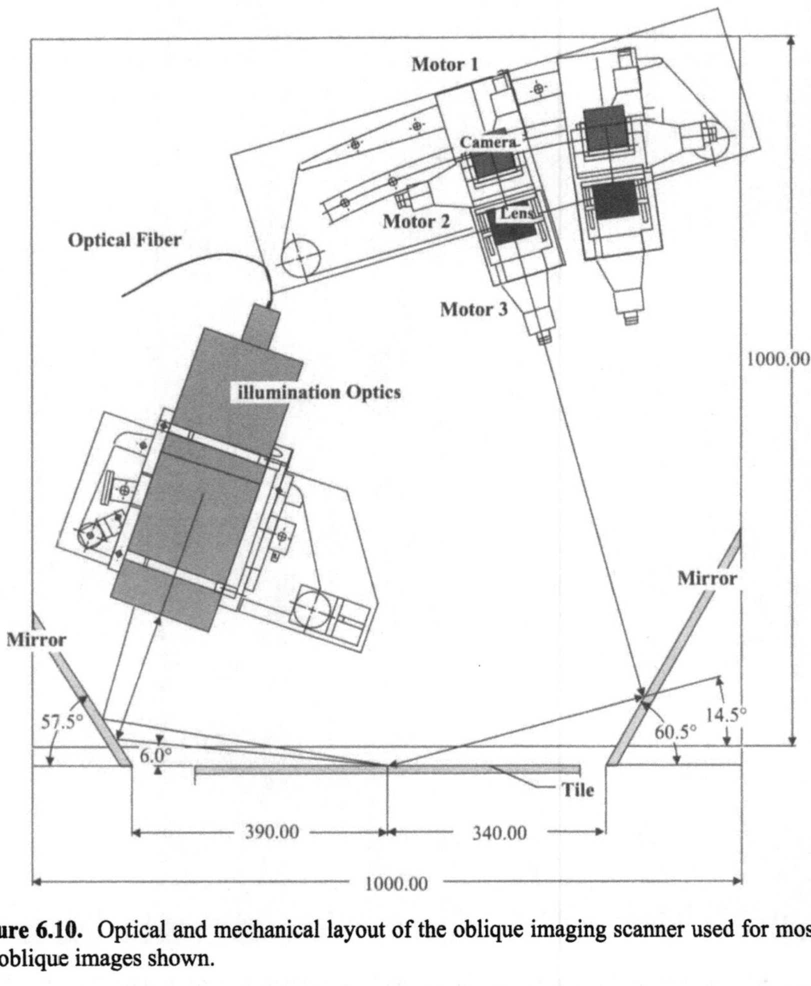

Figure 6.10. Optical and mechanical layout of the oblique imaging scanner used for most of the oblique images shown.

The surface is illuminated by laser. Gas lasers, although high performance, tend to be expensive and delicate. For this reason a diode laser of 300mW at a wavelength of 0.8μm driving an optical fibre and beam expander was used. The diode laser, which needs a thermo-electric cooler, could thus be separated from the rest of the optics allowing much greater convenience and the use of a cooling fan. The fibre also 'scrambles' the beam, which removes some problems of interference fringes on the object. The wavelength is just outside the visible range, which makes eye safety a particular problem. The beam is not visible but is focused effectively by the eye. This wavelength is, however, close to the peak responsiveness of CCD detectors used in the system. Diode lasers naturally produce an elliptical beam, which suits this application. To take advantage of this the diode laser would need to be mounted directly on the beam.

In theory only a strip of surface less than a millimetre wide needs to be illuminated. In practice a few millimetre is required to take up tolerances in the mechanics and optics of the system. The prototype system produced a beam approximately 5mm wide and 10cm across. This is produced by judicious use of a cylindrical lens in the beam expander. The collimator works better if the output beam is in fact not quite collimated but focused at a point just beyond the farthest edge of the surface. This allows the detector optics to go as close as possible to the illuminating beam without actually intersecting any specularly reflected light. The limit to how close the specular beam can come to the detector is determined by the flatness of the surface and the accuracy with which the tile is presented to the optical system. This last consideration is often the limiting factor.

The whole optical system is separated from the surface under inspection by a sheet of glass. This isolates the optics from the dusty industrial environment. The glass window needs to be very clean and preferably anti-reflection coated.

6.3 Obtaining More Information: Flying Spot Scanner

6.3.1 Introduction

Oblique imaging has the advantage of being relatively simple and giving very good results in simple cases, however, if the surface is not flat or two or more images from different angles are needed then the system becomes cumbersome. Image rectification must be accomplished very accurately and multiple linescan cameras are required. In cases where there are is a very large range of surfaces or where very detailed surface information is required it is probably better to use a completely different imaging system.

The flying spot imaging system is completely different in character from normal imaging with an area detector such as a CCD camera. The surface is unlit apart from a very small area which is raster scanned across the surface. This is usually done in one dimension only, the second dimension being produced by moving the object. Light scattered or reflected from the surface is detected by a single point detector. Rather than good imaging for the detector the imaging effort must be placed on scanning a small spot of light across the surface. More than one detector can be used to intercept light coming from the surface at different angles or polarisations. There is essentially no limit to the number of these detectors so that it is easy to generate pseudo-colour images with this technique. Processing the large amount of data can become a major problem. In most cases this technique can reveal surface details or characteristics that are very difficult or impossible to obtain in other ways. At least as important is that it produces data that would require multiple optical systems to acquire with normal imaging. For example data that would be obtained from normal and oblique images of a ceramic surface can be obtained with one scan.

6.3.2 Optical Layout

There are two areas of optical design required. The first is to illuminate a very small spot on the surface at the same angle for all points on that surface. The size of this spot will determine the spatial resolution of the system. The second is to make the single point detector equally sensitive to light scattered from all points along the scan line. The detector should also be sensitive to the same angular dispersion of light at all surface points.

The light projection system would ideally produce a spot of diffraction-limited size impinging on the surface from the same angle at every point on the scan. It is relatively easy to produce a very small spot from a laser source because of its high luminosity and single wavelength. A single movable mirror can be used to scan the spot in a line across the surface. There are two problems with this method. The first is that the length between focusing lens and surface changes through the scan, making perfect focus impossible. The second is that the beam impinges on the surface at different angles at different parts of the scan. The system has the very great advantages that the optical system is simple and uses a low number of small optical elements. The spot size is ultimately limited by the diffraction limit of the focusing lens. In practical cases this usually only requires a lens of 10–20mm diameter. Because the beam is so narrow the steering mirror can also be small. This allows rapid scanning which is a requirement in most uses. For relatively low-performance systems this arrangement is quite adequate.

For high-performance systems it is important that the spot size stay small and constant over the whole scan. The only practicable method of achieving this is to have an optical element, lens or mirror, that has a diameter as large as the scan length. The system in Figure 6.12 [5] can then be used. A mirror is normally used because large mirrors are cheaper and easier to make than the equivalent lenses and they do not suffer from chromatic aberration. It is also relatively simple to make a parabolic mirror, which has theoretically perfect imaging on-axis. Non-spherical lenses are much harder to make.

The large mirror is expensive but does not require very good optical quality. Only a small area of the mirror is illuminated at any one time so that as long as this area has a quality of $\lambda/4$ the spot size will be diffraction limited approximately the same as if a lens of the size of the illuminated mirror patch was used. As in the previous case the illuminated patch is unlikely to be larger than 10–20mm. The moving scanning mirror also needs to be no larger than this. The drawback of this system is that the imaging at the edge of the scan will be worse than in the centre. Photocopiers normally use an optical system of this type but include aspheric lens/mirror combinations to improve image quality across the scan. The optics are usually high-quality moulded plastic. The optical arrangement used for Figures 6.15 and 6.16 uses an alternative method of ensuring equal optical quality across the scan, which is the subject of a possible patent application. With these methods it is possible to produce very small spot sizes across the whole scan giving resolutions of 3000 and up. Some film scanners approach a resolution of 50,000.

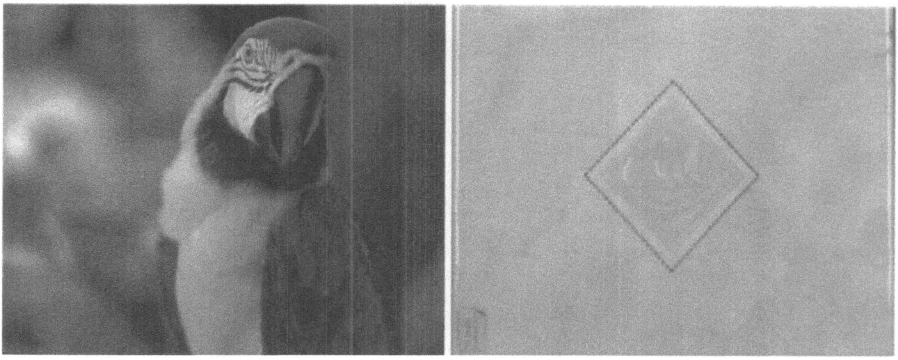

Figure 6.11. On the left is a normal colour picture of a parrot. This is about as multi-coloured as normal images get. On the right is a typical ceramic tile, which has little colour or indeed intensity contrast. Compare these to the images and histograms in Figures 6.15 and 6.17.

Both of these scanning techniques require a moving mirror to scan the beam. There are two basic ways to accomplish this. The first is to use a galvanometer-type movement suspending a light mirror. This can be scanned by changing the current in the driving coils. This has the advantage that the beam can be scanned in any direction at any time. It is not limited to a linear scan. The drawback is that the scanning is quite slow. Speed depends on the mass of the scanner mirror so very high speeds are impossible. One way used to give higher speeds is to make the mirror and its sprung suspension resonant. The system will then produce only a sinusoidal scan at one frequency but this frequency can be quite large.

The second technique is to use a rotating prism, usually with ten or more aluminised faces. The only limitation to speed is how fast this prism can be rotated. For very high speeds the prism needs to be in a vacuum but even television scan rates (approximately 16kHz) can be obtained with a relatively simple system operating at normal pressure. The system is of course limited to a linear scan. The figures in this chapter were obtained with a scanner of this type.

Having illuminated the surface with a small spot there are numerous ways to steer the reflected or scattered light into a detector. One method is to use the same optical path as the illuminating beam. The scanner is fed from a laser through a 45° semi-reflecting mirror. The detector therefore receives light back through the illuminating optics. The advantage of this system is that all the optics used to deliver the beam to the surface in an accurate manner is also used for the detector. The system therefore has very good consistency across the scan. There are two major drawbacks to this system. The optics are very wasteful of light. The semi-reflective mirror cuts the theoretical light gathering maximum to 25% and because the scanning system has a relatively small aperture the effective *f* number is probably above 40. The second disadvantage is that only one scattering angle is sampled. It is difficult to get multiple scattering angles back through the optical system.

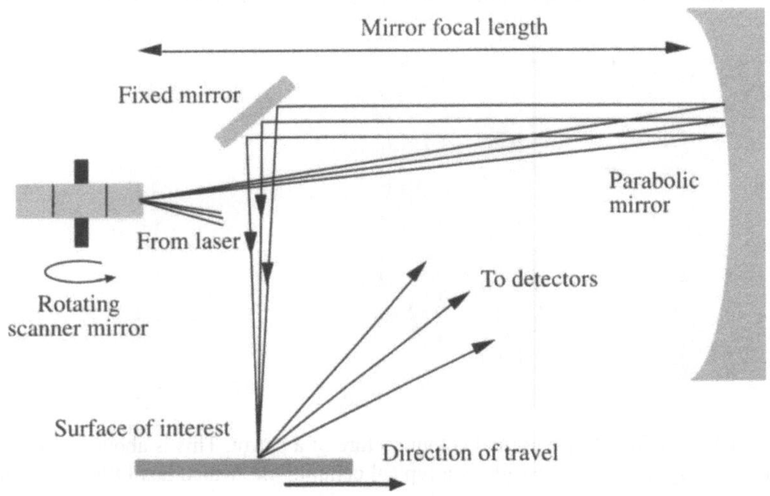

Figure 6.12. Optical arrangement of scanning system. Line of scan is in and out of the plane of the diagram.

An alternative is to use a light collector such as a perspex rod, which is placed beside the scanning line. A detector on the end of the rod receives scattered light at all points of the scan through internal reflections along the rod. This system has much different sensitivity at different parts of the scan. The images in this chapter were obtained by imaging the scan line onto a detector from various positions. This has the advantage of approximately equal sensitivity for all of the scan line, good *f* number and hence light gathering power and allows the use of multiple detectors to image light scattered at different angles. The method used to collect scattered light from the surface depends very much on the information required and the characteristics of that particular system. There is a very great variety of possible systems.

6.3.3 Detector Requirements

The detector requirements for a flying spot system are quite stringent. It is difficult to get good light collection efficiency from the optics. The high resolution and fast scanning speed imply very high pixel rates so the detector must be fast and efficient. The following is the expression for the signal to noise ratio in a photodiode:

$$S:N = P = \frac{S\sqrt{t}}{\sqrt{S(hc/\lambda) + D/e(hc/\lambda)^2 + N^2}} \quad (6.6)$$

where:

S = signal power in *watts*
D = leakage current in *amps*
N = noise power in *watts* $Hz^{-1/2}$

The second two terms in the denominator should be made as small as possible. The middle term is the contribution of the leakage current through the diode. This is set by the characteristics of the device and the temperature. The right-hand term is the noise power produced either by the amplifier to which the diode is attached, or the high-value resistor used in the feedback of the transimpedance amplifier (see later). It is difficult to make photodiodes with low leakage because the incoming photons must bounce electrons across the junction, so the junction must be made narrow to have reasonable sensitivity. High sensitivity and low leakage are therefore competing design aims. The right-hand noise term is determined by the quality of amplifier available and the required gain to obtain signal above the amplifier noise. Bigger gains require bigger feedback resistors but these produce more thermal noise. There is therefore a maximum performance that can be obtained which is likely to be limiting in this application. Both noise terms are significant for flying spot scanner detectors working at high bandwidths. The only sure way of reducing their effect is to use a more powerful light source.

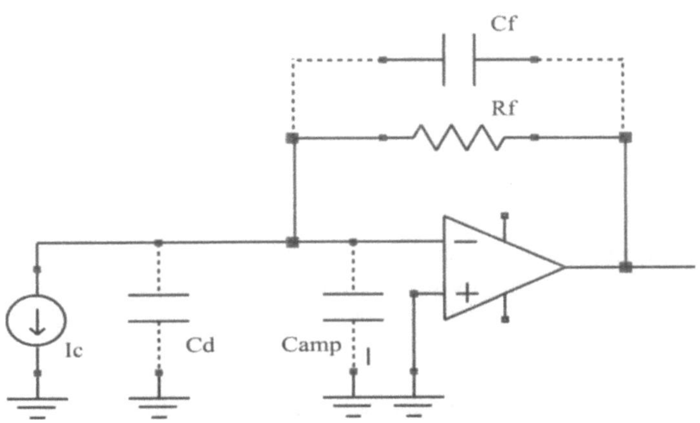

Figure 6.13. Transimepedance amplifier showing the parasitic capacitance around R_f, C_f and the parasitic input capacitances of the diode and the amplifier input.

The problems with photodiodes do not end with noise performance. It is also very difficult to build a suitable amplifier with the required bandwidth. The basic circuit for the transimpedance amplifier, which is preferred for this application, is shown in Figure 6.13. Current produced in the photodiode with zero bias is turned into a voltage output with low source impedance at the amplifier output. Zero biasing the photodiode allows very good linearity from the system and by careful construction can prevent electrical interference from entering the signal path.

The bandwidth of the system is limited by two effects. The first is reasonably obvious, that is that the leakage capacitance around the feedback resistor turns the circuit into a low-pass filter. Increasing the value of the feedback resistor increases the effective gain of the amplifier reducing noise problems but also reducing bandwidth. The second effect is much less obvious. The photodiode and amplifier input also have a leakage capacitance. At high frequencies the input capacitance and feedback capacitance dominate the circuit response. Because the input capacitance is larger than the feedback capacitance the effective gain increases with frequency. Any thermal or other noise at the input is therefore strongly amplified at high frequencies. This prevents the amplifier being used for signals at these frequencies (Figure 6.14).

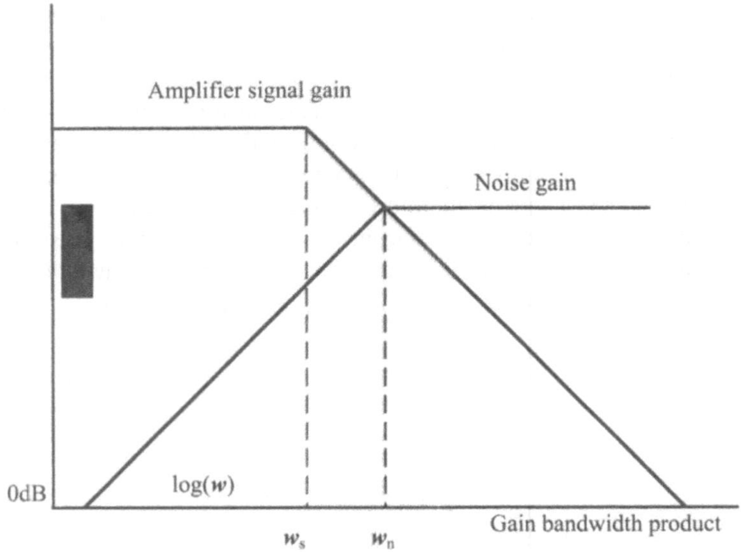

Figure 6.14. Performance of the transimpedance amplifier showing the effect of the parasitic capacitances. Noise gain corner frequency must be higher than signal gain *ws* corner frequency to prevent continuous high-frequency noise output from the amplifier detector combination w_n.

Details of the amplifier response and possible circuit modifications are complex [6,7] but by clever circuit design it is possible to make some improvements to these problems, however, with current components a bandwidth of 1–2MHz is the best that can be achieved for small signals. This is not enough for many flying spot scanner applications.

The consequence of these problems is that the choice of detector is a very important issue for the flying spot scanner. A detector with significant intrinsic gain solves all of the above problems. The images in this chapter were made with photomultiplier detectors with a 50 ohm load impedance and amplified by high-performance op-amp amplifiers. The photomultiplier can have very high gain, 100,000 in this case. The bandwidth is limited only by tube geometry. Leakage currents can be extremely low, down to a few photons a second, or better with cooling. Low leakage devices are not needed in this application because of the relatively high signal levels. The drawback is the need for a high voltage-biasing source. The quantum efficiency of these devices is also a little lower than semiconductor devices. An alternative halfway house detector would be the avalanche photodiode (APD). By biasing the APD an intrinsic gain of about 100 can be obtained. Quantum efficiency is very high but the devices need thermo-electric cooling to deliver their best performance.

6.4 Image Processing of Multi-channels

The multi-channel images that result from scanning even seemingly simple surfaces can be quite dramatic (Figure 6.15). These are three channel scans with detector angles arranged to emphasise texture and surface contour. It is easy to obtain more information by using more detectors at different angles; see figure 6.12. With multiple channels and high spatial resolution the amount of data obtained can be enormous. To reduce this to a more manageable level Principal Value Decomposition (PVD) also known as Karhunen-Loeve decomposition can be used. This is a statistical technique, which expresses the data by rotating the co-ordinate frame of the data such that the axis along which there is most variance is aligned with the first axis. The orthogonal direction with the next most variance is along axis two, etc. No data is destroyed in the process and it is reversible. The results is therefore expressed in the statistically most sensible way with the most "important" data along axis one. In the case of a normal colour image consisting of red green and blue channels the first axis almost always lies along the line of intensity, that is equal amounts of red green and blue. The result is therefore a channel of intensity data and two channels of colour data. From a statistical point of view most information in normal images is in the intensity. This is the basis of the HLS (Hue Lightness Saturation) colour imaging system [8,9] and the way in which analogue colour television pictures are transmitted and decoded.

More transmission bandwidth is given to the intensity data that contains most of the information.

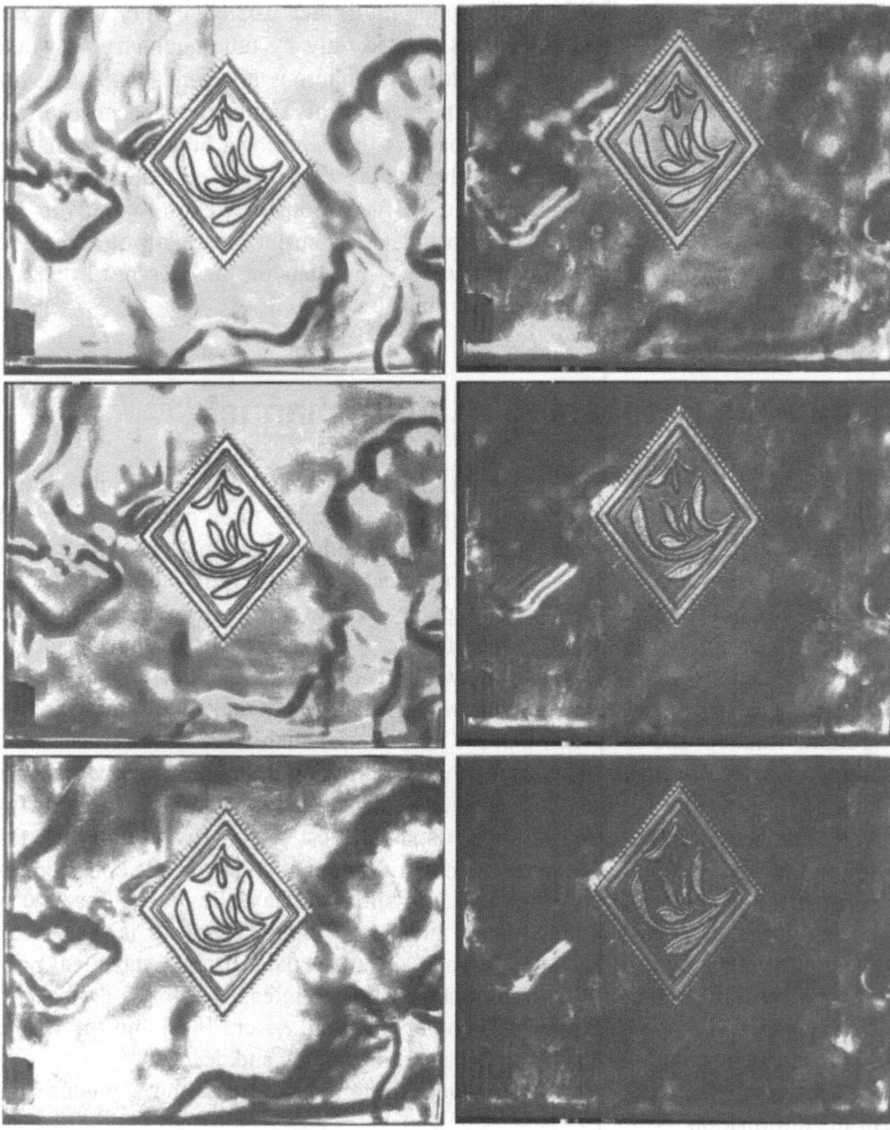

Figure 6.15. The same tile as in Figure 6.11 but images taken by a flying spot scanner at two sets of scattering angles, one to emphasise the specular components (on the left) and the other to emphasise surface detail. The amount of information exposed is huge. By producing pseudo-colour images from these frames subtle surface features are easily apparent.

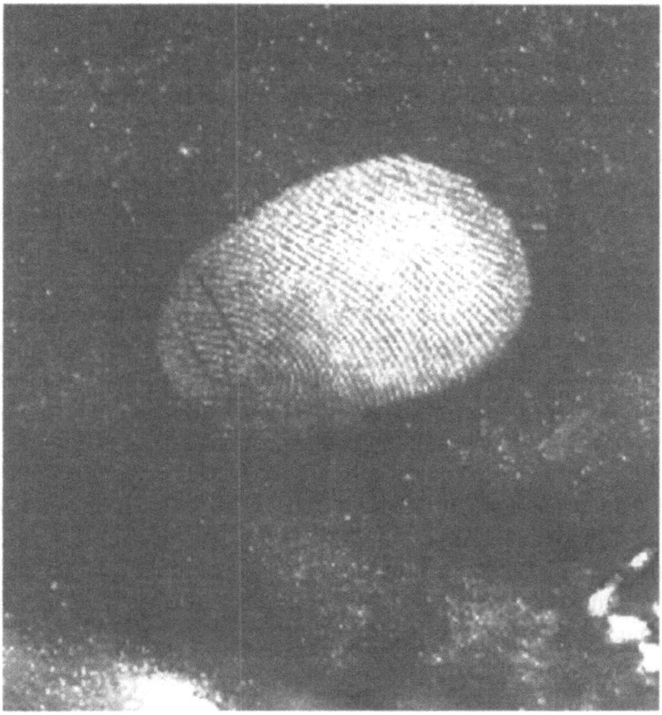

Figure 6.16. This is the same tile and scan parameters as Figure 6.15 but shown at the full scanning resolution. The original image is about 5000 pixels square. Obviously very fine detail on the surface is being seen.

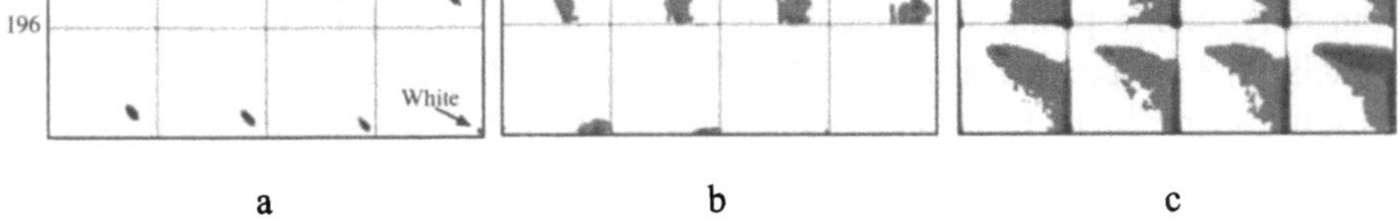

Figure 6.17. A representation of three colour histograms. 16 slices through the RGB histogram are taken at increasing levels of blue. For each of these slices green increases from left to right and red from top to bottom. Darker is more pixels of that colour: a. the histogram of the tile in Figure 6.11; b. the histogram of the parrot in Figure 6.11; c. a histogram of the pseudo-colour image produced from the left-hand images in Figure 6.15. Note the almost filled colour histogram produced by the scanning system.

For images with multiple channels, especially more than three, the PVD process can result in the least significant channel(s) containing only noise. These channels can be discarded reducing the quantity of data that need to be further processed. Alternatively some of the least statistically significant channels can show up subtle effects clearly by removing most of the normal image information [10] which is masking the more subtle data. This technique has been used for some time on imaging from satellites but also works well on complex surfaces.

An interesting feature of flying spot multi-channel images is that they fill the histogram of possible 'colours' much better than normal images. Even the most colourful image of a scene such as a parrot against foliage (Figures 6.11) can use as little as 10% of the space in the histogram. That is of the available colours only 10% are used. This is a property commonly used to compress and display images on the computer. The flying spot images however use virtually all of the available colour space. Colour image processing techniques may therefore have to be specially developed for flying spot multi-channel images.

It may be possible to separate depth and surface information from flying spot images. By taking the amount of light scattered at different angles from a single point on the surface an estimate of the angle of the surface can be made. The surface can have only one height at any point so integration of the surface slopes with suitable constraints may yield the surface topology. This is a possibility, which is being investigated at Imperial College.

6.5 Conclusion

The ceramic surfaces considered here have very complex characteristics. Simple images of these surfaces have no chance of revealing the complex behaviour of the surface. By imaging the surface in a different way and combining this with image processing algorithms designed for the task a whole new range of possibilities is opened up.

Oblique imaging allows the surface characteristics to be almost completely dissociated from the bulk characteristics of the tile. Tile pattern and colour is suppressed in favour of surface detail. The technique can be a little too effective showing up dust and finger marks which are of little interest. To detect scratches the only image processing needed is thresholding and counting the remaining pixels. The image processing has therefore changed from a very difficult and high-resolution problem to an extremely simple algorithm by the use of optics and lighting. As is usually the case special image processing is needed for a different optical system. Image rectification and intensity normalisation in this case. The technique is an extremely effective and cheap method of solving a relatively simple problem, which if approached by conventional image processing techniques would have been impossible or at the least extremely expensive to implement.

The flying spot scanner opens up completely new forms of imaging which give data that would be impossible to obtain by normal imaging. It is possible to compare the information obtained by the scanner with what could theoretically be obtained from viewing a surface. One could envisage obtaining the scattering distribution at all wavelengths and polarisations for every point on the surface. The flying spot scanner can approach if not actually reach this perfect performance. Conventional imaging cannot hope to compete with this. The images from these systems can open up a huge range of applications and completely new areas of image processing. This is surely one way in which research can make progress in the next few years.

Image processing and the way in which the images are obtained cannot and should not be separated. A broader view of imaging is the way to produce new data and techniques.

6.6 References

[1] Forrest A.K. (1999) Oblique imaging, magnification and image intensity, laser Metrology and Inspection, *J.Opt.A, Pure Applied Opt. A*, Vol 1, pp. 697–701.

[2] Forrest A.K. (1991) Improvement relating to an optical measuring system, PCT/GB90/09110, Describes a moiré fringe measuring system using a single grid.

[3] Bakolias C. (1996) Oblique imaging of scattered light for surface inspection PhD Thesis, Imperial College Mechanical Engineering Department.

[4] Forrest A.K., Bakolias C. (1996) Oblique imaging inspection system, PCT/GB96/19670.4. Oblique imaging and illumination system.

[5] Sureshbabu S.K., Forrest A.K. (1999) A real-time surface inspection technique using a multichannel telecentric flying spot scan system, Laser Metrology and Inspection, Proceedings of SPIE, Vol 3823.

[6] Graeme J. (1995) *Photodiode Amplifiers (op amp solutions)*, McGraw-Hill. ISBN 0-07-024247-X

[7] Winder S. (1997) *Filter design,* Newnes. ISBN 0-7506-2814-6

[8] Kang H.R. (1996) *Color Technology for Imaging Devices*, SPIE – The International Society for Optical Engineering. ISBN 0-8194-2108-1.

[9] McCollum A.J., Forrest A.K. (1999) A colour metric for production quality control: polarization and colour techniques in industrial inspection, *Proceedings of SPIE*, Vol. 3826.

[10] Bakolias C., Forrest A.K. (1997) Dark field, Scheimflug imaging for surface inspection, machine vision applications in Industrial Inspection V, *Proceedings of SPIE*, Vol. 3029, pp. 57–68.

Chapter 7

Editorial Introduction

Human beings cannot concentrate for long periods on repetitive inspection tasks and find it difficult to maintain constant standards when subtle judgements are required. For example, people do not perform well at grading fruit moving at high speed (10 fruit/second) on a conveyor belt. Motivated by this fact, automated visual inspection systems were developed some years ago for examining fruit on the basis of size, shape (silhouette) and colour. Although they are just as important in determining the value of fruit, detecting surface blemishes has not been addressed successfully until recently. As often happens in Machine Vision, it is beneficial to plan an integrated processing/inspection system from the outset, rather than retro-fitting a system into an existing plant. This is particularly important for this application because the mechnical handling sub-system is complicated. The whole surface of the fruit must be examined. To achieve a good all-round view of the fruit, some previous authors have used wire-frame cups. An alternative arrangement is to use some mechansim for turning the fruit in front of the camera. In order to count surface blemishes properly, some means must be found to recognise the fact that some blemishes are actually seen more than once. Moreover, the stalk and calyx are likely to be confused with blemishes, so sophisticated techniques are needed to identify them. Tasks of this type can be approached in two ways: make the image processing more intelligent (and hence more complicated), or modify the lighting–viewing arrangement. By studying the reflection characteristics carefully, it was possible to identify two different mechanisms that give an apple its distinctive appearance. Light at the blue end of the VIS spectrum is reflected from its surface, while light at longer wavelengths penetrates into the body of an apple and re-emerges as a diffuse glow. Optical filters and an HSI (hue, saturation and intensity) camera can help to separate these components. It is also possible to alter the geometry of the light source, making the dimple around the stalk and calyx more obvious to the image processor. Several cameras can be used to provide stereo viewing of the object surface. Detecting the stalk and calyx is eventually assigned to a heuristic procedure. Counting blemishes may seem to be straightforward but the uncalibrated and unpredictable rotation of the apple may lead to overcounting. In view of the unpredictable performance of non-algorithmic programs, rigorous testing is essential.

Chapter 7

On-line Automated Visual Grading of Fruit: Practical Challenges

P. Ngan, D. Penman and C. Bowman

7.1 Introduction

Huge quantities of fruit are grown and packed each year across the world. Most markets demand high quality and top-quality fruit can command premium prices. Hence, accurate grading to market requirements is highly desirable. Visual properties, such as size, shape, colour and surface blemishes, are important in determining fruit quality as perceived by the consumer. Consequently, visual grading of some sort is standard in most pack houses.

The sheer volume of fruit processed by modern pack houses poses considerable difficulties for human inspectors. Typically, teams of six or more inspectors will view rows of fruit passing by on a conveyor and endeavour to sort defective fruit. This is a challenging task—not just because of the sheer volume of produce, but also because objective grading criteria need to be applied based on complete fruit surface information and it is difficult to view all parts of a fruit at one time.

For these reasons, there is strong demand for automated fruit grading equipment and a number of companies now manufacture vision systems specifically for this task. These are usually integrated with handling and sorting equipment and other types of sensors, e.g., load-cells for weighing each fruit. Systems for visual sizing and colour grading are now fairly well established. However, there are still very few surface-defect grading systems used in pack houses. Those that do exist are rudimentary and are applied to fruit, such as citrus, with uniform surface and shape characteristics.

There remain a number of challenges to overcome before a cost-effective automated fruit-grading system can approach the performance of a human inspector:

- The complete surface of each fruit must be examined. This poses challenges for both handling and imaging systems.
- Sophisticated machine vision segmentation and classification algorithms are required to distinguish defects on a natural and varying product adequately.

- Fruit such as apples always have a stem and a calyx, which are difficult to distinguish from true defects.
- Production rates dictate that fruit need to be graded at 10 fruit a second or faster.

This chapter focuses on two of these challenges: complete surface imaging and stem and calyx discrimination. The following sections describe the various approaches taken by researchers to address these challenges and describe in more detail the techniques developed by the authors. For a review of photonics in fruit and vegetable quality assessment, the reader is referred to the paper by Tao [1].

7.2 Complete Surface Imaging

7.2.1 Introduction

Inspection systems employing a range of mechanical conveying, camera, and lighting components have been built in an attempt to perform complete surface inspection. These systems fall into two categories of approach.

The first approach is to secure the fruit in a holder that allows almost complete visibility of the fruit surface, and to capture multiple images of the fruit simultaneously. For example, Laykin *et al.* [2] constructed a prototype which uses a wire-frame cup to present the fruit to three cameras. The primary drawback of this system is the inability of the wire-frame cup to carry fruit of various shapes and sizes without replacement.

The second approach, now widely used by the industry and adopted by the authors, is to use a mechanism that rotates individual fruit as they move along the conveyor. Typically these conveyors use bi-cone rollers [3] to rotate the fruit, as shown in Figure 7.1. Cameras and lighting are housed in an inspection cell mounted above the conveyor and acquire a rapid sequence of images where each fruit is imaged in a succession of orientations as it moves past the cameras. Such systems have been commercially successful in meeting most of the requirements outlined in the previous section for modest visual inspection tasks such as sizing, shape measurement, and rudimentary colour grading.

Figure 7.1. Bi-cone rollers used to rotate single fruit on a conveyor (courtesy of Compacting Equipment: source: www.compacsort.com).

7.2.2 Surface Feature Tracking

Acquisition of multiple overlapping views of a fruit as it moves along the conveyor increases the difficulty of counting the unique features on the surface of a fruit. The ability to count surface defects is a key function of any inspection system because the feature count is often a key determinant of fruit quality.

A unique feature count is difficult to measure because a point on the surface of a fruit is imaged multiple times during an image sequence. In particular, a given point will appear in successive images until it rotates into the occluded region, when it will be absent from the sequence. Eventually the point may or may not rotate back into the visible region before the fruit disappears from the camera view. While a highly repeated data set minimises the chance of not collecting information about a part of the surface, it introduces the need to identify repeated instances of unique information.

We have developed a 3D surface-tracking algorithm to identify repeated data in the image sequence. The algorithm, whose components are illustrated in Figure 7.2, begins by acquiring a sequence of stereo images of a fruit as it moves along the conveyor. At each instant in the sequence, image processing operations are applied to the stereo pair of images to determine the boundary of the fruit and a set of 2D points to be tracked. These points of interest are processed using geometry-based techniques, which produce a 3D point set representing the 3D location of surface features visible in at least one of the views. The rotational offset of this 3D point set relative to a global 3D point set is estimated, and finally the global point set is updated using the new data.

Figure 7.2. System diagram for the 3D tracking algorithm.

Image Sequence Acquisition

At least two camera viewpoints are necessary to obtain adequate coverage of the fruit surface. On these conveyors, fruit often rotate about a fixed axis, and a single view will not always view the entire surface region—at least at an adequate resolution for inspection purposes. Two views are also required to calculate 3D co-ordinates of the surface points visible in the stereo overlap region. These 3D points are calculated using a geometric spatial intersection technique from calibrated views. When a surface point is visible in only one view, a geometric technique analogous to puncturing a sphere with a ray is used to calculate the corresponding 3D point. Both two and one-view 3D reconstruction techniques are described in Section 7.2.2.

Figure 7.3 illustrates typical stereo images captured by the image acquisition rig. The cameras are placed on either side of the conveyor at equal angles from the axis formed by the conveyor rollers. The images of a stereo pair are captured at the same instant in time. This is a necessary property when applying two-view geometry techniques to dynamic scenes.

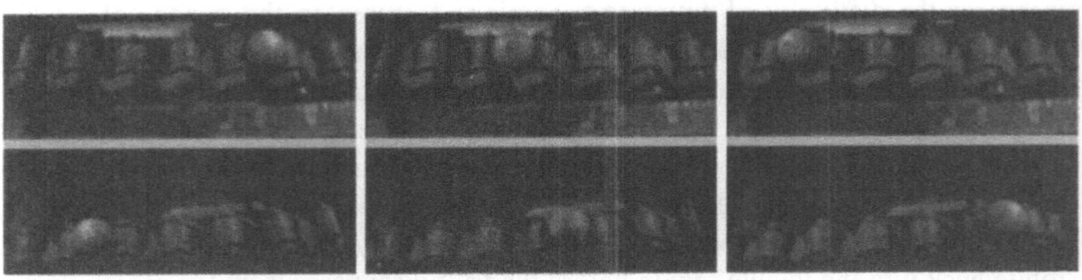

Figure 7.3. Stereo images excerpted from the beginning, middle, and end of an image sequence.

Boundary and Surface Measurement

The purpose of this step is to calculate a circle fitted[1] to the fruit boundary and extract a set of 2D points representing each surface feature to be tracked; an example is shown in Figure 7.4. The circle fitting procedure involves segmenting the fruit from the background scene, finding the pixels on the periphery of the region representing the fruit, and fitting a circle to those pixels.

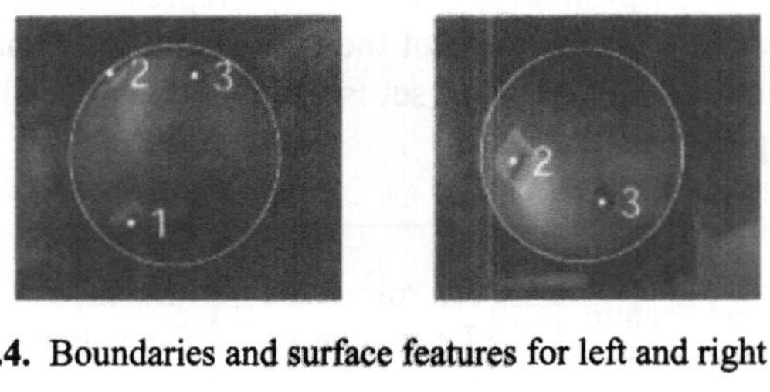

Figure 7.4. Boundaries and surface features for left and right fruit views.

---

[1] Even though a circle is fitted to the fruit boundary, the algorithm does not assume that the fruit is perfectly spherical; only that is it near spherical, e.g., a granny smith apple, or an orange. However, the algorithm will perform better for spherically shaped fruit.

Image segmentation refers to the process of categorising pixels of an acquired image into one of three groups: surface features on the fruit, the rest of the fruit, and the remainder of the image. In our experimentation, each pixel was categorised according to its Hue-Saturation-Intensity (HSI) colour components. The partition boundaries in HSI space were determined by manual inspection of the colour values present in a representative set of sample images.

Image regions of low intensity lack information useful for segmentation purposes. Segmentation of dark regions in an image tends to produce unreliable results, even when segmentation is based on colour, as intensity is a component of colour. Unfortunately when imaging a spherical object, such as some types of fruit, its visual perimeter is generally darker than the rest of the object because of the tangential viewing angle. Hence special care is needed to provide a light source that surrounds the fruit to illuminate the visual perimeter. While expansive light sources help illuminate the object's perimeter, complications arise owing to shadows cast onto the fruit by adjacent fruit, the conveyor, and bi-cone fruit support. Since these shadows are difficult to eliminate, we identify regions of low intensity on the fruit perimeter and exclude these pixels from the data set supplied to the circle estimation procedure.

The choice of surface features used for tracking may or may not include the features that concern the grading criteria, e.g., surface blemishes. In practice, grading-related features often make poor features to track. Features suitable for tracking have a stable appearance during tracking (i.e., should not move on the surface or change shape or size), and are sufficiently plentiful but not too numerous (e.g., 5–10) on all fruit irrespective of grade or size. Blemishes obviously are a poor choice because a perfect fruit will have no blemishes. Once a feature-type has been chosen and a suitable image processing algorithm developed for their extraction, each extracted feature must be represented in the tracking algorithm as a 2D point. We used a centroid operator to calculate the required 2D point representation.

3D Point Set Estimation

The purpose of the third step indicated in Figure 7.2 is to reconstruct the 3D location of every feature of interest present in one or both views, at a given instant of the image sequence. A list of 3D locations is called a 3D point set. The 3D co-ordinates are relative to a co-ordinate system fixed in the world. A 3D point can be reconstructed using image data originating from either one or two views. This section describes procedures for calculating 3D points for the two-view case and the single-view case.

A technique known as spatial intersection is used to calculate a 3D point in the world from a pair of corresponding points taken from each stereo view. Rays are back projected through the image points into 3D space, and the point at which these rays intersect is considered to be the 3D location of the physical point. The mapping from a point in an image to a 3D ray requires knowledge of various camera parameters including its orientation, position, scaling, and focal length. The values of these parameters are determined beforehand by using a technique commonly called camera calibration, which is described elsewhere [4, 5].

The spatial intersection algorithm will happily compute an "intersection" result for any pair of 2D points whether or not they belong to the same surface feature. Therefore it is important to have a way of identifying pairs of image points that belong to the same surface feature. The method of identifying corresponding image points is as follows. Initially all possible pairs of points are considered to be candidate correspondences. Then a series of heuristics based on geometric reasoning are applied to remove implausible pairs. The remaining pairs are then considered genuine matching pairs and their implied 3D locations form the 3D point set.

The first, and most powerful, heuristic is based on the fact that rays passing from the view-centres through image-points corresponding to the same surface feature must pass arbitrarily close to each other; that is they must notionally intersect. This heuristic eliminates almost all implausible point pairs, and is so useful that it is given a name: the *epipolar constraint* [4, 5]. Nonetheless this heuristic does not eliminate all implausible pairs because (a) the proximity threshold for intersecting rays must be set sufficiently high to allow for the uncertainty in the localisation of the image points; and (b), the epipolar constraint is satisfied by any two image points that together with the view centres, reside on the same plane.

The second heuristic is based on the fact that a genuine 3D point must be located arbitrarily close to the surface of the fruit. Points that satisfy the epipolar geometry but originate from different 3D features are likely to reconstruct to a 3D point distant from the fruit surface. For this purpose, the surface of the fruit is modelled by a sphere reconstructed from the circular approximations to the fruit boundaries obtained from each of the pair of views. The centre of the sphere is estimated by determining where the rays projected through the two circle centres pass closest to each other. The distance from an arbitrary point on the circle perimeter to the ray through the centre is measured in both views and averaged to give a diameter estimate.

The third, and last, heuristic is based on the fact that a point in one view cannot correspond to multiple points in the other view. If a point satisfies the first two heuristics and is matched to more than one other point, then the candidate point that reconstructs the 3D point closest to the sphere surface is regarded as the true match. The other candidate matches are discarded.

Single viewpoints are used to derive 3D co-ordinates by back projecting them into space and calculating the point at which they intersect the surface sphere. Of the two solutions, only the visible intersection point is kept, and the other is discarded.

Pose Estimation

An update of the global model requires the 3D point set described above to be translated and rotated so that it is superimposed onto the global model. The pose of the global model is defined as the pose of the 3D point set at the first instant of the image sequence. The translational component is the vector difference between the sphere centres of the 3D point set and the global model. This translational component is subtracted from the 3D point set so that its centre point coincides with the centre point of the global model.

Once the translation component has been removed, the corresponding points of the 3D point set and global model can be superimposed onto each other using a rotation operator. Conversely, knowing the correspondences between points on the 3D point set and the global model allows the estimation of a rotation operator that superimposes the points onto each other in a least squares sense. Calculation of rotation and the search for correspondences are co-dependent and form two steps of an iterative procedure. First, a candidate rotation is applied to the 3D point set and then nearest-neighbour matches between the two point sets are found. Using these matches the initial rotation estimate is refined and the matching is repeated. These two steps are iterated until no new matches are found. The initial candidate rotation is calculated using a rotation-tracking algorithm, which assumes a constant rate of rotation between frames.

Update Global Model

Once the corresponding points in the 3D point set and the global point set have been identified, and the rotation between them has been estimated, the points on the global model can be updated using the new observations. Three possibilities exist. First, when a 3D point matches a global model point, the co-ordinates of the global model point is adjusted using a weighted sum with the corresponding rotated 3D point. The weight assigned to a rotated point is determined by whether the point was derived from two views or one view. A result derived from two views is inherently more accurate than one derived from a single view and so is assigned a higher weight. Each point in the global model also has a weight, which is the sum of the weights of all contributing points. The second possibility occurs when a rotated point does not match a global point. In this case, the unmatched rotated point seeds a new point in the global model. The third possibility is that a global model point does not match any point in the rotated point set. Such points in the global model are left unchanged.

Algorithm Demonstration and Evaluation

A target object was simulated using a 70mm diameter plastic ball which had five colour patches fixed onto its surface. A test sequence of 59 frames was captured of the ball translating and rotating under the camera rig. Figure 7.5a shows the tracks formed by lines connecting point positions in adjacent frames. The tracks appear circular because the translation component has been removed for clarity. Each surface feature is represented by a different marker symbol. The tracks do not form complete circles because the features rotate out of both camera views. Figure 7.5b shows one track in detail and illustrates that this surface feature disappears from view at frame 12 and reappears at frame 27. The fact that the feature was uniquely identified throughout the sequence, despite an extended period of occlusion, suggests that the algorithm could be used to prevent double counting of surface features on a freely rotating object.

To validate the algorithm under more realistic conditions, the plastic ball was replaced by an orange, upon which between 1 and 20 surface patches were placed. Twelve experimental runs were conducted per feature number. Figure 7.6 plots the relationship between the actual number of surface features against the quartile results of the 12 runs returned by the tracking algorithm. Overcounting occurs for

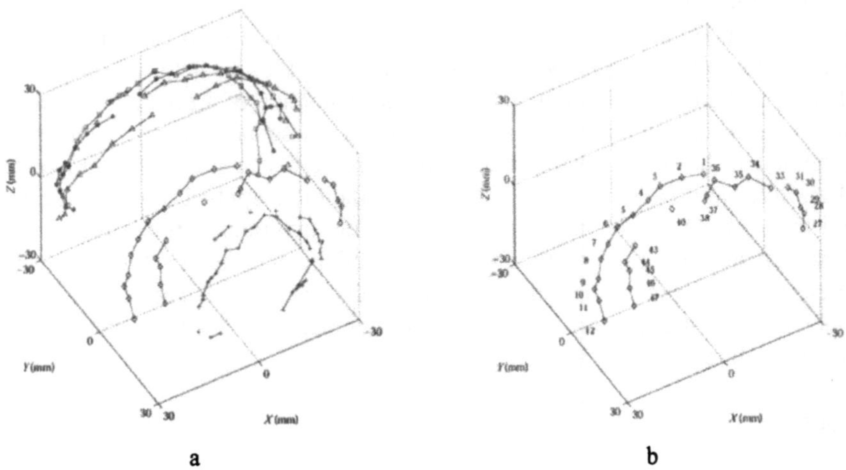

Figure 7.5. (a) point tracks for all five features on the test object; (b) point track for the '♦' feature; each point is labelled with the frame number

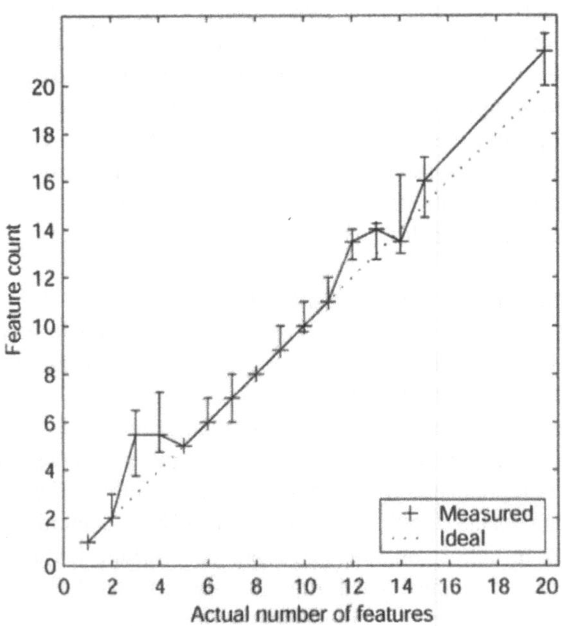

Figure 7.6. Influence of number of surface features on counting performance.

low number of features (< 5) because many frames contain insufficient surface features to reliably estimate rotation. In such cases, rotation is determined using extrapolation schemes and consequently these estimates become unreliable after a small number of frames. In practice, when the tracking algorithm returns unreliable rotation estimates, alternative sources of information may be employed such as the rotation of adjacent fruit and the rotational speed of the conveyor.

Overcounting occurs for high number of features (> 11) because the images are overcrowded with features and incorrect matches occur. Both cases of over counting could be reduced by introducing additional information to the feature matching operation, such as texture and shape based descriptors. These avenues of improvements were not tested because the purpose of the investigation was to develop a generic approach.

7.3 Stem/Calyx Discrimination

Automatic surface blemish detection systems for apples can be confused by the stem, calyx, and concavities. If the stem and calyx are incorrectly classified as blemishes, a false grade will be assigned to a fruit. This section is concerned with this issue and presents two approaches for solving the problem.

Various approaches to solve the problem have been investigated by other workers. One approach is to use a mechanical device to orient the fruit so that problematic regions, such as the stem and calyx are in known positions. Such systems have been extensively studied [6, 7, 8]. Campins *et al.* [9] concluded that it was not possible to mechanically orient all apples, and they sought other approaches to solving this problem. Outline views have been used to determine the orientation of oranges, blueberries and peppers [10, 11], but this technique can only be applied to certain types of fruit, and even then correspondence between different views may need to be established in order to identify the stem or calyx location in all views. Other workers have identified stem and calyx concavities from the gradients of reflected light intensity [12], but this approach must necessarily make assumptions regarding the natural colour and shading variations.

A further technique has been developed using mid-infrared (MIR) images [13]. In this technique, the surface of previously cooled fruit is rapidly heated with brushes. This results in temperature differentials due to the irregular structure of the apple, and these are detected in the MIR images.

Structured light techniques can be used to directly indicate the location of concavities, independently in each view and essentially for any orientation of the apple. We present here two techniques, both of which can be classed as structured-light systems, but which use markedly different underlying techniques for image formation. The first technique utilises the diffuse reflection of light stripes projected onto the surface of the fruit, whereas the second technique [14] utilises the surface (or specular) reflection of a small set of unfocused light sources. Consideration of these two techniques demonstrates the breadth of approaches that are often required in machine vision tasks. The second technique was built on an understanding of aspects of the material behaviour rather than merely treating it as an image processing task. We also hope that the presentation of these two techniques might stimulate others to explore their application to other problems.

7.3.1 Concavity Detection Using Light Stripes

This section comprises two parts. The first part describes the general approach of using projected light stripes to detect concavities, and reviews applications of this approach developed by other researchers. The second part describes techniques we have developed to specifically address practical difficulties in interpreting light stripe images for the purpose of concavity detection.

Overview of the Technique

Projected light stripes and a camera can be used to measure the 3D shape of objects that have diffuse reflecting surfaces. Focused stripes of light form planes when projected into 3D space. The intersection of these planes with the surface of an inspected object creates a profile whose shape is determined by the curvature of the object's surface. The shape of the profile is apparent from viewpoints off the light plane. In theory this curvature is best seen from viewpoints perpendicular to the light plane. In practice, the magnitude of the viewing angle is balanced against the need to minimize object occlusion and the ability to image multiple profiles from a single viewpoint. Examples of light stripes on apples are shown in Figure 7.7, where the viewing angle is offset from the central stripe plane by 30°.

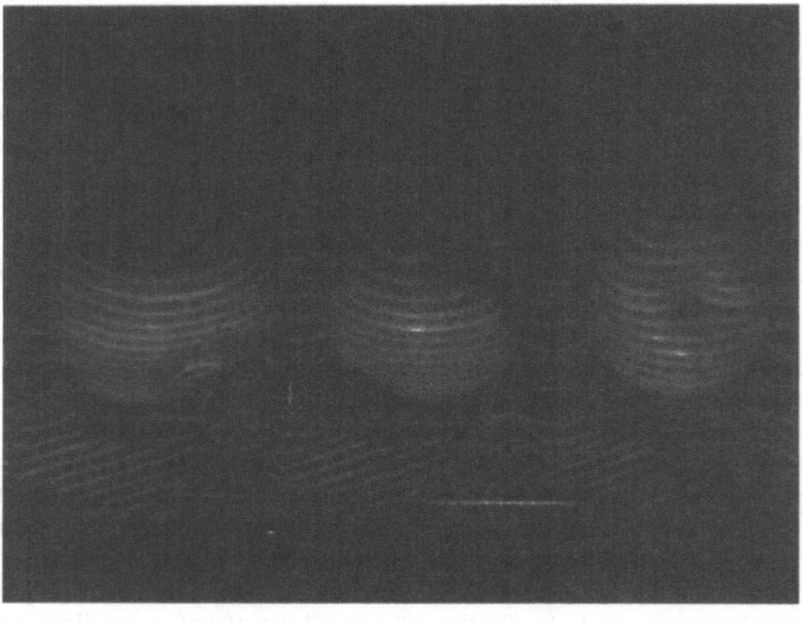

Figure 7.7. Light stripes on the surface of Braeburn apples. The centre apple presents no stem or calyx, while the outer apples present stems to the camera.

We used a light stripe projector consisting of a 50mW laser diode and an output diffraction grating. This projector produced 33 parallel light stripes, with a stripe-to-stripe angle of 0.38° and an end-to-end angle of 60°. The laser emits light in the near infrared (NIR) spectrum, and at this wavelength (780 nm) the surface reflectance of apples is generally uniform. The resultant images possessed well-defined stripes and no clutter caused by variations in the fruit's surface colouration. The camera was equipped with a matching optical band-pass filter.

Related Approaches

The use of projected light stripes as an approach to infer 3D surface information for apples has been extensively investigated by others. Yang [15] identified candidate regions on a fruit by using a watershed segmentation algorithm to delineate regions of low intensity. The stripe patterns within these regions were analysed for properties such as curvature. Finally, image-based features and stripe-related features were passed into a neural network classifier, which calculated the likelihood of the region to be a stem/calyx or a surface blemish.

Crowe and Delwiche [16, 17] deliberately set out to develop a fruit inspection system to address the issues of complete surface coverage, blemish detection, stem/calyx detection, and practical throughput rates. Concavity detection was performed by analysing the curvature of six laser stripes projected onto the fruit per view. Each line nominally ran vertically through the image and the curvature was analysed over a span of eight vertical pixels applying a combination of pixel offset counting and a one-dimensional convolution. This technique was implemented on a pipeline processor, which returned results extremely rapidly.

Stem/Calyx Detection Algorithm

Figure 7.7 illustrates how imaged inter-stripe distance is affected by surface orientation. In convex surface regions, and with this camera placement, the inter-stripe distances decrease monotonically when traversing the image from top to bottom. The proposed algorithm detects regions which break this trend of decreasing inter-stripe distances. In doing so, the algorithm acts as a detector of non-convex regions. The algorithm could be applied to the raw acquired images, such as the one shown in Figure 7.7. However, pre-processing each fruit image with a Cartesian to polar transformation greatly improves the accuracy and robustness of the inter-stripe analysis. This section describes the stripe segmentation process, then the inter-stripe analysis, and finally the adjunct pre-processing step.

Extracting the Individual Fruit from the Background Scene

The process of segmentation mostly concerns calculating separate binary masks for each fruit present in an acquired image. Firstly, a stripe-free image is calculated by applying a grey-scale minimum filter followed by a maximum filter to the acquired image. The kernel size for both filters must be identical. In our system, the fruit reflects the ambient NIR light more strongly than the conveyor, or any other parts of the image. Based on this observation, we compute a binary image representing fruit and background by thresholding the stripe-free image at a single intensity value. The binary image is then divided into three separate image masks, one for each fruit, using a connected components labelling routine. Each mask is used to extract an individual fruit from the acquired image. Finally, the pixel intensities of the fruit surface are suppressed by subtracting the stripe-free pixel intensities from the acquired pixel intensities, which results in images like those shown in the left-hand column of Figure 7.8.

Stripe Analysis

The algorithm is based on the observation that in convex regions of the fruit, the spacing between stripes steadily decreases when following down a column of pixels. A concave region causes the change in inter-stripe spacing to fall outside an expected range because:

- the surface orientation is concave and so causes the inter-stripe spacing to increase instead of decrease.
- the surface orientation is excessively convex (which occurs when the surface enters a concavity) and so causes an abrupt decrease in inter-stripe distance.
- surface occlusions cause stripes to be broken.
- tangential surface viewing angle suppresses the reflected intensity of the stripes.

The stripe analysis starts at the upper edge of the fruit and traverses down a column of pixels. A simple algorithm continuously maintains the values of previous two inter-stripe distances, and extrapolates these values to give an expected range of values for the next measured inter-stripe distance. The output of the analysis is a binary value that indicates whether the distance of the previous inter-stripe region falls within the expected range.

This inter-stripe analysis is repeated in the reverse direction by traversing the image from bottom to top. Reverse processing is necessary because the analysis is asymmetric since a stem/calyx region may seem convex when entering it, and only manifests itself as concave when exiting. So the complete result is calculated by processing the data from both directions and taking the union of results.

Each column is analysed independently, and the final result is produced by stacking the one-dimensional results to form an image. This image is post-processed to remove concave regions that are too small to be stem/calyx regions.

Cartesian to Polar Image Co-ordinate Transformation

The extremities of the stripes shown in Figure 7.7 have a property that reduces the effectiveness of the inter-stripe analysis technique. Each stripe tends to curl at its left and right extremities. The stripe analysis algorithm described above is most effective when the stripes are nominally horizontal in the image, since the inter-stripe distance is measured in a vertical direction. This shortcoming can be largely addressed by spatially transforming the pixel locations from Cartesian co-ordinates to polar co-ordinates, prior to performing the inter-stripe analysis. Careful choice of the centre point of the polar transform is needed. Figure 7.8 illustrates the intermediate and final results for processing the fruit shown in Figure 7.7. The centre point for the polar transform is calculated to be mid-way horizontally across the image of the individual fruit, and vertically above the fruit at a distance related to the width and height of the image. The analysis algorithm was configured to perform row traversal, instead of column traversal, for these transformed images.

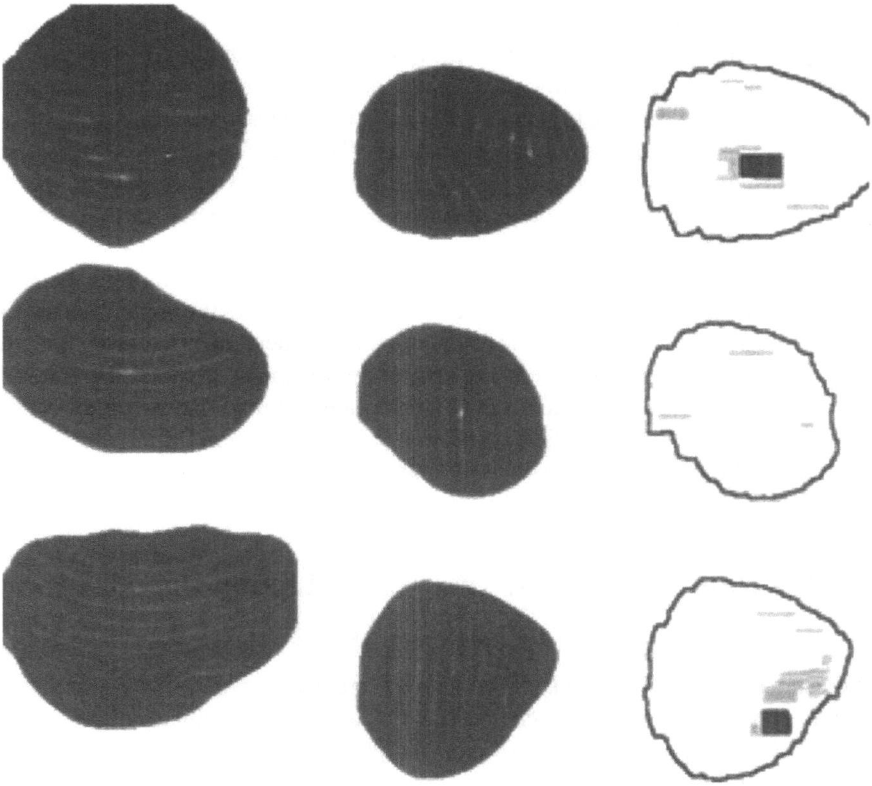

Figure 7.8. Fruit sub-images (left column); polar co-ordinate images (middle column); concavity regions (right column). Grey regions indicate small concavities that were eliminated by size filtering; black regions indicate significant concavities. The outline is shown for clarity.

The Cartesian–polar transformation significantly reduces the generation of spurious concavities especially at the regions near left and right extremities of the fruit.

Discussion

A number of practical trade-offs must be made in an implementation of this algorithm:

- *Stripe intensity versus stripe density*. Stripes of high intensity are desirable because they are easily segmented, even at surface points viewed nearly tangentially. Stripe density must be adequate to detect a minimum concavity size because at least one stripe must fall in a concavity before it can be detected. However, for a given laser, stripe intensity is traded off against density because the radiation is divided equally over all stripes.
- *Boundary delineation versus stripe delineation*. Segmentation of the fruit from the background is performed by an intensity threshold operation. This

difference in intensity exists because fruit reflects the NIR light more strongly than the background. For the fruit to have a higher intensity, a certain amount of NIR light must be present in the ambient illumination. However the intensity of the ambient NIR should be low enough to maintain a contrast between the fruit and the stripes on the fruit.

- *Viewing volume versus 3D information*. The useful working volume of any structured light system for measuring 3D shape is defined to be the intersection of the volume illuminated by the projector and the volume viewable by the camera. The greatest intersection is achieved when the projector and camera are coincident, but this configuration provides no 3D information. The maximum 3D information is achieved when the relative viewing directions of the projector and camera is set to 90°, but this configuration also offers the minimum working volume. The trade-off between viewing volume and 3D information is important when attempting to achieve whole surface inspection, where being able to inspect as much of the fruit surface as possible from one image is a highly desirable property.

7.3.2 A New Approach to Structured Lighting

Overview of Technique

Increasing processing speeds of computers are making more inspection tasks viable. However, if advantage can be taken of the natural attributes of a situation, the task of the processing stages can be substantially reduced. The technique presented in this section arose out of a careful consideration of the characteristics of the apple surface geometry and surface characteristics. The reflection of a non-diffuse light source from the surface of an apple is dependent on the shape of the apple. This suggests that such reflections might be used to detect major non-uniformities such as the stem and calyx concavities. The technique exploits a few basic opportunities presented by the overall task of inspecting apples for surface defects.

Firstly, we take advantage of the basic principles of the interaction of light with fruit [18–22]. Reflection from a dielectric material results from both surface reflection (sometimes referred to as regular or specular reflection), and body reflection, the light that re-emerges through the surface of the material. We have devised a system that utilises body reflection for defect detection and specular reflection for stem/calyx detection, separating the two on the basis of illumination colour.

Surface reflection is the mirror-like reflection of the light source that occurs right at the surface. For most practical purposes, we can consider that the light reflected at a point on the surface has the same colour or spectral content as the incident light, and is independent of the colour of the material. We can also say that this reflected light will be in one direction, dependent on the position of the light source, and the local surface orientation. Such reflection is defined by the Fresnel equations, which define the dependency on the refractive index of materials, the angle of incidence, and the polarisation of the light. Although the refractive index is a function of wavelength, it is commonly assumed to be constant over the visible spectrum, leading to the above assumptions.

Conversely, both the spectrum of the light source and the pigments of the material influence the spectral content of the body reflection. The body reflection will normally be diffuse, as the light having been absorbed, reflected and refracted by the pigments in the skin and internal material, will exit in random directions. It is the body reflection that is primarily responsible for our perception of colour in most situations, although the total reflection that we observe is the sum of the two components.

We apply these basic principles to the formation of an apple image when a non-diffuse light source is used to illuminate an apple. At particular locations, dictated by the geometry of the apple surface, the light source and the sensor, the surface reflection component will dominate, and will have a spectral distribution governed principally by the spectrum of the light source. The diffuse reflection that is observed from the remainder of the apple will have a spectrum that is greatly influenced by the colour of the apple. The dominant body reflectance of an apple is at wavelengths longer than 500 nm [22, 20, 23], and this is the portion of the spectrum that is most useful for defect detection, using a diffuse illumination system. This enables the remaining blue end of the spectrum to be used for other purposes.

If a non-diffuse light source is chosen with a wavelength of about 450 nm, the body reflection will be very low, and the only significant reflection at this wavelength will be the surface reflection. The shape of this reflection will, as stated above, relate to the shape of the fruit. This subdivision of the spectral properties of fruit is well matched with the spectral response of the three channels of a standard colour video camera, the blue channel being largely superfluous for defect detection. Using a suitable arrangement of 450 nm light sources, the stem and calyx concavities provide sufficient deformation of the surface to enable this technique to detect them using the spare capacity of a standard colour video camera.

Illumination System

The geometrical arrangements of illumination sources that have been tested for this purpose can be grouped into four generic categories

- an array of sources, each of small spatial extent;
- a single linear source;
- intersecting linear sources;
- parallel linear sources.

A single small source produces a simple reflection pattern, but does not cater for all apple orientations. Although an array of small sources can be devised to cater for a sufficient range of apple orientations, the ambiguity of the reflection pattern increases with the number of sources and it becomes increasingly difficult to associate portions of the pattern with a specific light source, and to interpret the pattern.

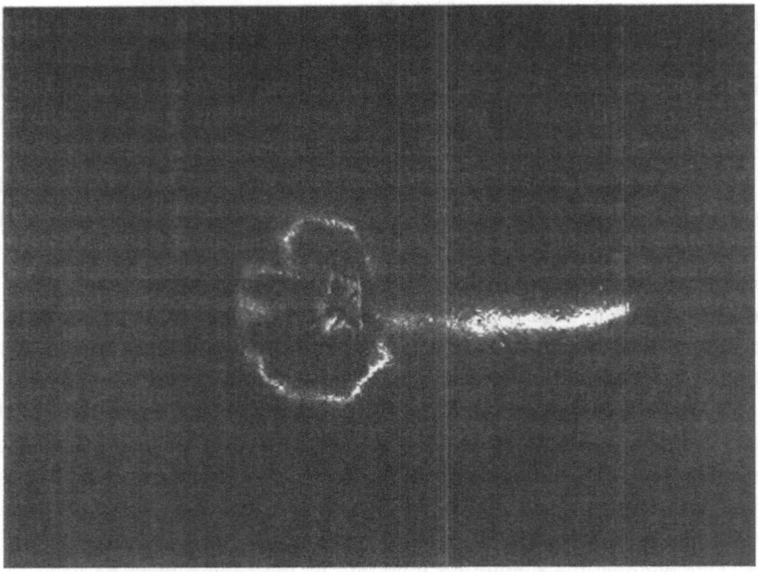

Figure 7.9. Example reflection pattern obtained using a single linear source.

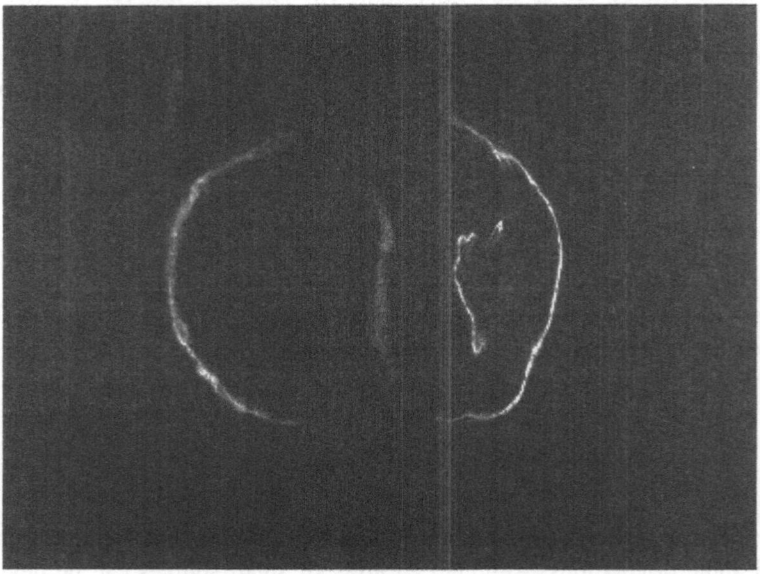

Figure 7.10. Example reflection pattern obtained using three parallel illumination sources.

An example of a reflection pattern from a single linear source is shown in Figure 7.9. Although only suitable for a limited range of apple orientations, the patterns generated are much less fragmented than those using an array of small sources and interpretation is considerably easier. Using an array of intersecting sources improves the coverage, but the patterns generated are much more complex, and are highly dependent on apple position.

By arranging three parallel blue fluorescent sources, one above the conveyor centreline and one to each side below the level of the conveyor, reflection patterns are produced that provide information relating to a wide range of stem and calyx positions. Figure 7.10 shows a typical image of an apple illuminated in this way and viewed from above.

It is interesting to compare the characteristics of this illumination with the projected light-stripe system described in Section 7.3.1.

Fundamentally, the projected light-stripe technique relies on a sufficient density of stripes to ensure that some are projected into a concavity, and that the body reflectance is adequate in magnitude and uniformity, for the formation of an image. The deformation of the light stripes in the image is discernible for a wide range of concavity depths. In contrast, provided a concavity constitutes a sufficient deformation of the surface, the technique described in this section will yield some information regarding the location of the concavity, does not place any requirement on body reflectance, but does require adequate surface reflectance. The NIR portion of the spectrum is suitable for the projected light-stripe technique because the NIR reflectance is substantially independent of fruit colour, and does not interfere with the use of the visible spectrum for defect detection. However, this can interfere with bruise detection systems using NIR. Because the stem and calyx position must ultimately be registered with the defect detection results, it is an advantage if the sensors are integrated. While special purpose multi-spectral cameras capable of working in the NIR region are available [25], the technique described here can utilise the spare capacity of a standard video camera.

Analysis of Reflected Light Patterns

Six views of a single apple in different orientations are shown in Figure 7.11. These show some of the characteristic stripe patterns reflected from the illumination system. Figure 7.11a demonstrates that curvature is not by itself significant when determining the location of concavities, the high curvature at the left of the apple being due merely to the general shape of the apple in this orientation. With the exception of determining the apple outline, the entire analysis is performed with the blue channel image.

In the discussion that follows, each separate portion of the reflection pattern is called a stripe. Stripes were divided into two classes, major and minor stripes as described later. The analysis may be divided into the following sequence of steps:

- extraction of apple outline from red channel image;
- extraction of stripes from blue channel image;
- measurement of main geometric parameters of stripes;
- identification of major stripes;
- determination of stem and calyx concavity location.

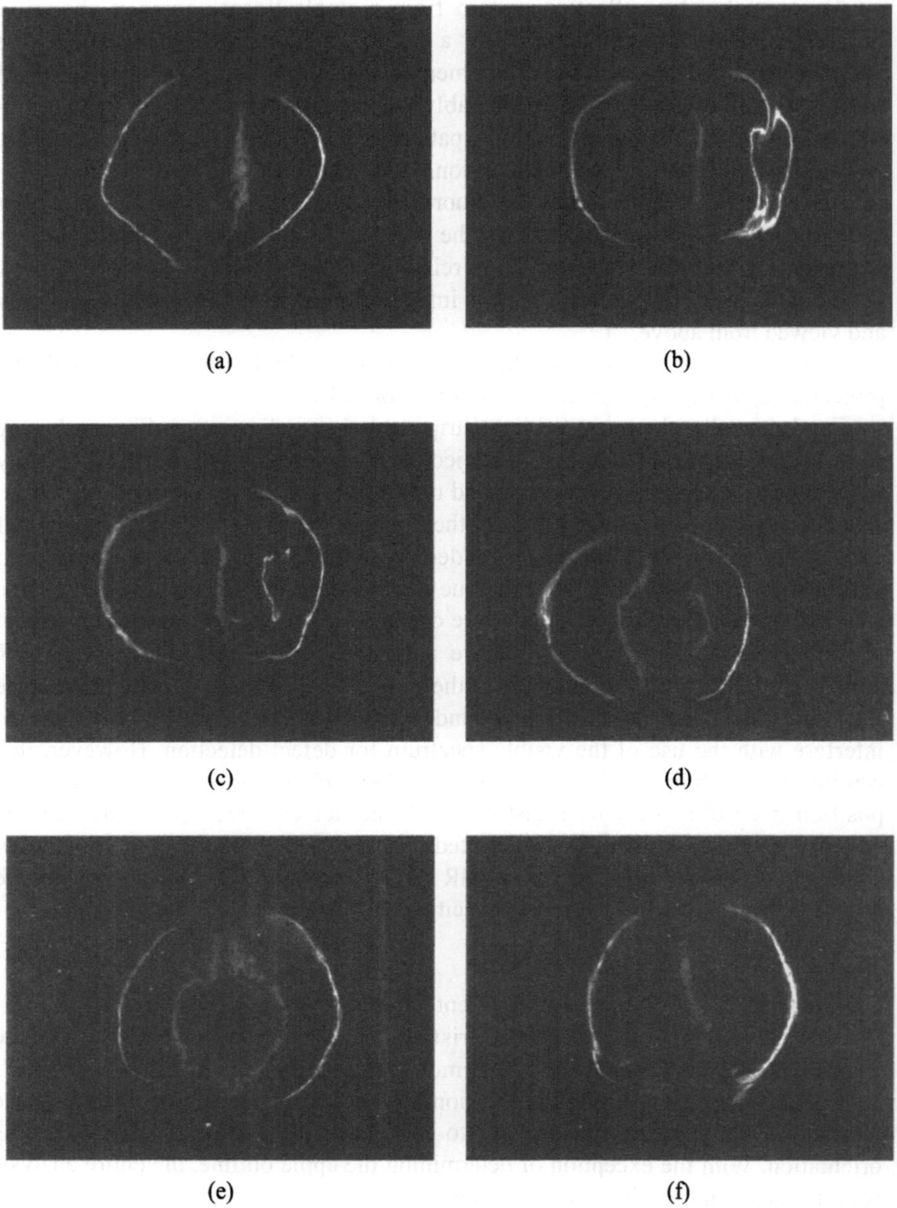

Figure 7.11. Reflection patterns of an apple in six different orientations, obtained using three parallel illumination sources.

Providing the mechanical components of apple grading inspection systems are blue, the segmentation of apples from other objects is readily achieved using the red channel image. Similarly the extraction of the stripes within the image is readily achieved using the blue channel image.

The image of each apple has a few dominant stripes, related to the dominant shape of apples, a deformed sphere. The left and right illumination sources produce long curved reflections near the sides of the fruit. The central source produces either a central stripe as in Figure 7.11a, or a more complex stripe pattern extending around both sides of the concavity, as one continuous stripe, or two distinct stripes as in Figure 7.11e. Rotating the concavity slightly away from the centre, as in Figure 7.11d, results in one part of the central stripe pattern being dominant (the left in this case), the other being much smaller. Visual examination of many images indicated that an analysis based on five major stripes was appropriate. The major stripes have been called, Left, Centre Left, Centre, Centre Right and Right major stripes. A stripe is only considered to be a Centre Left or Centre Right stripe if the other member of the pair is present.

To find the major stripes, the stripes in an image are considered in decreasing order of length until either all stripes have been considered, or Left and Right major stripes are found, and a Centre stripe or a Centre Left-Right pair has been detected. The longest stripe that is close to the left or right of the fruit is called a Left or Right stripe. The longest stripes between these two are called Centre Left and Centre Right if a pair is present, and Centre if there is only one. Any remaining stripes are re-examined by the algorithm to see if they might be a detached segment of a major stripe. The simple structure of the illumination sources aids this process, as major stripes have well-defined characteristics. To be considered an extension of a major stripe, the end of a remaining stripe must be close to a major stripe, and must not overlap (e.g., run alongside) the major stripe. A pair of stripes that are close to each other, but have significant overlap are not coalesced because they indicate a concavity between them. In many cases, a major stripe will branch, but no attempt is made to identify the main portion of each major stripe. Any stripes that are not part of a major stripe are called minor stripes.

To locate the stem and calyx regions, the stripe patterns are analysed using four classes. These are based on whether the stem or calyx location is determined using

- a single major stripe;
- a major stripe and minor stripes;
- forked major stripes and minor stripes;
- minor stripes only.

The main characteristic that indicates the presence of a concavity is that it is bounded on both left and right by portions of stripes. A single major stripe will only bound a region on both left and right if there is a concavity in that region, as in Figure 7.11e. Similarly a minor stripe will only exist if there is a concavity in that region, as in Figure 7.11c.

The first class includes cases such as Figure 7.11b and Figure 7.11e where the Centre Left and Centre Right stripes are considered together as a single major Centre stripe. In the case of Figure 7.11b, the Right major stripe encircles the concavity and in the case of Figure 7.11e the Centre stripe encircles the concavity. Sometimes a major stripe may have a small hook on it or appear to enclose a group of pixels, resulting in a false determination of a stem or calyx concavity location. The treatment of this is discussed below.

The second class (major stripe and minor stripes) is typified by Figure 7.11c in which the calyx was rotated closer to the camera than in Figure 7.11b. This causes the two branches of the Right major stripe to separate, forming major and minor stripes. The appropriate major stripe to include in the analysis is determined from the direction of curvature of the minor stripe. The concavity is always in the direction of the centre of curvature of the minor stripe.

The third class (forked major stripes and minor stripes) arises from the limited extent of the reflection when the stem or calyx is located near the top or bottom of the image, as in Figure 711f. (The forks are near the bottom of the image in this case.) Other cases occur when the concavity is located near the perimeter of the fruit image and the only indications of its presence are isolated minor stripes.

The fourth class (minor stripes only—not illustrated in Figure 7.11) encompasses those in which the location can be determined entirely from the minor stripes. These occur when one or more minor stripes in a single region of the image, bound a group of pixels on both the left and the right, identifying the position of the concavity.

Images that do not fit one of the above categories, such as Figure 711a are deemed to not have a stem or calyx concavity in the view of the camera.

In some situations, using the above analysis, more than one concavity is identified in a single image. In Figure 711f, the hook formed by the upper branch at the bottom of the Right major stripe, identifies one possibility (a case of the first class). An analysis of the forked ends of the major stripes together with the minor stripe between them, results in the second and correct possibility (a case of the third class). These situations are resolved by determining the area enclosed by the portions of stripes that are used to identify the concavity and using it as a quality measure. The greater the area enclosed, the higher the probability of it being a true location.

7.4 Conclusion

A method for tracking the orientation of fruit to prevent multiple counting of surface features, and two methods for stem/calyx detection have been presented. The motivation for the development of stem/calyx detection methods is to provide means to distinguish stem and calyx from actual surface blemishes in automated defect grading systems.

The most important synergy between the methods exists in the way that information derived from one method can be used to reinforce the computation of the other method.

In one direction, localisation of stem/calyx can be used to reinforce the tracking algorithm. Figure 7.6 illustrates the tendency for the tracking algorithm to over count features when either a low number of tracked features or a high number of features exist. Although overcounting occurs in these situations for different underlying reasons, the algorithm commits the same error in both cases—a feature appearing around the occlusion boundary is not matched to its instance in the global model. But consider the situation where a feature is not only localised but is also identified as a stem or calyx. Since a fruit has two such features (a stem and a calyx), even a coin toss will yield an even chance that a stem/calyx feature will be correctly matched to its instance in the global model. But with the addition of pose information, the stem/calyx feature should almost always be correctly matched. When the fruit is of a type that has no calyx, the chances of obtaining a correct match to its instance in the global model increases to 100% (even without tracking!). One reliable match into the global model constrains the unknown fruit rotation to one degree of freedom, which is much simpler to solve than a problem with three degrees of freedom.

In the other direction, if concavities are included in the set of features to be tracked, then the tracking algorithm can provide the expected location of the concavities, before they are located by the light-stripe localization techniques. This prior information of concavity location can be used to advantage by both light-stripe techniques presented. For example, the light-stripe algorithm of Section 7.3.1 can be modified so that the inter-stripe analysis is performed only on the part of the fruit surface that is likely to contain the concavity, rather than the whole fruit surface. The reduction in computation correlates with the certainty to which the location of a concavity can be predicted. The new approach to structured lighting described in Section 7.3.2 can also benefit from the tracking system. This technique examines a range of possible light reflectance patterns to determine concavity location. The information provided by the tracking algorithm can be used to order the sequence that these light patterns are examined, so that likely patterns are examined first, thus reducing the computation required.

This chapter has outlined the challenges that are preventing full-surface fully automated fruit grading from being practically realised. It has also described in detail several techniques that have been developed by the authors and their colleagues in an attempt to overcome these challenges. While real-time pack house implementations have yet to be realised, the techniques we have demonstrated in the laboratory show considerable promise.

7.5 Acknowledgements

The authors gratefully acknowledge the support of the New Zealand Foundation for Research Science and Technology (FRST) in funding this work.

7.6 References

[1] Tao Y., (2001) Photonics in fruit and vegetable quality assessment, *Optics in Agriculture, A Critical Review 1990–2000,* James A. DeShazer and George E. Meyer eds. SPIE Vol. CR80, pp. 64–100.

[2] Laykin S., Edan Y., Alchanatis V., Regev R., Gross F., Grinshpun J., Bar-Lev E., Fallik E., and Alkalai S. (1999) Development of a quality sorting machine using machine vision and impact, ASAE/CSAE-SCGR Annual International Meeting, Toronto, Ontario, Canada, July 1999.

[3] Crowe T.G. and Delwiche M.J. Real-time defect detection in fruit (Part I): Design concepts and development of prototype hardware, *Trans. ASAE*, vol. 39, No. 6, pp. 2299–2308.

[4] Hartley R.I. and Zisserman A. (2000) Multiple View Geometry, Cambridge University Press.

[5] Faugeras O. (1993) *Three-dimensional Computer Vision: a Geometric Viewpoint,* MIT Press, Cambridge, Ma.

[6] Miller W.M. and Drouillard G.P. (1997) On-line blemish, color and shape analysis for Florida citrus, Proc. Sensors for Nondestructive Testing: Measuring the quality of fresh fruit and vegetables, vol. 97, pp. 249–260.

[7] Davenel A., Guizard C.H., Labarre T. and Sevila F. (1988) Automatic detection of surface defects on fruit by using a vision system, *J. Agricultural Engineering Research*, vol. 41, pp. 1–9.

[8] Throop J.A., Aneshansley D.J. and Upchurch B.L. (1997) Apple orientation on automatic sorting equipment, Proc. Sensors for Nondestructive Testing: Measuring the quality of fresh fruit and vegetables, Orlando Florida, vol. 97, pp. 328–342.

[9] Campins J., Throop J.A. and Aneshansley D.J. (1997) Apple stem and calyx identification for automatic sorting, ASAE Paper 973079, Proc. of the 1997 ASAE Annual International Meeting, Minneapolis, MN, USA, August.

[10] Pla F. and Juste F. (1995) A thinning-based algorithm to characterize fruit stems from profile images, *Computers and Electronics in Agriculture*, vol. 13, no. 4, pp. 301–314.

[11] Wolfe R.R. and Sandler W.E. (1985) An algorithm for stem detection using digital image analysis, *Trans. ASAE*, vol. 28, No. 2, pp. 641–644.

[12] Miller B.K. and Delwiche M.J. (1991) Peach defect detection with machine vision, *Trans. ASAE*, vol. 34, no. 6, pp. 2588–2597.

[13] Wen Z. and Tao Y. (2000) Dual-camera {NIR/MIR} imaging for stem-end/calyx identification in apple defect sorting, *Trans. ASAE*, vol. 43, no. 2, pp. 449–452.

[14] Penman D.W. (2001) Determination of stem and calyx location on apples using automatic visual inspection, *Computers and Electronics in Agriculture*, vol. 33, No. 1, pp. 7–18.

[15] Yang Q. (1996) Apple stem and calyx identification with machine vision, *J. Agricultural Engineering Research*, vol. 63, no. 3, pp. 229–236.

[16] Crowe T.G. and Delwiche M.J. (1996) Real-time defect detection in fruit –(Part I): Design concepts and development of prototype hardware, *Trans. ASAE*, vol. 39, n. 6, pp. 2299–2308.

[17] Crowe T.G. and Delwiche M.J. (1996) Real-time defect detection in fruit –(Part II): An algorithm and performance of a prototype system, *Trans. ASAE*, vol. 39, no. 6, pp. 2309–2317.

[18] Butler W.L. (1962) Absorption of light by turbid materials, *J. Optical Society of America*, vol. 52, no. 3, pp. 292–299.

[19] Birth G.S. and Zachariah G.L. (1973) Spectrophotometry of agricultural products, *Trans. ASAE*, vol. 16, no. 3, pp. 548–552.

[20] Birth G.S. (1978) The light scattering properties of foods, *J. Food Science*, vol. 43, no. 3, pp. 916–925.

[21] Gunasekaran S., Paulsen M.R. and Shove G.C. (1985) Optical methods for non-destructive quality evaluation of agriculture and biological materials, *Journal of Agricultural Engineering Research*, vol. 32, no. 3, pp. 209–241.

[22] Gudrun J., Klinker S., Shafer A. and Kanade T. (1988) The measurement of highlights in color images, *International J. Computer Vision*, vol. 2, pp. 7–32.

[23] Reid W.S. (1976) Optical detection of apple skin, bruise, flesh, stem and calyx, *J. Agricultural Engineering Research*, vol. 21, pp. 291–295.

[24] Upchurch B.L., Affeldt H.A., Hruschka W.R., Norris K.H. and Throop J.A. (1990) Spectrophotometric study of bruises on whole Red Delicious apples, *Trans. ASAE*, Vol. 34, No. 3, pp. 1004–1009.

[25] Graydon O. (1999) Light scatter checks tissue health, *Opto and Laser Europe*, January, p. 11.

Chapter 8

Editorial Introduction

It is a plausible conjecture that the primary purpose of human colour vision is to enable us to locate and make quality judgments about food items. Until recently in human history, the ability of the human senses to detect unsafe food was critical for our very survival. For this reason, we can approach the study of machine vision systems for this purpose with some optimism. Nowadays, bacteriological assay techniques have taken over the primary responsibility for food safety but there remains the question of judging aesthetic appearance, which is important for commercial reasons. Undesirable features, such as stains, discolouration, tears, missing parts, distortions of form, etc., present a challenge to the vision engineer, because these often require fine discriminations, based on subtle changes of colour or texture. Grading criteria for food products, such as chickens, fruit, etc., is often expressed in the form of rules, tabulated by government or international bodies, specifically to guide human inspectors. Trying to represent these rules in a form that is suitable for computer implementation requires the use of techniques such as Fuzzy Logic, Artificial Neural Networks and other Pattern Recognition procedures. To convince would-be users of automated grading systems, it is often necessary to compare machines and human graders, neither of which is "perfect". We must first appreciate how unreliable human beings are on these tasks. It is only by correlating the results from different human inspectors and comparing individual graders with their performance on exactly the same task on a different occasion, that we are in a position to judge a machine properly. It is a regrettable truth that managers are often blissfully unaware of the unreliability of human inspectors but are unwilling to commit themselves to purchase a system that is clearly superior, when judged by objective, quantifiable criteria, but is still manifestly imperfect.

Chapter 8

Vision-based Quality Control in Poultry Processing

W. Daley and D. Britton

"Where there is no vision the people perish..."

Proverbs 29:18

Heard in the plant. "The problem is always the people, but so too is the solution."

Daley 2001

"It is deeper than we thought"

Britton, 1999

8.1 Introduction

The theory has been suggested that the primary purpose of human color vision is to enable us to both locate and make quality discriminations about food products. For example, we can tell whether or not something is green, ripe or spoilt using visual input alone. So it continues today.

Most of what are typically known as first world countries now have the ability to produce food and its allied products at very rapid rates. A line for processing chicken for example is capable of running at about 180 birds per minute, and fruit processing lines will run at rates of ten pieces per second. Even with these advancements, however, most of the inspection and quality control tasks are conducted manually.

The main reason for this state of affairs is the variability of the product and the lack of adequate software to accommodate the great deal of subjectiveness associated with the decision-making process. Currently existing computer systems are not able to incorporate this kind of flexibility while achieving the desired throughput rates. Visual sensing for machine guidance is a key area of need. Many material handling systems require the location of specific points on the product for manipulation. This is also problematic at times for non-uniform products. In this chapter we will describe some approaches being taken to address some of these issues.

Post-chill grading is one of the activities that is ubiquitous to most poultry processing operations. In the more advanced operations automated sorting is conducted after the grading. The grading tasks at this point are not for health or safety issues, but relate more to total product yield in terms of quality and appearance. In the typical American poultry plant the factors of interest include: missing parts, bruising and discolorations, gall stains, feathers, skin tears and size. In some respects automating the grading process is not currently a critical issue, because a manual sorting operation occurs right after the chiller. However, as labour becomes more of an issue it is likely that more automation will be introduced. Then, the ability to accurately and automatically sort product will be of more direct importance. A block diagram of the poultry production process is shown in Figure 8.1.

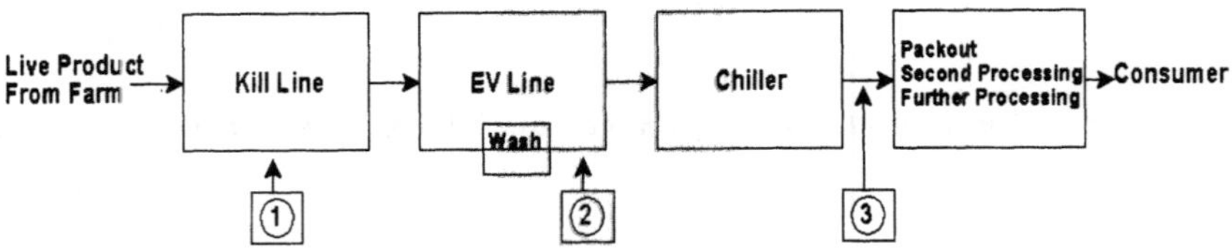

Figure 8.1. Block diagram of the poultry production process.

Live product enters the production process on the kill line, where the birds are killed and defeathered. After being transferred to an eviscerator (EV) line and having the inner organs removed they are inspected for wholesomeness. They then enter the chiller, where they are cooled to about 34°F to retard the growth of bacteria. The final stage involves any further processing necessary before delivery to the consumer. The positions labelled 1, 2 and 3 identify key quality checkpoints in the process.

In this chapter we consider the design of an automated system to do grading after the chill stage of the poultry production process. For simplicity, we will constrain the grade evaluation to be based solely on the level of bruising on the carcass. However, the principles can be extended to other kinds of defects such as skin tears, gall stains, missing parts or other plant-specific concerns.

8.2 Poultry Grading Application

A section of the United States Department of Agriculture's (USDA) guidelines that establishes grades based on bruising, is shown in Table 8.1. This conveys the rules for determining the grade of a whole bird carcass. From this table we deduce that birds with an average weight of between 2 and 6lbs color. with a lightly shaded discoloration greater than one inch in diameter on the breast or legs will be downgraded; similar rules exist for other grades. What happens if the bruise is 1.1 inches in diameter? Most reasonable graders would allow this carcass to pass as an

A grade, unless they have a tremendous dislike for the shift supervisor. Most computer systems would behave as if they had an ongoing feud with said supervisor and downgrade the product. Another question also arises; what does lightly shaded mean? As you can tell, this is not a typical programming statement. This is an example of a problem that would require special treatment. Techniques utilising what are called "soft computing" approaches now exist to address these sorts of problems.

These methodologies are investigated as a way to program the decision-making process on the back end of machine vision systems. The algorithm development based on adaptive Acceptance Quality Levels is presented within the framework of the poultry grading application to decide whether a chicken is accepted, partially accepted, or rejected. The results are then compared with a more common technique called a minimum distance classifier, to assess qualitatively the performance of this approach.

Table 8.1 Excerpt from USDA grading summary of specifications for maximum allowable discoloration of a grade A carcass.

Grade A Carcass: Max Allowable Discoloration	**Lightly Shaded**		**Moderately Shaded**	
	Breast and Legs	Elsewhere on Carcass	Hock of Leg	Elsewhere on Carcass
2 lbs or less	$^3/_4$ inch	1 $^1/_4$ inches	$^1/_4$ inch	5/8 inch
Over 2 lbs – 6 lbs	1 inch	2 inches	$^1/_2$ inch	1 inch
Over 6 lbs – 16 lbs	1$^1/_2$ inches	2$^1/_2$ inches	$^3/_4$ inch	1$^1/_4$ inches
Over 16 lbs	2 inches	3 inches	1 inch	1$^1/_2$ inches

8.2.1 Soft Computing: Fuzzy Logic and Neural Networks

"Soft computing" is a technique that attempts to humanize machines. In most of our day-to-day functions we do not operate with an extreme level of precision, and yet in most situations we are able to perform our tasks quite effectively. As another illustration, think of how you decide on a gratuity. Most of us have some rules of thumb by which we operate. For example: *if the food is great give the server a good tip*, or *if the food is great and the service is atrocious give the server a poor tip*. Implementing soft computing methods allows one to program and compute based on concepts such as these.

A brief digression is necessary to summarize the approaches used in this study. As mentioned earlier, "soft computing" is the name that has been given to the area of research geared towards programming and making decisions with imprecise concepts. Also called intelligent computing, it refers to the ability to do symbolic manipulation and pattern recognition [1]. This includes the areas of expert systems, fuzzy logic, and neural networks. All of these fields use computational models that are based on inferred mechanisms about the operation of the brain and human reasoning. This approach is needed to address the many problems that cannot be precisely formulated mathematically, or where this formulation would be prohibitively expensive. Examples of the application of these approaches to natural products can be found in areas as varied as cork inspection to computer aided tomography [2–8]. Two of the more useful approaches have been neural networks and fuzzy logic. We will briefly explain the concepts behind the use of these tools.

8.2.2 Fuzzy Logic

Fuzzy logic is considered a superset of conventional logic extended to handle the concept of partial truth. This idea was proposed and developed by Lotfi Zadeh in 1965 [2]. It was not immediately popular in America, however, the Japanese more readily embraced it. The main idea is that the decision is no longer a binary 0 or 1, but it can lie somewhere in between. This then leads to the concept of a fuzzy variable and membership functions. Using this representation, objects (in our example chickens) are members of classes to some level or degree. The word "fuzzy" has probably been an unfortunate choice, as it is not that the concept is fuzzy, but rather the methodology presents a way of reasoning and calculating with imprecise concepts and definitions. Examples would be TALL and OLD; how do we describe these concepts? One way is shown in Figure 8.2 and Figure 8.3 where it is not after a certain height that someone is called tall or after a certain age that someone is referred to as being old. It is all a matter of degree. Zadeh describes these as "linguistic variables".

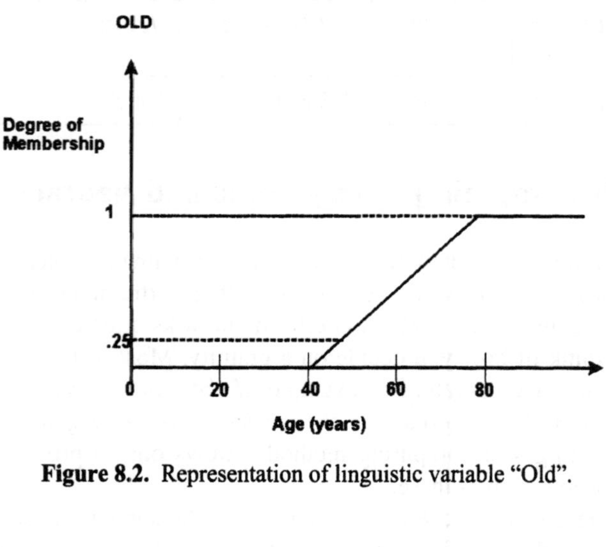

Figure 8.2. Representation of linguistic variable "Old".

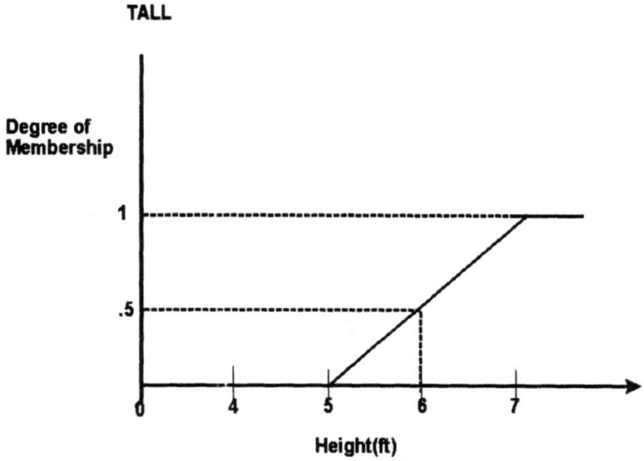

Figure 8.3. Representation of linguistic variable "Tall".

Using the definitions of OLD and TALL, we can assign degrees of truth to different statements. For example if Frank is 4'9", a statement such as "Frank is TALL" would yield an output value of 0.0. If Tom is 6'0" however, a statement such as "Tom is TALL" would yield an output value of 0.5. In addition if Tom is 50 years old, then "Tom is TALL and OLD" would equal 0.25. The benefit of these representations is that we now have a mechanism for reasoning with these imprecise concepts. Additionally, we have a mathematical representation to compute with; it is no longer purely semantic as was the case with early expert systems. With this foundation, one can implement what are called fuzzy expert systems using fuzzy membership functions and rules (the choice of the membership function is user and problem dependent). We extend these concepts to compute quality-grade parameters.

8.2.3 Neural Networks

Neural networks are computational engines patterned after the operation of the human brain. The similarity stems from the fact that both consist of many simple computational elements that are connected together. The influence of each of these elements on the output is determined by weights at the connections. The system learns through training, which adjusts the weights at the connections to obtain the desired outputs.

Conceptually, it is easy to see how these devices might work. Looking at Figure 8.4, we see that with a proper choice of weights, w(i), we are able to get a different output signal, y, for different combinations of inputs. One of the important operations in using these devices is choosing these weights to allow the device to make the right decisions. The process of choosing these weights is called learning, and many algorithms have been developed to implement this activity including supervised, unsupervised, and reinforcement learning. These techniques are used primarily in modelling, clustering, function approximation, and pattern recognition.

A comprehensive review of the concepts and models used in this field can be found in [1].

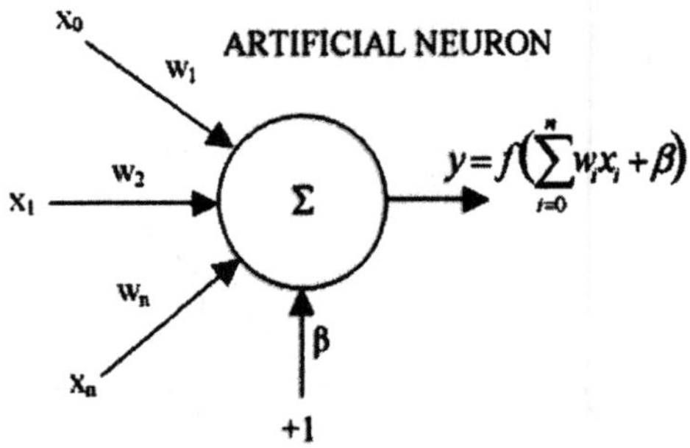

Figure 8.4. Sketch of artificial neuron.

Researchers have shown that with three layers of neurons you are able to classify regions with arbitrary shape [1]. Another area that has become more prevalent is data mining, where we look for hidden relationships in the data. This sort of activity has grown in importance with the increasing amount of data being stored on computer systems [9]. Research has demonstrated the ability of neural network systems to extract the relationship between input and output data without the benefit of an explicit model [9].

8.3 Algorithm Development

In the poultry grading system, images are acquired of the birds on a shackle line using a three-chip colour camera and a standard frame grabber. These digital images are then subdivided into five zones and processed to identify candidate areas with defects. An example of the image regions is shown in Figure 8.6. These zones are labelled 'B' for the breast zone, 'LL' for the left leg zone, 'RL' for the right leg zone, 'LW' for the left wing zone, and 'RW' for the right wing zone. The processing is done using extracted image features and a trained neural network. The candidate defect areas classified by zone and defect type are then used as inputs to the final grade decision-making algorithm. The output of this algorithm should be a single number, such that a final decision could be made based on its value. This number should also be able to communicate an overall "feel" concerning the chicken's final disposition and how reasonable this is based on the governing quality standards.

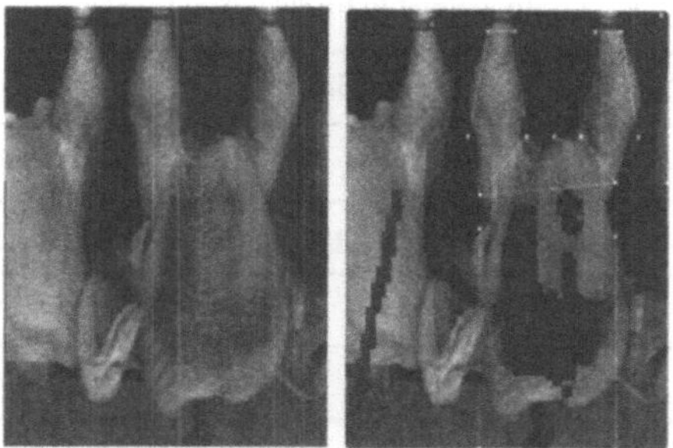

BR	LL	RL	LW	RW	WB
1.00	0.01	0.01	0.15	0.15	0.51

Figure 8.5. Decision outputs for a bruised bird.

8.4 Bruise Detection

The actual areas detected as bruise are found by training a neural net to extract the colour features. Once this has been accomplished identifiers that take into account the presence of defect neighbours are used to identify bruised regions. A brief explanation of terminology is necessary at this point. Zones refer to specific areas on the bird, such as leg or breast zones. Blocks refer to the smallest processing area used by the computer. Regions will refer to the defect area as found by the computer while area will represent the size of the defect. In a similar vein, letter grades A, B, etc. will be the plant grade while the corresponding numbers 0, 1,etc. will represent the machine grade.

The approach for detecting bruises consisted of first identifying 15×15 pixel blocks in the image. This represented the size of the smallest area that would be of interest. The average RGB inputs from these regions were then used as inputs to a trained neural network to identify bruised blocks. Neighbourhood relationships were then used to confirm the decision that there was a bruised block. Once the presence of a bruised region has been confirmed then a decision on its effect on the quality grade is determined using the fuzzy logic approach.

8.5 Fuzzy Logic Approach

As was previously mentioned, in order to use the fuzzy logic approach we need to establish reasonable membership functions. We decided to do this by modeling, and then using sample data provided by a plant for which birds were pre-graded. Sample data from these birds are shown in Table 8.2.

Table 8.2 Grade determinations based on sizes of defects.

OUTPUT DEFINITION	Lower Limit				Upper Limit			
	Nominal		Real		Nominal		Real	
Grade A	Value	Units	Value	Units	Value	Units	Value	Units
Bruise Breast	0	in	0.0	Regions	1	in	2.7	Regions
Bruise Leg	0	in	0.0	Regions	0.75	in	1.5	Regions
Bruise Wing	0	in	0.0	Regions	1.5	in	6.1	Regions
Grade B								
Bruise Breast	1	in	2.7	Regions		in	5.4	Regions
Bruise Leg	0.75	in	1.5	Regions		in	3.1	Regions
Bruise Wing	1.5	in	6.1	Regions		in	12.3	Regions

Using this data, we constructed membership functions by assuming normal distributions centred around each grade as shown in Table 8.3. Using this approach we were able to generate membership functions to describe grades A through H for each of the zones (Breast, Legs, and Wings).

Table 8.3 Sample of membership functions using normal distributions.

	Breast		
	Grade A	**Grade B**	**Grade C**
Mean	0.3	3.5	7.1
Std.Dev.	0.65	0.75	0.90
# obs.	213	27	12
Mean+3σ	2.25	5.75	9.8
Mean-3σ	-1.65	1.25	4.4
	Legs		
	Grade A	**Grade B**	**Grade C**
Mean	0.1	2.5	4.0
Std.Dev.	0.30	0.51	0.44
# obs.	483	28	5
Mean+3σ	1	4.03	5.32
Mean-3σ	-0.8	0.97	2.68

Table 8.3 (continued) Sample of membership functions using normal distributions.

	Wings		
	Grade A	Grade B	Grade C
Mean	0.4	5.0	13.5
Std.Dev.	0.99	2.76	0.71
# obs.	524	34	2
Mean+3σ	3.37	13.28	15.63
Mean-3σ	-2.57	-3.28	11.37

The strategy used in this evaluation was first to grade each region (B, LL, etc.) and then combine these results to establish the grade of the whole carcass. This final grade assignment was done using rules based on the number of defects in each zone. An example evaluation for the breast zone is shown in Figure 8.6.

The process for the whole bird is done in a similar manner as illustrated in Figure 8.7. In this evaluation, the influence of each zone is weighted differently with the breast being the most significant.

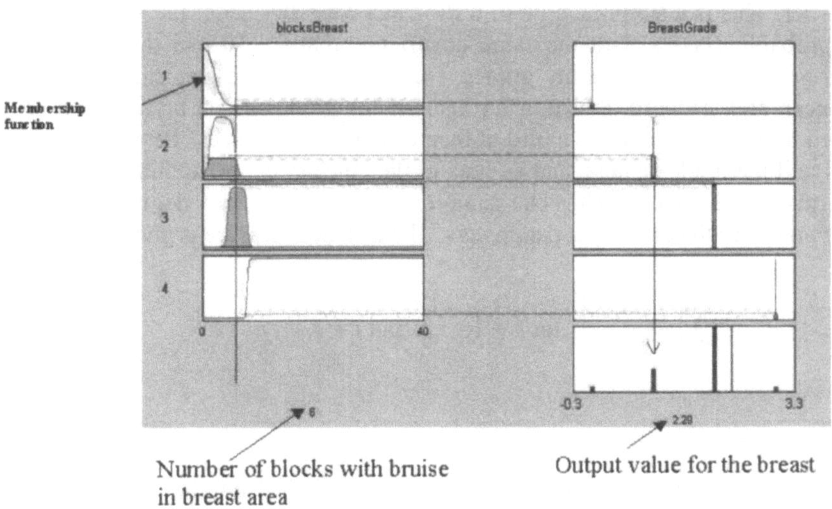

Figure 8.6. Sample evaluation for breast zone grade.

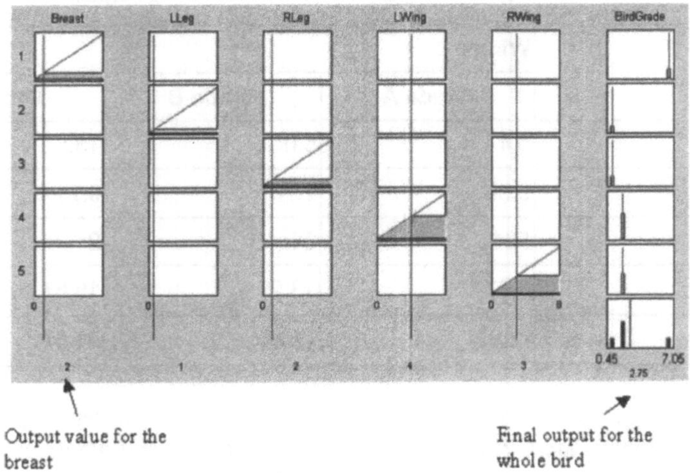

Figure 8.7. Whole bird sample grade evaluation for breast region.

8.6 The Minimum Distance Classifier

To test the approach, we implemented the more traditional minimum distance classifier. This is a simplification of a Bayesian classifier under the assumptions of equiprobable classes with the same covariance matrix [10]. In this method the users predetermine the quality grades. During training, the algorithm determines the mean and standard deviation for each of the grades based on a known set of graded birds. For an unknown bird, it then calculates "distances" between the given input and the grade centers, and assigns to the chicken the grade that is the closest in terms of "distance" [10]. The sample mean and standard deviation for each input/grade are the only input parameters. The distance measure is given by

$$z_{ik} = \sqrt{\sum_{j=1}^{n} \left[\frac{F_{ij} - o_{kj}}{v_{kj}} \right]^2} \quad \text{for } k = 1,\ldots,m \text{ and } j = 1,\ldots, n$$

where:

o_{kj} is the mean for jth input feature of the numerical training data for the kth grade;
v_{kj} is the standard deviation for jth input feature of the numerical training data for the kth grade;
z_{ik} is the weighted distance of the pattern vector F_i from the kth grade

The weight $1/v_{kj}$ is used to compensate for the variance of the classes so that a feature with higher variance has less significance in characterising a class. The output is a discrete value between 0 and 7; 0 is the best and 7 the worst. The features here correspond to the amount of bruising on each part of the carcass. Each carcass is then assigned a class based on the minimum distance.

8.7 Comparing the Fuzzy Logic to the Minimum Distance Classifier Approach

The results of using this approach are shown in Table 8.4, Figure 8.10 and Figure 8.11. The values in the first five columns of Table 8.4 represent the number of defect areas found in each zone. The columns denoted "Method 1" and "Method 2" show the output grade for the fuzzy logic and minimum distance classifier approaches respectively. In this case 0 denotes the best grade and 7 the worst. It can be seen qualitatively that both approaches exhibit reasonable behaviour and are strongly correlated (0.8201) with each other. The fuzzy logic output in Figure 8.10 better represents a continuum allowing more flexibility in assigning classes. When compared with the output in Figure 8.1.

Table 8.4 Sample inputs and outputs using fuzzy and nearest neighbourhood techniques.

Breast	Lleg	Rleg	Lwing	Rwing	Method 1	Method 2
1	5	12	0	3	1.0	1
4	3	9	5	4	1.7	1
0	11	7	19	9	1.8	1
8	13	8	4	13	2.6	3
17	0	10	8	17	3.2	6
10	11	6	6	4	4.1	3
4	10	5	11	3	1.6	1
14	6	13	17	12	3.2	5

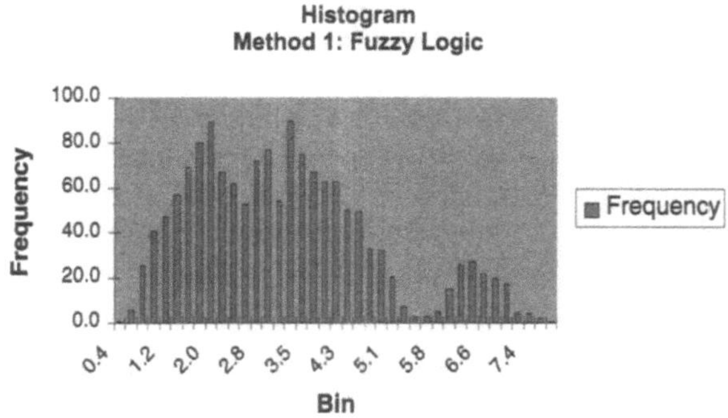

Figure 8.8. Whole bird grades using fuzzy approach (500 samples).

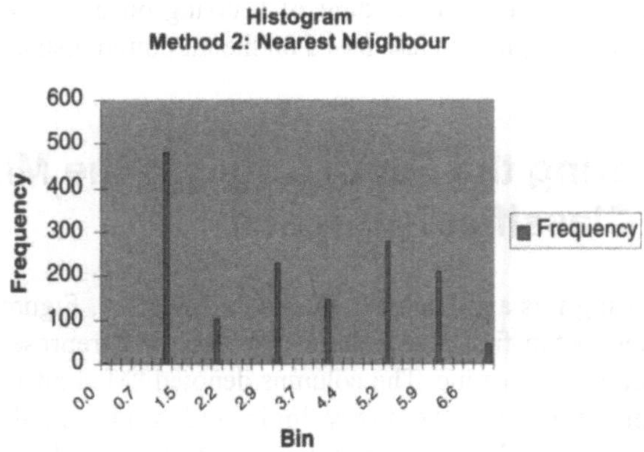

Figure 8.9. Whole bird grades further development.

8.8 Comparison with Human Operators

In order to test the effectiveness of these soft computing concepts for the poultry quality grading application a series of tests were conducted using off-line historical image data. Three human graders were asked to evaluate 105 images of bruised and good birds. These images were also processed using fuzzy logic, and the results were compared with those of the human graders

Experiment 1: Whole Bird Binary Decision - Grade A or Downgrade
The objective of this experiment was to compare performances for the simple binary grade A/downgrade product decision.

Table 8.5 shows the correlation between the human graders and the fuzzy logic process. However, since the decision was binary an arbitrary threshold was needed to convert the fuzzy values into a good/bad decision. This threshold was determined experimentally, to maximize the correlation between the graders and the fuzzy logic process and was chosen to be 0.3.

Table 8.5 Whole bird, grade A/downgrade correlation.

	Fuzzy	Grader 1	Grader 2	Grader 3
Fuzzy	1.00	0.23	0.21	0.32
Grader 1		1.00	0.54	0.65
Grader 2			1.00	0.44
Grader 3				1.00

From the results, it does not appear that a very high correlation exists between each of the graders and the fuzzy logic process. While the correlation between the graders is higher, even these values are quite low suggesting that there exists much more diversity of opinion on what is considered USDA Grade A.

Experiment 2: Overall Whole Bird Grading on Scale 0–7

In this test we asked the graders to make a quantitative assessment of the birds on a scale of 0 to 7, where 0 was a perfect bird and 7 the opposite extreme (worst-case bird).

Table 8.6 shows the correlation between the fuzzy logic process and the graders. The average correlation between all the graders and the fuzzy logic process is 0.33.

Table 8.6 Whole bird, 0–7 clipped grading correlation.

	Fuzzy	Grader 1	Grader 2	Grader 3
Fuzzy	1.00	0.35	0.16	0.36
Grader 1		1.00	0.70	0.69
Grader 2			1.00	0.45
Grader 3				1.00

On the average the correlation between the fuzzy logic outputs and the human graders is better using a graduated scale rather than a pass/fail criterion as used in the previous test.

Experiment #3: Reverse Testing

In this test, the three professional graders were asked to review the grade assigned to a particular bird by the fuzzy logic system. They were then asked to declare if they strongly agree, agree, borderline/questionable, disagree, or strongly disagree with the fuzzy logic value or grade. The objective was to try and determine how often the graders were in agreement with the fuzzy logic assessment of the bird quality.

Table 8.7 shows in terms of percentages how often the graders agreed with the fuzzy assessment. Notice that one grader, number 3 was asked to perform the test twice without the grader knowing that the data were repeated.

Table 8.7 Whole bird, reverse test agreement of graders with fuzzy grade.

	Strongly Agree	Agree	Borderline/ Questionable	Disagree	Strongly Disagree
Grader 1	17.14%	38.10%	5.71%	23.81%	15.24%
Grader 2	11.43%	38.10%	19.05%	20.95%	10.48%
Grader 3a	0.00%	73.33%	5.71%	17.14%	3.81%
Grader 3b	0.00%	69.52%	8.57%	17.14%	4.76%
Average	7.14%	54.76%	9.76%	19.76%	8.57%

Table 8.8 shows the correlation between the graders on their opinions of the fuzzy logic performance for the individual birds.

Table 8.8 Whole bird, fuzzy logic performance assessment - between grader correlation.

	Grader 1	Grader 2	Grader 3a	Grader 3b
Grader 1	1.00	0.03	0.03	-0.13
Grader 2		1.00	0.13	0.40
Grader 3a			1.00	0.46
Grader 3b				1.00

It is interesting to note that the graders agreed with the fuzzy logic assessment of the bird between 50 and 73% of the time. On average they agreed with the fuzzy grading 63% of the time. Except for one grader they very rarely thought that the fuzzy decision was borderline/questionable in either direction. What is even more interesting is that when you start looking at the opinions of the fuzzy grade for each of the individual birds, the graders are not at all in agreement. This is indicated by the poor correlation results given in Table 8.7. Even between the same grader, the opinion of the same bird was not consistent. Over the entire test data set however, the same grader achieved very similar results. This means that within the data set, opinions were not very consistent, but on the average over a large set, distribution was consistent. This might indicate some bias towards unconsciously needing to achieve a specific classification distribution.

While the correlation between the fuzzy logic approach and the human graders were not high, neither was the correlation amongst the individual graders. This would be expected however, as the fuzzy logic system could just be considered another human grader. The last experiment shows that for the most part the graders could live with the decisions made by the fuzzy logic system. The one factor that is of significance but not measured in this study is that of consistency. The graders got extremely fatigued during these evaluations and in some cases took more time than they would normally have. The fuzzy logic system would not exhibit these tendencies and thus overall would tend to be more consistent and reliable.

8.9 The Future

Continued advancement in PC capabilities will serve the machine vision community well. Using the PC as a machine vision platform is therefore a reasonable strategy for developers of machine systems as it enables them to ride the wave of technological developments geared towards the consumer and professional markets. This will result in systems that are lower in cost while exhibiting improved performance.

Specific developments that are of interest are the technologies for handling digital images especially for transmitting, handling and storing this form of data so as to meet the needs of the Internet generation. Many of these capabilities are

directly related to our ability to solve machine vision problems; with this in mind it is prudent to allow most of the applications to reside in software adding functionality as the processors and hardware provide the needed horsepower.
These benefits not only translate to computing hardware, but also to the development in camera technology making digital cameras more available at lower costs. The existence of USB and IEEE 1394 cameras are examples of this kind of process. Consumer demands have driven the imaging industry before as evidenced by the existing CCTV standards, which were adopted by the machine vision community; these same forces are likely to play a significant role in the digital age.

Software development continues to be the most significant cost in the development and deployment of any machine vision solution. Most solutions today depend on experience, heuristics, and rules of thumb and more fundamental approaches could prove to be useful and cost effective. One approach is to look at using features derived from human models of vision. Further work on the development of these techniques could prove to be beneficial.

8.10 Conclusion

It will be necessary for us to continue to produce food for the general populace. In order to realise improved efficiencies in the production process it will be necessary to utilise more automation. As these systems become more sophisticated the need for automated quality control processing and sensing will be even more necessary. We will need to continue to develop and refine techniques that will allow us to develop more cost effective and reliable systems for quality control in food production systems.

Visual techniques are still the main approach for conducting quality control operations in food processing. This is inherently a difficult task for most people, as it plays to one of our shortcomings, our inability to focus for extended periods of time on repetitive tasks [11]. Automated systems would go a long way towards increasing productivity, improving performance along with the work environment. The development of workable systems is still a challenge mainly because of the product variability and the subjectiveness involved in making some of the final determinations. Advances being made to computer hardware as well as the development of programming techniques and tools to specifically address these kinds of problems will result in more systems being delivered that perform acceptably in the production environments in the near future.

8.11 References

[1] Jain A.K., Jianchange M. (1996) Artificial neural networks: a tutorial, *IEEE Computer*, March, pp. 31–44.

[2] Chang J., Han G., Valverde J., Griswold N., Carrillo Francisco D., Sánchez-Sinencio E. (1997) Cork quality classification system using a unified image

processing and fuzzy-neural network methodology, *IEEE Transactions on Neural Networks*, vol.8, no.4.

[3] Li H., and Chung Lin J. (1994) Using fuzzy logic to detect dimple defects of polished wafer surfaces, *IEEE Transactions on Industry Applications*, vol.3, no.2.

[4] Piironen T., Kantola P., Kontio P., Kiuru E. (1994) Design considerations for a metal strip surface inspection system, *SPIE* vol.2353.

[5] Sarkodie-Gyan T., Chun-Wah Lam, Hong D, Campbell A. (1996) A fuzzy clustering method for efficient 2-D object recognition, Proceedings of the Fifth IEEE International Conference on Fuzzy Systems, 1996, pg 1400–1406, vol.2.

[6] Lu N., Tredgold A., Fielding E. (March 1995) The use of machine vision and fuzzy sets to classify soft fruit, SPIE Vol.2620/663.

[7] Brown M., McNitt-Gray M., Mankovich N., Golding J., Hiller J., Wilson L., Aberle D. (1997) Method for segmenting chest CTImage data using an anatomical model: preliminary results., *IEEE Transactions on Medical Imaging*, Vol.16, No.6.

[8] Ching-Teng L., Lee C.S. (1996) *Neural Fuzzy Systems-A Neuro Fuzzy Synergism to Intelligent Systems*, Prentice Hall, p.3.

[9] Bigus, J.P. (1996) *Data Mining with Neural Networks: Solving Business Problems—from Application Development to Decision Support*, McGraw-Hill.

[10] Theodonidis, S., Koutrombas, K. (1999) *Pattern Recognition*, Academic Press.

[11] Drury C.G., Fox J.C. (1975) *Human Reliability in Quality Control*, Halstead Press, New York.

Chapter 9

Editorial Introduction

The inspection of wood is commercially important for two reasons: a. to ensure that timber load-bearing structures (e.g., in buildings) are safe and b. to detect unsightly blemishes on decorative surfaces (e.g., panelling, parquet flooring and furniture). Illuminating smooth, planed and polished wooden surfaces is straightforward. However, lighting for a rough-cut surfaces should be multi-directional, to avoid casting shadows, which can be mistaken for cracks. Ideally, timber defects should be classified, since the nature, size and location of a flaw affects the value of a piece of timber. Maximising the value of a stock of timber samples, by judicious cutting requires a high level of intelligence. The computational processes required are similar to those required for packing (Chapter 11). However, an essential prerequisite for these high-level functions is the ability to detect defects cracks and splits. These must be distinguished from the benign features forming the grain. Of course, other defects are important as well, particularly in decorative timber. Advanced image processing and data analysis techniques are required to optimise the inspection algorithm. Of particular note here is the use of genetic algorithms, which achieve results that would probably not have been found by interactive experimentation. A vision engineer finds it very difficult indeed to formulate an effective inspection algorithm when he is given a large number of sample images containing subtle feature changes. This is just the type of situation in which the use of Pattern Recognition methods is most appropriate. A large set of images must be available to provide a representative sample of both the "good surface" and defects. Another large set of images must be used to test the procedure independently of the design data. Obtaining and particularly classifying, a set of images that properly covers all likely situations that can occur is often difficult and costly.

Chapter 9

Quality Classification of Wooden Surfaces Using Gabor Filters and Genetic Feature Optimisation

W. Pölzleitner

9.1 Introduction

In this chapter we report on the design of a new system for the inspection of quality features on wooden surfaces. Recently we have focused on the problem area of profiled board inspection [1], textured parquet tiles [2], and lumber [3]. These applications have varying demands in terms of the complexity required for imaging. We will focus on two specific areas of inspection: texture inspection, and defect inspection. The first, texture inspection, deals with the overall appearance of the board depending on its two-dimensional distribution of grain lines. Some examples are given in Figure 9.1, which shows three different quality classes. The second, defect detection, deals with local disturbances of the surface causes by cracks, knots, resin galls, bark pockets, worm holes and the like. These have to be detected, classified, and graded using a set of specification rules. Digital image analysis also has to take into account the different surface properties of wood. Profiled board inspection and parquet tiles have a reasonably smooth, planed surface, resulting in a very even reflecting surface, which is comparatively easy to illuminate. This is opposed to the rough sawn surfaces of lumber, where even the slightest directionality in illumination will cause shadow effects on the surface, and give rise to artefacts (like illusionary cracks and holes) in the digital images.

The underlying task requires the characterisation of surface textures on wooden boards in terms of several texture classes, which can be sorted in terms of decreasing statistical homogeneity. Typically, textures are mixed on a single surface and only local deviations (disturbances) of statistical parameters will cause them to be sorted in a different class. Thus the task also involves texture segmentation and requires the identification of these disturbing textured areas.

(a)

(b)

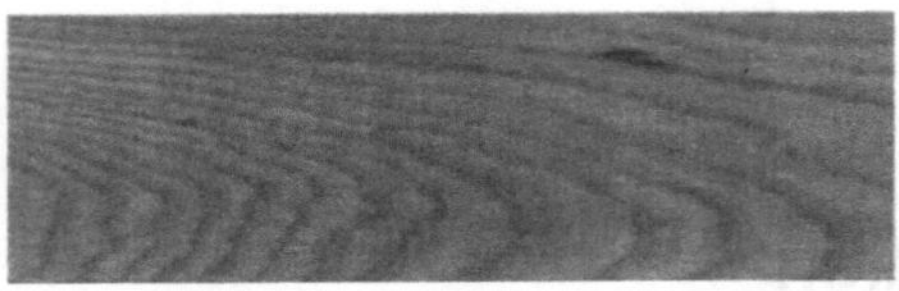

(c)

(d)

(e)

(f)

Figure 9.1. Examples of different grades of textures on wooden surfaces. These are the major classes: a. c001; b. c008; c. r006; d. r008; e. r010; f. s001. Final classification will also be influenced by local deviations and combination of these texture classes.

Any procedure to identify these candidate disturbances will be subject to false alarms. We will study the applicability of filter fusion concepts to reduce the number of false alarms.

In this chapter we describe the overall processing chain we use in processing this type of image. The major philosophy is to use a generic stream of processing with modules characterised by a set of parameters. These parameters, in conjunction with the desired classification performance, are then subject to an optimisation scheme using ideas from genetic programming. We give an example of this type of optimisation showing how the detection of cracks can be optimised. A similar approach can be used for the other areas of processing as well.

9.1.1 Problem statement

Cracks on wooden surfaces are hard to detect for two reasons. Due to price constraints the system is limited to a particular geometric resolution. We use about 0.25mm $\times$ 0.25mm per pixel. This resolution defines the minimum width of cracks we will be able to detect. Grain lines are of the same width, and also appear in the same direction as cracks, as in Figure 9.2 for example. Any detection algorithm will cause false alarms when dark grain lines are present, and will miss thin cracks when this false alarm rate is kept small. The task is to find an optimal point between crack detection and false alarms.

9.1.2 Algorithmic Approach

For the described task, we found that the application of Gabor filters as feature detectors in combination with morphological processing was a major enhancement of our previously implemented systems. Features computed by convolution of the image with Gabor kernels provide great flexibility, because they can be tuned to be highly selective in terms of orientation and shape of the objects. Processing consists of the following steps:

- Filter with a set of Gabor filters. Every filter in this set is described by a filtering kernel, which can be described analytically by the parameters of the Gabor filter function.
- Combination of the filter outputs either by linear combination, yielding so-called Macro Gabor Filters, or by non-linear operations like shifting and taking the maximum, local maximum detection, etc.
- Fusion parameters. The parameters for fusion of the filter outputs need to be chosen. Here we can start to combine them in a binary fashion using only 0.0 and 1.0 as the multiplier for fusion. This is in essence a feature selection process. By allowing the full range between 0.0 and 1.0 as multipliers, we would drastically increase the complexity of the optimisation, and we decided to leave this for later investigations.

(a)

(b)

(c)

(d)

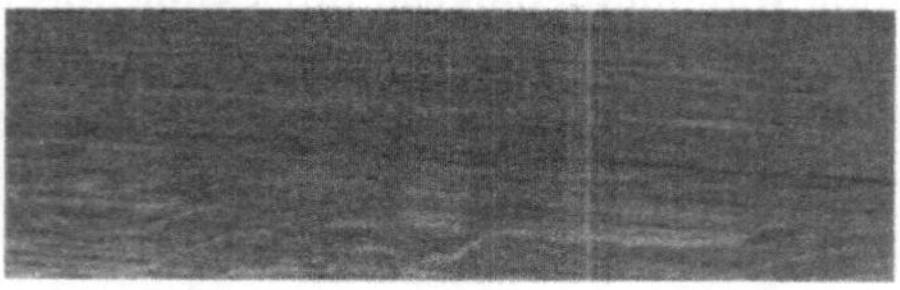

(e)

(f)

Figure 9.2. Examples of cracks on wooden surfaces: a. c001; b. c002; c. c006; d. c007; e. c008; f. c010. Notice the similar directionality and shape appearance of cracks and grain lines.

- After this fusion step, the fusion result is improved by morphological operations. The purpose is to improve the connectivity of the blob detector in the next step, and to decrease noise in the detection result.
- Finally we detect blobs and connected components, which are then analysed in a straightforward manner using feature vectors and statistical pattern classification.

9.1.3 Trade-offs

As in every pattern recognition problem we have to trade-off false alarms and correct classification. In the above processing chain we have to find optimum trade-offs in the following parts:

- Gabor filters: the choice of the filtering kernels directly effects the detectability of features. Whilst the basic structure of the filter can be selected *a priori* (e.g., the orientation of the filter if we use an edge detector type of kernel; or the blob size parameters a and b), the frequency parameter is tightly connected to the selectivity between cracks and grain lines;
- morphological processing: a suitable process to remove noise in the feature detection process, while at the same time improving the connectivity of objects, is a sequence of dilation and erosion steps. Here we can choose the shape and size of the structuring element, as well as the number of dilation and erosion steps.

In the following section we review the underlying formalism of Gabor filters.

9.2 Gabor Filters

The advantages of Gabor filter methods as feature detectors for the present application lie in several areas.

The Complexity Viewpoint

Today's digital signal processors (DSPs) supply fast implementations of finite impulse response (FIR) filters. This occurs mainly because the filter coefficients are known *a priori* and the number of multiplications necessary can be optimised. When Gabor filter kernels are used, the implementation could be in terms of dyadic scales, where the processing complexity is $O(n)$, but care must be taken that the architecture is truly shift-invariant [4]. Another interesting type of implementation is via optical processing [5–8].

Biological Evidence

For the special case of Gabor features, biological arguments and digital filters exhibiting similar behaviour have been brought about for support. [9–13].

Texture Processing

We will extend the work reported here to the complete processing of wooden surfaces. Particularly we will use the Gabor filters described later as pre-processing steps for texture processing. This will facilitate multiple use of the computed features.

9.2.1 Gabor Wavelet Functions

The Gabor wavelet function in 2-D is defined as

$$G_n(x, y) = C_n\, exp[x^2/a^2 + y^2/b^2]\, .\, exp[j2\, \pi \varpi r\, cos(\delta - \phi)] + B_n \tag{9.1}$$

It consists of parameters

$$n = (a, b, \delta, \phi, \varpi) \tag{9.2}$$

and the constants C_n and B_n .

When applied to an input image f , it yields a 2-D analogue output (the correlation $f * G_n$) whose value at each pixel location indicates the amount of data present in each local region (given by the Gaussian widths a, b) at a radial spatial frequency ϖ at an orientation ϕ. Different choices of n and different Gabor functions result in different 2-D outputs.

The Gabor filter (GF) is complex-valued in the image domain. For image representation and compression, the real part of the GF is sufficient and is significant for detection. By tuning the parameters a and b of the Gabor filter, it can be made an ideal detector for blob-shaped objects. As detailed elsewhere [14] for a blob-shaped object of width l (at an angle ϕ) in a background of width of at least $l/2$ around the object, an optimal choice is $\varpi = 1/(2l)$. To derive the GF set for the case of the images shown in Figures 9.1 and 9.2 the parameters are summarized in Table 9.1 For the detection of blob-like objects, the real part $\mathrm{Re}[G_n]$is an excellent filter, whereas the imaginary part $\mathrm{Im}[G_n]$ detects edges.

All filters are normalised to have zero DC output (to achieve independence on local intensity variation) and unity energy. We choose C_n and B_n such that

$$\int_{x_a}^{x_b}\int_{y_a}^{y_b} \mathrm{Re}[G_n(x,y)]^2\, dxdy = 1 \text{ and } \int_{x_a}^{x_b}\int_{y_a}^{y_b} \mathrm{Re}[G_n(x,y)]\, dxdy = 0 \tag{9.3}$$

Table 9.1 Examples of the various filter types used in the experiments. Rel and Iml are the real and imaginary parts of filter 1, respectively. They are used in four different orientations as shown in the table (see also Figure 9.3).

Parameter choice filter	**Width of blob = *l***	***a = l***	***b = l***	***ω=1/(2l)***	***ϕ***	**Discrete filter size**
Rel	3	3	3	1/6	0, $\pi/4$, $\pi/2$, $3\pi/4$	7×7
Im1	3	3	3	1/6	0, $\pi/4$, $\pi/2$, $3\pi/4$	7×7
Re2	2	2	2	1/4	0, $\pi/4$, $\pi/2$, $3\pi/4$	5×5
Im2	2	2	2	1/4	0, $\pi/4$, $\pi/2$, $3\pi/4$	5×5
Re3	1	1	1	1/2	0, $\pi/4$, $\pi/2$, $3\pi/4$	5×5
Im3	1	1	1	1/2	0, $\pi/4$, $\pi/2$, $3\pi/4$	5×5

9.3 Optimisation Using a Genetic Algorithm

The processing chain proposed above is characterised by the mutual dependence of the various processing elements. For instance when the threshold for combining the filter outputs is increased, we will see more instances of "crack" pixels detected. In this case, we should perform fewer dilation steps, otherwise too many objects will be linked together, and we will merge grain lines with cracks. Also if we use many dilation steps, very likely we will need more erosion steps afterwards. We can view the combination of all parameters in the processing chain as points in a multidimensional search space. The dimension is the number of parameters we have, the goal is to choose a parameter setting that minimises the total error.

In order to find a good set of parameters, we need a definition of a suitable error function. This is done by storing for each board in our test set the total length of cracks we should detect. Given a particular value of the parameter vector, we can compute the length of cracks detected by running the processing chain. The smaller the difference between the desired and the actual outcome, the better the parameter set. This difference is called a "fitness" function in our setting.

Genetic algorithms are a very useful approach to our problem for the following reasons [15]:

- We cannot compute any derivatives of the "fitness" function.
- Owing to the interdependence of parameters, we will very likely have many good choices. A standard search technique might fail owing to the resulting large number of local minima.

We approach the search for a good combination of parameters in the following steps:

Step 1 *Initial population:* this is a collection of some settings of parameters that were found by experimenting with a limited number of test samples.

Step 2 *Reproduction:* we run the classification on a set of boards and compute the fitness function for every sample. Let $j = 1, \ldots, J$ be the number of members in the population and $k = 1, \ldots, K$ be the index of the board tested. We then have a fitness function for every single board and member of the population, $F(j, k)$. The total fitness of a member j is

$$F_j = \sum_{j=1}^{J} F(j,k)$$

The value of F_j is used for reproduction: the higher F_j, the higher the probability of reproduction of member j. We assigned 100% to the highest fitness achieved, and 50% to the lowest value. After the reproduction step, we will have copied many "good" members from the original population, and fewer "bad" members.

Step 3 *Crossover:* using the newly generated population, we randomly select pairs of members, and exchange single parameter values in their parameter vectors. This results in a new population that undergoes a new run of testing and fitness function computation.

Step 4 *Mutation:* A small fraction of the better members resulting from Step 2 also undergoes mutation. Here we randomly change the value of a single parameter to form a new member.

To exemplify the procedure, we first summarise the actual parameters used in Table 9.2.

Table 9.2. Processing chain parameters requiring genetic optimisation.

A, b, ϖ, ϕ	parameters describing a specific Gabor filter kernel
S	the (symbolic) direction and amount of shifting the output of a filter "N", "S", "E", "W" indicate shifting of the output image in directions north, south, east, or west
W	the weights used for fusion of the filter outputs. $F_j = \sum_{j=1}^{J} F(j,k)$ where F_j is the output of the i th filter, in our case $i = 1, \ldots, 4$
S	the structuring element for morphological processing, e.g., a circle with 7_7 pixels diameter
N_d	the number of dilations
N_e	the number of erosions

0		0		-6		-6	
0	0	-1	0	-6	-6	-6	-6
0	0	-2	-1	-4	-6	-7	-6
0	0	0	-2	-3	-7	-8	-7
0	0	2	0	-4	-8	-7	-8
0	0	1	2	-6	-7	-6	-7
0	0	0	1	-6	-=6	-6	-=6
	0		0		-6		-6
-1		0		-6		-6	
-4		-4		-7		-7	
-11		-11		0		-11	
-16		0		11		-13	
-11		11		0		-11	
-4		4		-7		-7	
-1		0		-6		-6	

-2	0	-7	-4
-11	-11	-11	0
-32	-32	11	11
-46	0	40	17
-32	32	11	11
-11	11	-11	0
-2	0	-7	-4

0	0	-8	-3
0	-16	-13	11
0	-46	17	40
0	0	59	59
0	46	17	40
0	16	-13	11
0	0	-7	=3

2	0	-7	-4
11	-11	-11	0
32	-32	11	11
46	0	40	17
32	32	11	11
11	11	-11	0
2	0	-7	-4

1	0	-6	-6
4	-4	-7	-7
11	-11	0	-11
16	0	11	-13
11	11	0	-11
4	4	-7	-7
1	0	-6	-6

(a) (b) (c) (d)

Figure 9.3. The Gabor feature set used in the experiments is computed using these filter kernels. Columns of figures should be read from top to bottom before reading the last group at the top of the second column. Continued on next page.

0	0	-7	-7
0	-3	-6	-8
0	0	-3	-10
0	3	-6	-8
0	0	-7	-7
-3	0	-8	-6
-27	-27	-6	-6
-59	0	34	-6
-27	27	-6	-6
-3	0	-8	-6
0	0	-10	-3
0	-59	-6	34
0	0	82	82
0	59	-6	34
0	0	-10	-3
3	0	-8	-6
27	-27	-6	-6
59	0	34	-6
27	27	-6	-6
3	0	-8	-6
0	0	-7	-7
0	-3	-6	-8
0	0	-3	-10
0	3	-6	-8
0	0	-7	-7
e	f	g	h

Figure 9.3 (continued).

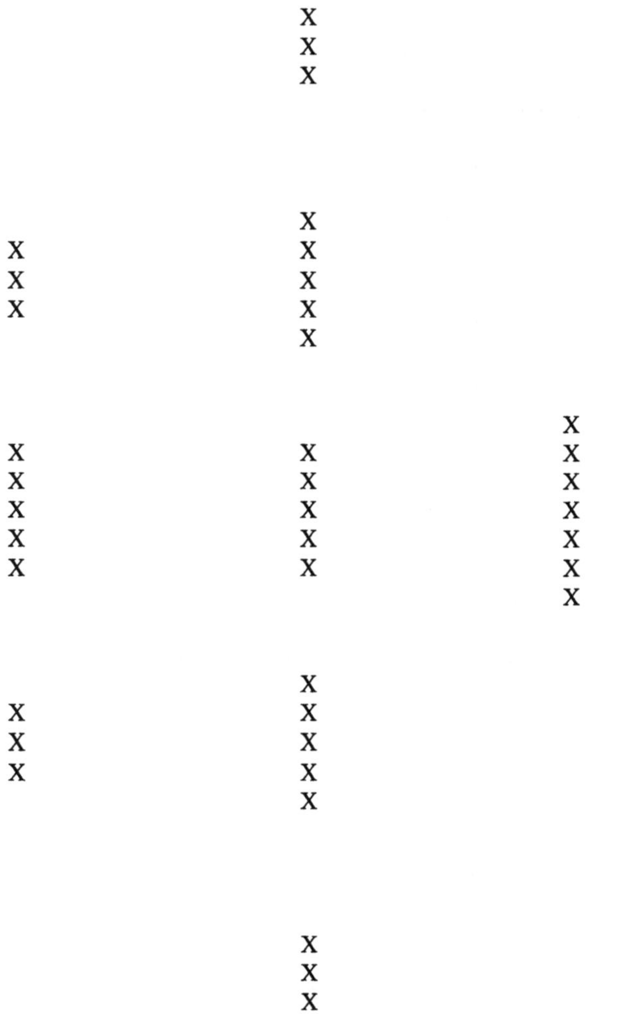

(a) (b) (c)

Figure 9.4. The three different structuring elements used for morphological processing: a. ellipse; b. circle; c. horizontal bar.

These are the parameters that form our "genetic" code. With actual values, the testing algorithm will perform in a particular way and yield a certain detection result. For illustration we show the processing steps, parameters describing them, and examples in Table 9.3:

Table 9.3 Examples of different parameters for each processing step.

Processing step	Parameters	Example
Gabor Filter 1	Type, a, b, ϖ, ϕ, s	Im73_0, "N'
Gabor Filter 2	Type, a, b, ϖ, ϕ, s	Im73_2, "S"
Gabor Filter 3	Type, a, b, ϖ, ϕ, s	Re73_0, -
Gabor Filter 4	Type, a, b, ϖ, ϕ, s	Re73_2, -
Fusion weights	ϖ_1, ϖ_2, ϖ_3, ϖ_4,	0.0, -1.0, 1.0, +1.0
Threshold on fusion output	T_j	80.0
Dilation	S, N_d	7×7, 3 times
Erosion	S, N_e	7×7, 2 times

The complete parameter vector for one member of the population reads as follows:

$$P_j = (a_1, b_1, \varpi_1, \phi_1, s_1, a_2, b_2, \varpi_2, \phi_2, s_2, a_3, b_3, \varpi_3, \phi_3, s_3, a_4, b_4, \varpi_4, \phi_4, s_4, w_1, w_2, w_3, w_4, T_f, N_d, N_e)$$

An example for such a vector is

$$P_j = (\text{Re}, 3, 3, 1/6, 0, \text{“W”}, \text{Re}, 3, 3, 1/6, \pi/4, \text{“N”}, \text{Im}, 3, 3, 1/6, \pi/2, \text{“E”}, \text{Im}, 3, 3, 1/6, 3\pi/4, \text{“S”}, 0.0, -1.0, -1.0, +1.0, 80.0, 3, 2)$$

We define $C_{jk} = C_k(P_j)$ as the resulting length of cracks found on testing a particular board k with the parameter set j. The dimensionality of the search space in the example above is 31.

The fitness function for this board is $F_{jk} = 1 - s \mid C_{ik} - L_k \mid$, where L_k is the stored, desired length of cracks for board k, and s is a normalising coefficient. F_{jk} is 1.0 for a parameter vector that detects the perfect length of cracks, e.g., for which $C_{ik} = L_k$.

9.4 Experiments

To test the proposed genetic algorithm we used a set of 300 tiles, where 50 of them contained cracks. For running the genetic algorithm we used 25 boards with cracks and 25 boards, which did not contain cracks, but grain lines that are bound to be confused with cracks. For the initial population we generated a random configuration of parameters. To limit the search space, we selected a set of eight Gabor filter kernels, three structuring elements (see Figure 9.4), and allowed the output of the Gabor filters to be shifted by at most one pixel position.

To illustrate the behaviour of the pre-processing and morphological processing steps, an example is shown in Figure 9.5. In Figure 9.5 two different filter outputs are shown separately ((b) and (c)), and in combination (d). In the case of (b) and (c), the positive and negative outputs of Gabor kernel Im73 were used, both with magnitude thresholding at 80.0. In (d)) the combination of both filters was computed and then thresholded at 80.0. From these images it should be visible that any one of these filter outputs emphasises different parts of the cracks. The filter combination in (d) emphasises the larger cracks, and loses some detail for finer cracks. The detail is kept better by b, with the downside of a higher sensitivity to false alarms. Notice that even with the same setting of the threshold value, different results are produced. This variability is of course even more obvious when we extend these filter results and apply the morphological cleaning and segmentation steps in Figure 9.6. Here only Figure 9.6)d_ is shown as input to the morphological steps. After several dilation and erosion steps applied to the filtered image, we could at best say we detected the desired number of significant cracks, but miscalculated the total length of cracks.

It should be clear by now that a small variation in one of the processing parameters will cause a change in the final result. In standard praxis, this means in industrial practice, an extensive fine tuning phase is necessary which may not converge to the optimum due to time and monetary constraints. Such an automatic "fine-tuning" phase is provided by our genetic algorithm. An additional advantage is that it can work off-line and use a much larger database of samples than can be handled by an operator trying to optimise the system.

A sample output of the genetic algorithm is shown in Figure 9.7. We show the output of the best member of the population in steps of 50 populations. Every population contains 10 members; the total number of populations was 300, which results in a total of 3000 parameter combinations tested. First, the choices taken by the genetic algorithm seemed to focus on the elliptically shaped structuring element. This is mostly due to the (random) situation that the initial population contained only two circles, and one horizontal bar in its parameter vectors. Notice the change of dilation end erosion steps, which went from three and two, to four and one, going from (a) to (e). We also limited the computational effort by allowing only two possibilities for shifting Gabor kernels: "N", and "S". The major reason for including shifted versions is the fact that the imaginary Gabor filters respond to the edges of cracks.

(a)

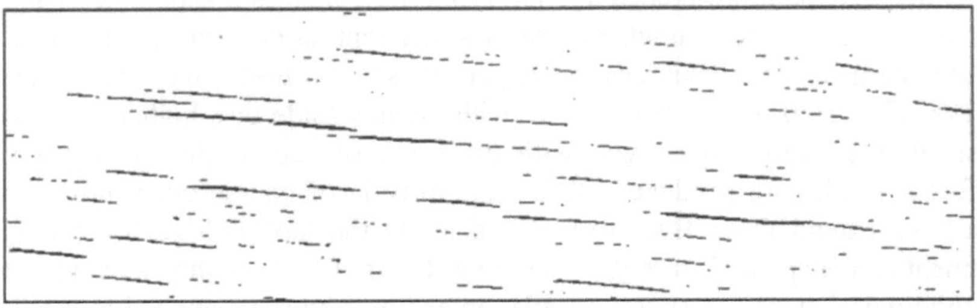

(b)

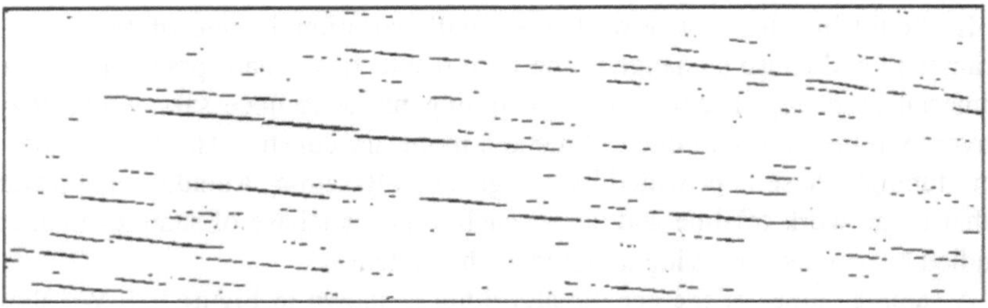

(c)

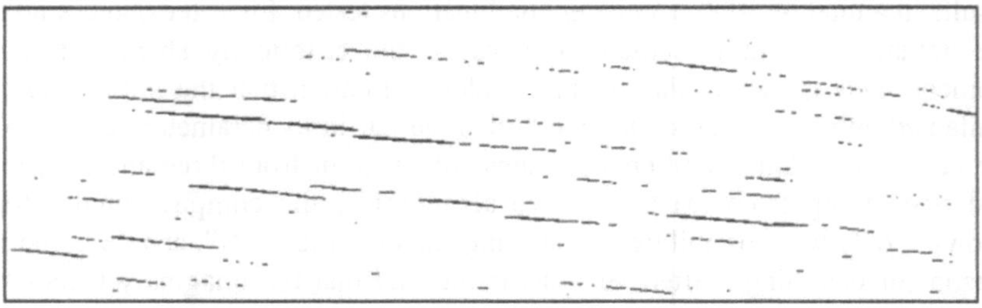

(d)

Figure 9.5. An example combination of steps: (a) the original input image; (b) output of Gabor filter Im73, negative, shifted north; (c) Gabor filter Im73, positive, shifted south; (d) sum of the two, thresholded at 80.0.

(a)

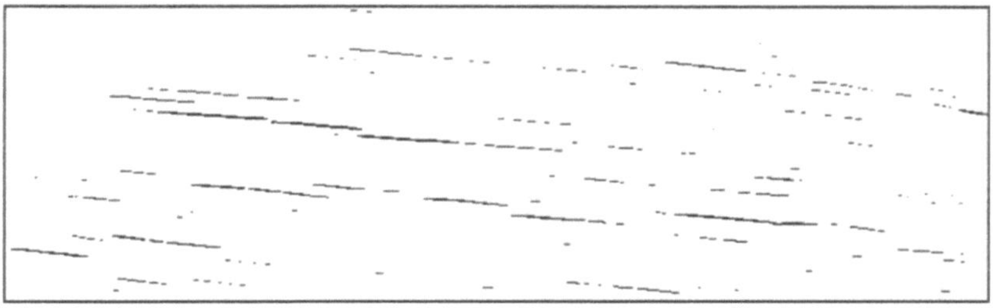

(b)

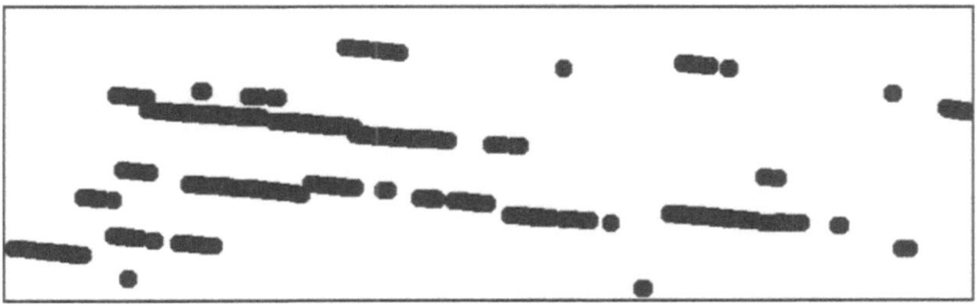

(c)

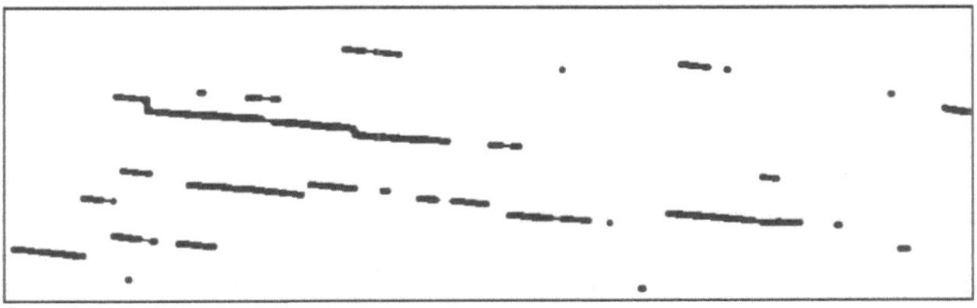

(d)

Figure 9.6. An example combination of morphological processing steps: (a) the original input image; (b) result of feature detection (see Figure 9.5); (c) after three dilations (d) after two erosions. In (d) we have the desired number of cracks detected, although their overall length differs from the desired one.

	Image after filtering and morphological operations	Threshold	Number of Dilations	Number of Erosions	Structuring Element
(a)	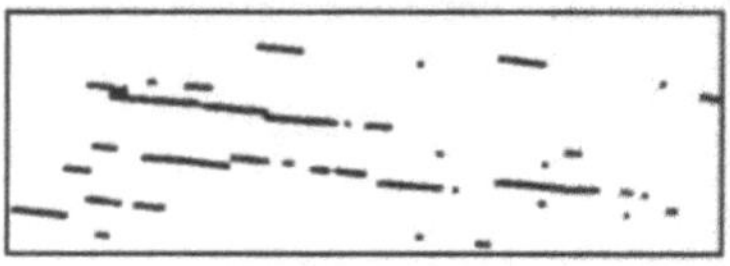	115.0	3	2	Ellipse
(b)	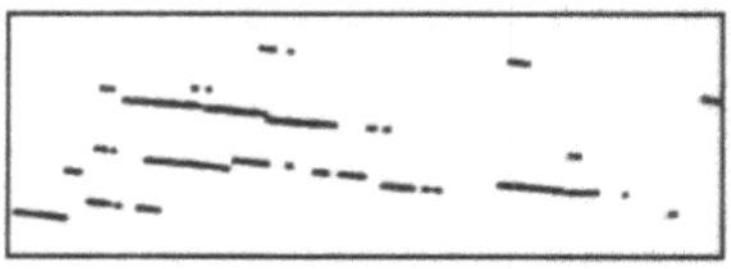	125.	3	2	Circle
(c)	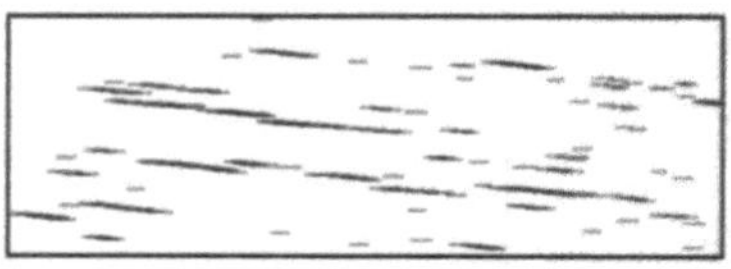	115.0	4	1	Horiz Bar
(d)	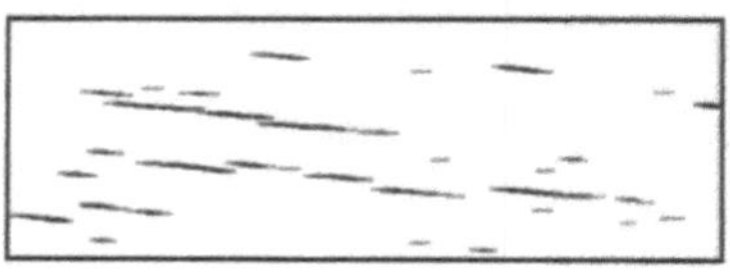	100.0	4	1	Horiz Bar
(e)	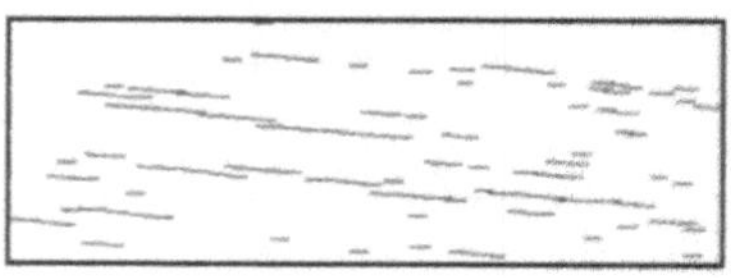	Skeleton for c			
(f)		Skeleton for d			

Figure 9.7. Output of the processing chain at various iteration steps. (a)–(d). the output for various combinations of parameters for one example board;. (e) and (f) the skeletonisation result of the last two samples (c) and (d), which is the final step before cracks are classified. It can be seen that the detection improves while the genetic algorithm proceeds. The final configuration is the one with the best fitness score, and the result for one board is shown in (f). Notice that (f) has the smallest number of fragmented lines, while keeping false alarms at a minimum.

By combining the top edge output and the bottom edge output, each of them shifted over the centre of the crack, both filters work together to produce an improved recognition result (see [16] for more details on this approach).

The final parameter vector, which yielded the best fitness score after 300 iterations, was

C_k = (*Im, 3, 3, 1/6, 0, "N", Im, 3, 3, 1/6, 0, "S", Re, 3, 2, 1/6, π/2, "-", Re, 3, 2, 1/6, π/2, "-", 1.0, -1.0, 1.0, 0.0, 100.0, 4, 1*)

Notice that from four possible Gabor filters, only three as 100.0, and four dilations and one erosions were found optimal.

The major result of our experiments was to find a set of parameters for our processing chain that indeed produced a minimum of disconnected and fragmented objects, while at the same time minimising the false alarm rate. The resulting parameter set is one that would not have been found by interactive experimentation. Overall, we found that the combined approach of using good features for detection, and optimising them with a genetic algorithm drastically alleviated the training step in our system. With the proposed algorithm, training of the system is fully automatic, typically runs off-line, and needs a minimum of operator interaction.

9.5 Conclusion

We proposed a method for detection of surface defects on wooden boards. This method uses a set of Gabor filters, whose output is combined and thresholded, followed by morphological processing for the detection of line like objects. We used a genetic algorithm to compute an optimal set of parameters for the various processing steps. The optimisation was done by generating a set of candidate parameter sets and changing them in an iterative manner such that the overall fitness function improved. Fitness was measured in terms of deviations from the desired output of the detection result. The method was found to be a feasible approach to the underlying training problem. With the genetic algorithm, the training step can now run with very little operator intervention. Future work in the proposed direction will focus on the application of a similar procedure to detect other surface defects on wooden boards, especially texture classification.

9.6 References

[1] Polzleitner W., Schwingshakl G. (1992) Real-time surface grading of profiled board, *Industrial Metrology*, Vol. 2, No. 3 & 4, pp. 283–298.

[2] Polzleitner W., Schwingshakl G. (1994) Argus: A flexible real-time system for 2d defect and texture classification of wooden material, *Optics in Agriculture, Forestry, and Biological Processing, Proc. SPIE*, Vol. 2345.

[3] Polzleitner W. (1993) Convex shape refinement using dynamic programming for surface defect classification on wooden materials, *Optical Tools for Manufacturing and Advanced Automation, Intelligent Robots and Computer Vision XII: Algorithms and Techniques, (Boston), Proc. SPIE* Vol. 2055.
[4] Shenoy R., Casasent D. (1995) Fast non-dyadic shift-invariant gabor wavelet, (Orlando, FL.), Proc. SPIE - Int. Soc. Opt. Eng. (USA).
[5] Casasent D., Ye A., Smokelin J.S., Schaefer R. (1994) Optical correlation filter fusion for object detection, Opt. Eng., vol. 33, no. 6, 1757–1766.
[6] Casasent D., Smokelin J.S. (1994) Real, imaginary, and clutter gabor filter fusion for detection with reduced false alarm, Opt. Eng., vol. 33, no. 7, 2255–2263.
[7] Casasent D., Smokelin J.S. (1994) "Neural net design of macro gabor wavelet filters for distortion-invariant object detection in clutter", Opt. Eng., vol. 33, no. 7, 2264–2271.
[8] Casasent D., Smokelin J.S., Ye A. (1992) Wavelet and gabor transforms for detection, Opt. Eng., vol. 31, no. 9, 1893–1898.
[9] Hubel D., Wiesel T. (1962) Receptive fields, binocular interaction, and functional architecture in the cats visual cortex, Journal of Physiology, London, vol. 160, 106–154.
[10] Campbell F., Cooper G., and Enroth-Cugell C. (1969) The spatial selectivity of the visual cells of the cat, Journal of Physiology, London, vol. 203, 223–235.
[11] Campbell F., Robson J. (1968) Application of fourier analysis to the visiblity of grating, Journal of Physiology, London, vol. 160, 551–566.
[12] Gabor D. (1946) Theory of communication, *J. IEE*, vol. 93, 429–457.
[13] Daugman J. (1980) Two-dimensional spectral analysis of cortical receptive field profiles, *Vision Research*, vol. 20, 847–856.
[14] Casasent D., Smokelin J.S. (1994) Real, imaginary, and clutter gabor filter fusion for detection with reduced false alarm, *Optical Engineering*, vol. 33, no. 7, 2255–2263.
[15] Goldberg D. (1989) *Genetic Algorithms in Search, Optimization, and Machine Learning*,. Reading, MA.: Addison-Wesley.
[16] Pölzleitner W., Casasent D. (1996) A unified approach to control point detection and stereo disparity computation, *Intelligent Robots and Computer Vision XV: Algorithms, Techniques, Active Vision, and Materials Handling*, vol. 2904–24, (Boston), SPIE.

Chapter 10

Editorial Introduction

The inspection of woven fabric and knitwear has been the subject of research for many years. Inspecting plain fabric still presents a considerable challenge, on account of the variable nature of the weave. The task is made even more difficult if there is a strong surface pattern inherent in the weave, or applied later by printing. Frequency (Fourier transform) and sequency (Walsh/Haar transform) methods present possible ways to characterise the weave, although wavelets are used in this study. Whichever method is chosen, analysing the resulting data is far more difficult, since we have a surfeit of low-grade information. In such a situation, there is no obvious and specific clue that can help us to calculate the decision that we need. This has to be "teased out" of a large mass of data by whatever cunning we can devise. We can simplify the problem somewhat, by applying a series of IF_THEN rules. These attempt to encapsulate a human inspector's experience in a simple and convenient form. In this kind of application, the decision rules are best based upon Fuzzy Logic combinations of ill-defined concepts such as "large", "similar to", "near", etc. It is necessary to resort to such an *ad hoc* approach, because there is no possibility of our ever being able to derive analytic solutions for this type of task. The reason is that we are aiming for an uncertain and ill-defined goal that can only be approximated by even the best human inspector. The situation is in fact worse than playing football in the fog, as nobody knows where the goal posts are, nor how wide they are. We cannot even be sure whether any real goal posts exist at all! Pattern Recognition techniques, such as learning and cluster analysis, are also appropriate in this type of situation. Performance indicators are needed to guide and evaluate learning and have to be defined, using our instinct and experience. In situations such as this, we have to stretch human intelligence, to try to build a machine that is "intelligent". This machine has to work in conjunction with a modern high-speed loom. For this reason, it is necessary to pay detailed attention to the implementation of the heuristic procedures that we devise.

Chapter 10

An Intelligent Approach to Fabric Defect Detection in Textile Processes

M. Mufti, G. Vachtsevanos and L. Dorrity

10.1 Introduction

Since the beginning of fabric weaving, manufacturers have relied upon visual inspection of fabrics. It is well known [1] that it is often less expensive to do 100% inspection than it is to do statistical sampling. Juran states that: "Actually, 100% inspection is often less expensive than sampling if the average incoming quality is poor enough to cause 40% or more of submitted lots to be rejected, since the expenses of administration of the sampling plan and a double handling of rejected lots are eliminated." This has been the case in textile production. Although quality levels have been greatly improved with new yarn and weaving equipment, most weavers still find it necessary to perform 100% inspection because customer expectations have also increased and the risk of shipping inferior quality fabrics without inspection is not acceptable.

Consistent objective evaluation of fabric quality has always been a problem. This task requires constant full attention to the fabric passing before the inspector so that all defects may be seen and recorded. Humans are distracted by many things and to expect one to concentrate for hours at a time on the fabric is expecting a great deal. It is also common that the inspection supervisor or other plant authority might observe to the operators that the quality level from the cloth room has decreased and might even question whether the inspectors have become too stringent. This type of subtle "hint" could be taken as a mandate to "loosen up" on the standards for inspection. Suddenly, the quality of the plant increases though the process may still be the same as before. It would be better to give consistent objective evaluation of quality and either improve the process or segregate according to quality and sell to customers according to their requirements. The best possibility of objective and consistent evaluation is through the application of an automatic system.

Inspection has been historically done by removing the rolls of cloth from the weaving machines and running them over an inspection board with either transmitted or reflected light or both. This is a costly process and does not significantly improve quality of production. Only a small amount of off-quality may be prevented in the event that a continuing defect is found and the weaving machine could be stopped and the defective condition corrected. As production speeds increased and roll sizes increased, the time delays became unacceptable.

To address this problem, weavers installed some means at each machine so that most running defects could be seen by the weaver or by an inspector who would roam the room and look for defects. In some cases, a light box is employed and in other cases ambient light is sufficient. This process is somewhat more effective in eliminating the obvious running defects; it is less effective in noting repeating filling (cross machine) defects. It also requires persons on each shift to be inspecting fabric on the weaving machines. Most companies would allow for 30% or less inspection and would therefore allow some off-quality production to continue until the roving inspector or the weaver noticed it and stopped the machine.

A minimum of defective fabric would be produced before the system would stop the machine to allow a technician to correct the condition. The cost of doing this with persons is prohibitive and until recently the cost of doing it automatically was as well. The cost of doing this with human inspection is prohibitive and until recently the cost of doing it automatically was also prohibitive. With the development machine the ideal is to have 100% inspection on the weaving which with the reduction in costs of low-price vision system components such as CCD arrays, gate arrays and DSP chips, has become affordable. The problem then becomes one of adequate software algorithms.

Product characterisation (defect detection) can be closely mapped to a feature space of that product, where feature processing is intended to extract only those characteristic signatures or attributes in the image that represent adequately the required product information. The failure detection problem is in fact a problem of classifying features into different categories. It may be viewed as a mapping from feature space to decision space. One of the most popular fuzzy classification routines has been the Fuzzy C-Means (FCM) algorithm derived from its crisp version called ISODATA [2]. Pal and Majunder [3] have used similarity measures of vague features with known patterns for classification. These approaches work on the assumption that the fuzzy classes are fully understood by the user and there exists sufficient knowledge of the associated features. They do not allow the classes to be self-generated or to evolve over time. Hence, they lack the element of learning that would enable the system to work independently without user assistance.

Binaghi *et al.* [4] have suggested a multi-level architecture for feature classification based on fuzzy logic. The highest level is the application level, the middle level provides the definition of a language for the specification while reasoning, and the lowest level contains the elementary data structure definition and operators for classification and reasoning. Other popular methods for classification use a fuzzy rule-base [5], fuzzy decision hypercube [6], fuzzy relational matrix [7], and fuzzy associative memories (FAM) [8]. All these techniques rely upon the user to provide the expert knowledge for the inference engine. Unfortunately, the generation of a fuzzy decision hypercube or FAM is not very simple for most applications.

Most of the intelligent techniques employ a learning mechanism (unsupervised or supervised), which uses information from an expert, historical data, extrinsic conditions, etc. The learning procedure, in most cases, is cast as an optimisation problem [9] which adjusts the parameters of the detection algorithm, modifies the knowledge-base, initiates mode switching, etc. Loskiewicz-Buczak [10] use learning to determine the optimum weights for aggregation of information from different sources for vibration monitoring. Neural network based techniques [2] use learning to adjust the weights of individual neurons while Fuzzy Associative Memories [11] employ learning to design the inferencing hypercube.

The Fuzzy Wavelet Analysis (FWA) technique detailed in the following sections attempts to accommodate some of the shortcomings of the existing methods of feature extraction and failure detection and identification for effective implementation of a higher level control loop. This technique uses wavelet transforms for feature extraction purposes and allows one to extract features from various signals and to optimise them for use in the knowledge base. The method requires both on-line and off-line learning as described in the sequel. Finally, an application example relating to textile fabric inspection is used to demonstrate the validity and effectiveness of the approach.

10.2 Architecture

An intelligent feature extraction, defect detection and identification algorithm generally works at two levels of abstraction: the first one is the detection level which decides upon the presence of any one of the possible detects. Once the presence of a targeted detect has been established by the detection algorithm, the latter triggers the identification routine which attempts to classify the defect into one of the anticipated defect classes. Both units are similar in nature but the identification part is much more comprehensive and capable of intensive mathematical manipulations. The detection part is deliberately kept simple to fulfil timing constraints dictated by the dynamics of the system under observation. For the identification part, the data can be buffered and processed off-line in some applications. The identification process is triggered only when a defect is detected.

The intelligent identification approach utilises three types of units, based on the way the data is handled. Firstly, the main units (Type I) that follow the general flow of information from input to output. These units function on-line and include a preprocessing element, feature extraction, and intelligent decision making. The second types of processing elements (Type II) are those that are not involved in the mainstream flow of information but perform auxiliary functions. They include performance assessment and on-line learning. They also work in real time and provide feedback information for adding robustness to the system. The third kind of unit (Type III) work off-line, and do not participate in the main information flow. These include the learning mechanism that generates the rule base using performance assessment tools. The core of the operation is the on-line knowledge base that controls the inferencing mechanism. An intelligent feature extraction, detection and identification system, in general, consists of the following architectural components:

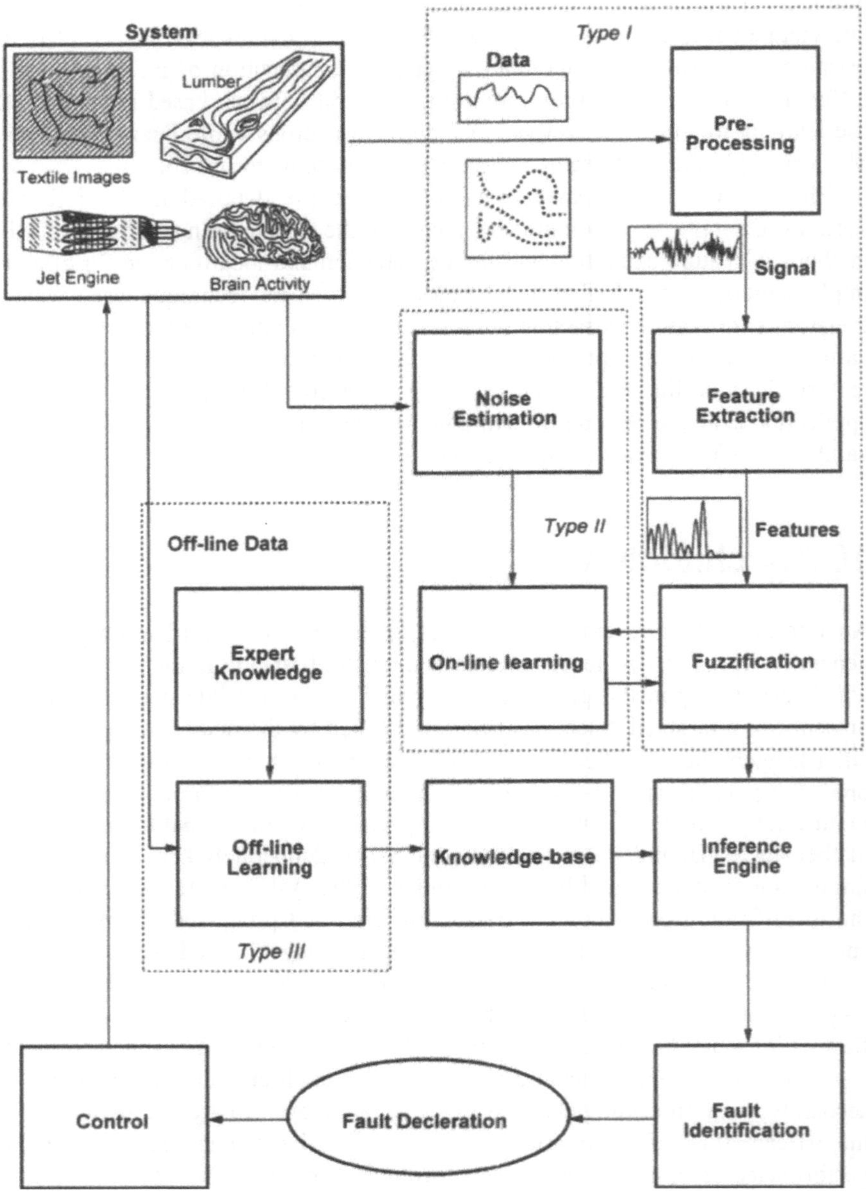

Figure 10.1. General architecture of an intelligent identification and control scheme.

Pre-processing. Raw data derived from sensors possess signal and measurement noise. This noise can be either random or systematic. The latter type can be filtered to increase the usability of the data. Noise can be further characterised as high-frequency Gaussian noise, DC bias, etc. Proper signal processing is applied at this stage to decrease random noise levels. Preprocessing also involves data format conversion, sampling, quantisation, digitisation, etc. In order to reduce the computational effort and the amount of data processing, high-dimensional data is simplified wherever possible; for example, two-dimensional images can be scanned to produce a one-dimensional data stream. This operation reduces exponentially the amount of computations required for 2-D image analysis.

Feature Extraction. Features or signatures are characteristic signals representing a particular operating mode of a system. Feature extraction is accomplished by processing the input data so that the presence of the operating modes is accentuated. An effective feature extraction methodology ensures a more robust identification process. However, feature extraction can be the most difficult part of any identification scheme. This is due to the fact that normal operational modes and noise have signatures that are very similar to feature signatures. Different processing methods yield different results. A combination of different processing techniques is usually employed and features from different sources can be used in an intelligent inferencing scheme. Most popular feature extraction tools have exploited the frequency and time domains. Due to the dual and complementary nature of these domains, both are used simultaneously. This is a primary motivation for using techniques like the Short Time Fourier Transform, Gabor Transform and Wavelet Transform for feature extraction. Feature extraction using wavelet transform techniques is discussed in detail in the following sections. Intelligent identification schemes have the advantage of using qualitative features in addition to quantitative measures, prompting researchers to utilize features like *erratic, smooth*, etc. in signals and *dense, hollow*, etc. in images.

Fuzzification. Since fuzzy tools aim at accommodating uncertainty and are easy to interface in hierarchical expert system architecture, most intelligent identification techniques employ a fuzzification process. This also allows one to use several fuzzy logic tools already available. Fuzzification involves utilisation of a fuzzy membership function, which assigns a grade of membership to the input crisp feature set.

Inferencing. All intelligent identification schemes entail an inference engine that utilises knowledge about the system to provide decisions about the input data. The inference engine employs a knowledge base which is usually generated off-line and may be updated on-line as additional information about the system becomes available. Knowledge about the system can also be acquired from an expert or compiled off-line through various optimisation techniques. The inference engine (forward and backward reasoning) may consist of a Fuzzy Relational Matrix, Fuzzy Associative Memories (FAM), Neural Networks or other inferencing tools.

Learning. Intelligent identification systems continuously monitor their performance and decide upon necessary changes in the knowledge base or other system parameters. This is the process of learning through which the system attempts to improve its performance as the knowledge about the system increases.

Learning can be either supervised or unsupervised. Parametric changes and minor additions to the knowledge base are accomplished on-line, i.e., in an unsupervised mode, while changes that involve massive calculations and processing are made off-line, i.e., in a supervised mode. Supervised learning is usually carried out when the knowledge base has no prior information about the system.

Defect Declaration. An important part of an intelligent identification algorithm is decision making and interfacing with external units. The objective here is not only to provide user interface functions but also to assist in implementing control strategies on the basis of the results from the identification scheme. A decision about the presence of a particular defect and the type of the defect is reached in this unit. It also assigns a degree of certainty (DOC) to the decision-making process which is a measure of how closely the input signal matches the information in the knowledge base.

Control. Based on the results of the defect detection, the control unit takes necessary actions to rectify or at least minimise the defect/fault conditions. The control unit in such a scheme usually entails expert systems, artificial intelligence, fuzzy logic, neural networks, etc.

10.3 Fuzzy Wavelet Analysis

Fuzzy Wavelet Analysis (FWA) combines fuzzy tools [12], [13], [14] and wavelet transform techniques [15], [16] to provide a robust feature extraction and failure detection and identification scheme. The detailed description is given below:

The input signal $x(t)$ is in the form of a stream of 1-D data. Various preprocessing techniques can be applied which depend upon the application at hand. These may vary from simple averaging to various signal processing algorithms in order to reduce its noise content The input signal $x(t)$ is sampled at regular intervals to obtain its sampled version $x(l); l \subset N$. Preliminary adjustments are made at this stage to obtain a uniform signal mean and variance. This is helpful in suppressing external variations and sensor imperfections.

The FWA employs the wavelet transform (WT), with different wavelet functions, to extract features from the signal $x(l)$. This is illustrated in Figure 10.2. The wavelet transform generates wavelet coefficients which are employed by a fuzzy inference engine to seek a match between the current state of the wavelet coefficients and the templates stored in the knowledge base. A wavelet is represented by $_{a,b}(l)$, where _ is the mother wavelet, a is a scaling factor and b is the translation in time or space depending upon the context. $n \subset N$, the number of wavelet scales is chosen which produces the best results for the anticipated defects. The set of wavelets for some $b = b_j$ is given by $D = \{\psi_{a_1,b_j}, \psi_{a_2,b_j}, \ldots, \psi_{a_n,b_j}\}$. The wavelet scales a_i; $i = 1 \ldots n$, are obtained via an optimization process that works off-line. The wavelet coefficients for each wavelet ψ_{a_i,b_j} and the input signal $x(l)$, are calculated as

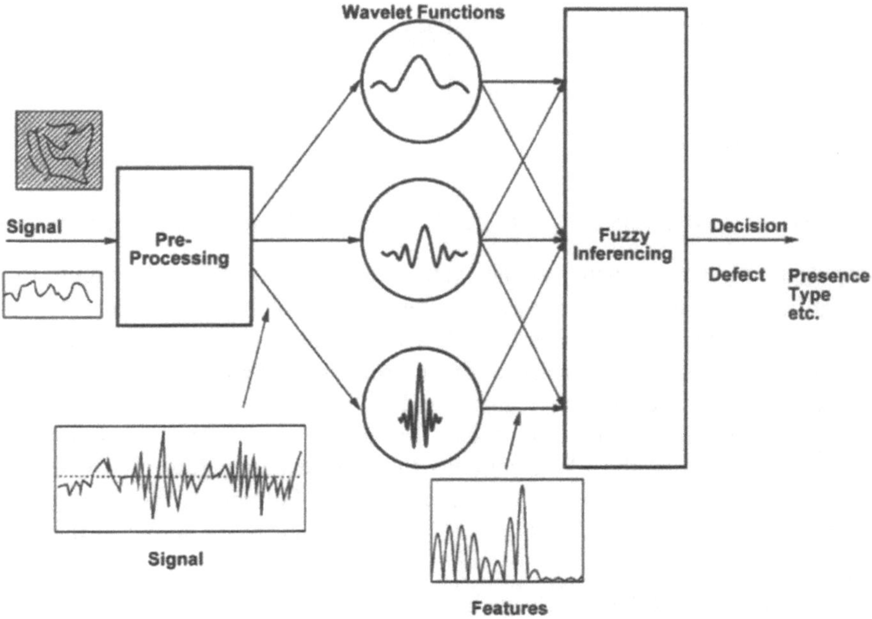

Figure 10.2. Feature extraction using wavelet coefficients.

$$c_{ij} = \sum_{l=0}^{N} x(l)\psi_{a_i,b_j}(l) \tag{10.1}$$

where N is the number of samples for which ψ_{a_i,b_j} is non-zero. Since most of the features produce a signature in a wide range of frequencies that is spread over a range of time (or space), $m < 0$ number of coefficients $c_{i,j}$ are buffered. The coefficients $c_{i,j}$ are stacked in a matrix arrangement that is referred to as the *Information Matrix* W, since it stores the information about all the features under examination.

The matrix W with elements $c_{i,j}$ has the following characteristics:

- for a fixed $j=u$, the c_{iu} give the frequency response of the input signal at a particular instant of time;
- for a fixed $i=v$, the c_{vj} give the relative level of a particular frequency over a period of time (or space);
- each row of the matrix W is referred to as w_i where $i=1,\ldots,n$, and is comparable to a band-passed version of the signal.

The matrix W is represented as follows:

$$W = \begin{bmatrix} c_{11} & c_{12} & \cdots & c_{1m} \\ c_{21} & c_{22} & \cdots & c_{2m} \\ \vdots & \vdots & \ddots & \vdots \\ c_{n1} & c_{n2} & \cdots & c_{nm} \end{bmatrix} \begin{matrix} \textit{frequency} \\ \downarrow \end{matrix}$$

Time / Space $\rightarrow$

The rows, w_i, where $w_i \subset U_i \subseteq \Re^m$ represent the features obtained from the input signal. The elements of the input space $\Re^m$ constitute the inputs to the identification algorithm and are vague representations of the failure modes.

The knowledge base stores the information employed in the decision-making process and attempts to organise all the information available about the system (the relationship between failure features and failure modes) through mathematical models, heuristics, input-output data or any other information source. The knowledge base of the FWA contains the representative features for different wavelet scales, for each one of the anticipated failures. It is represented by $\kappa \subseteq B = U \times V$, where $U = U_1 \times \ldots \times U_n$ is the input space and V is the output space.

Development of the knowledge base is carried out off-line. Let the coefficients $\_{i,k}$ represent the trend for scale a_i (frequency) for the kth feature (k=1...M), where M is the total number of anticipated failures. The $\_{i,k}$ are similar to w_i, but they are optimised through a learning process and are the representative signatures of the failure modes. The optimisation process chooses the best set of wavelet scales (a_i) and stores the template features for these scales. The knowledge base κ is generated via the following stepwise procedure:

i. *experimental analysis of known defects:* signal data representing features from the system under observation are collected and stored;

ii *initial guess of wavelet functions:* a finite number, n, of wavelet functions are chosen with arbitrary wavelet scales. The choice of n is initially based on heuristics, but if the FWA system fails to perform adequately after optimisation, its value can be increased;

iii. *formation of the information matrix:* the wavelet coefficients are calculated using the selected wavelet functions and are stored in the information matrix. The components of the information matrix, w_i, now represent the reference features and are called $\_{i,k}$ (kth reference feature with scale a_i);

iv. *optimising the wavelet scales:* the components, $\_{i,k}$, of the information matrix are optimised by adjusting the wavelet scales, a_i, to maximise certain detectability and identifiability measures, as described in the sequel;

v. *formation of the rule base:* the optimized $\_{i,k}$ are then fuzzified into corresponding fuzzy sets, F_i^k, using a similarity measure:

$$sim(X,Y) = \frac{1}{K}\sum_{i=1}^{K}\frac{1-|x_i - y_i|}{1+\alpha|x_i - y_i|} \tag{10.2}$$

where X, Y are two sets with elements $x_i \subset [0,1]$ and $y_i \subset [0,1]$, respectively, K is the number of elements in each set and $\alpha > -1$ is a predetermined constant. Thus,

$$F_i^k(w_i) = sim(\beta_{i,k}, w_i), \qquad k = 1 \ldots M, w_i \subset U_i \tag{10.3}$$

The F_i^k s (acting as templates for future reference) are stored in the rule base the in form of fuzzy rules. The output sets, G^k, are constructed as fuzzy singletons via:

$$\{0/1 + 0/2 + \ldots\ 0/(k\text{-}1) + 1/k + 0/(k{+}1) + \ldots\ 0/M\backslash\}$$

The kth component, R^k, of the fuzzy relation is constructed as follows:

$$R^k = F_1^k \wedge F_2^k \wedge \cdots \wedge F_n^k; \qquad k = 1, \ldots, M$$

where $\wedge$ is the fuzzy min operation [13]. The collection of the R^ks constitutes the knowledge base $\mathcal{K}$. M is the total number of IF-THEN rules.

The fuzzification unit of the FWA calculates the membership functions for the features w_i. The methodology suggested here uses a similarity measure, as defined below:

The membership function for the input fuzzy sets, $\overline{A} = (A_1, \ldots, A_n) \subset U$, is defined using the similarity function of Equation 10.3 as:

$$\mu_{A_i^k}(w_i) = sim(\beta_{i,k}, w_i) \qquad k = 1, \ldots, M \tag{10.4}$$

$\mu_{A_i^k}(w_i)$ is, therefore, the grade of membership of the vector w_i in F_i^k. This approach differs slightly from conventional techniques, as reported in [2,3,17], but is especially suited for the fuzzy interpretations of wavelet transforms.

10.4 Fuzzy Inferencing

Inferencing or decision-making is based on a set of IF-THEN rules. Let the input universe of discourse be $U = U_1 \times \ldots \times U_n$ and the output universe of discourse be V. The input (features) variables are $w_i \subset U_i$ and the output (decision) variable is $y \subset V$. w_i are the columns of the information matrix and F_i^k and G^k are fuzzy sets in U_i and V, respectively. The rule set is given by the following statements:

If w_1 is F_i^1 and w_2 is F_2^1 and … and w_n is F_n^1 then y is G^1

If w_1 is F_i^2 and w_2 is F_2^2 and … and w_n is F_n^{21} then y is G^2

⋮

If w_1 is F_i^L and w_2 is F_2^L and … and w_n is F_n^L then y is G^L

The output fuzzy set B in V is calculated as $B = \kappa \circ A$, where $\circ$ is the fuzzy composition [13]. Since κ is composed of a number of relational rules, each one generates an output $B^l = R^l \circ A$. The first step in achieving this is to calculate the premise part, that is,

$$\hat{A} : A_1 \ \ and \ \ A_2 \ \ and \ldots \ and \ \ A_n$$

This is accomplished as follows:

$$\hat{A} = A_1 \wedge A_2 \wedge \cdots \wedge A_n$$

The next step is to compare $\hat{A}$ with the rulebase and generate the output fuzzy set B^k,

$$\mu_{B^k}(y) = \vee_{w \subset U} [\mu_{R^k}(w, y) \wedge \mu_{\hat{A}}(w)]$$

The final output is obtained as follows:

$$\mu_B\ (y) = \mu_{B^1}(y) \wedge \mu_{B^2}(y) \wedge \cdots \wedge \mu_{B^k}(y)$$

The failure mode is identified by defuzzifying the final output fuzzy set B:

$$y = \arg\ \sup_{y \subset V} (\mu_{B^k}(y)) \tag{10.5}$$

and gives the decision that the yth fault or defect has been detected.

10.5 Performance Metrics

Once a decision about the fault/defect has been reached, the FWA assigns two performance metrics:

i. a Degree of Certainty (DOC) to the failure decision;
ii. a Reliability Index (RI) for the overall diagnostic process.

10.6 Degree of Certainty

The Degree of Certainty is a measure of confidence in the decision and is employed to account for the uncertainty inherent in the system under test. Under perfect recall conditions, the output of the inference engine, B^k, would be identical to the output association, G^k, in the training set. However, under normal operating conditions, B^k and G^k do not match exactly due to the uncertainty in the input. The DOC gives an indication of the closeness of the actual decision to the original training output

$$DOC_k = h(G^k, B^k), \quad k = 1,2,\ldots,M \tag{10.6}$$

where $h : [0,1] \rightarrow [0,1]$.

$$h = \frac{\mu_{B^k}\left(y^k\right)}{\mu_{G^k}\left(y^k\right)} \tag{10.7}$$

Since, in this case, G^k is defined as a singleton,

$$DOC_k = \mu_{B^k}\left(y^k\right)$$

A value of $DOC_k = 1$ indicates a perfect recall for the kth rule, while a value of DOC_k close to 0 implies that the belief in the occurrence of that particular failure is very small. The values of DOC are indicative of the robustness of the decision-making logic residing in the inference mechanism.

10.7 Reliability Index

While the DOC gives a measure of confidence in the decision-making process, the Reliability Index (RI) gives a measure of confidence in the performance of the overall diagnostic and classification routines. The reliability index is dependent on the entropy of the system, the DOC of each individual failure, and the probability of occurrence of a failure.

The DOC_k values are sorted and stored in descending order in a set m_i, where m_1 is the maximum DOC value, m_2 is the second highest, and so on. The probabilities of occurrence, p_i, of all anticipated defects are specified *a priori*. Thus, the RI can be written as:

$$RI = \frac{1-H}{p_{\max}} \sum_{i=0}^{M} p_i e^{-\sum_{j=1}^{M}\left(m_j - V_{ij}\right)} \tag{10.8}$$

where, $p_{\max} = \max_i\{p_i\}$, H is an entropy measure of the system, and v_{ij} is the expected value of m_j given that i number of defects/faults have occurred. Mathematically, v_{ij}, is given by

$$v_{ij} = \begin{cases} 1 & \forall j \leq i \\ 0 & \text{otherwise} \end{cases}$$

As the probability of occurrence of more than two defects simultaneously is very small, the equation for *RI* reduces to:

$$RI = \frac{1-H}{p_0}\Big(p_0 e^{-((m_1-v_{01})^2+(m_2-v_{02})^2)} + p_1 e^{-((m_1-v_{11})^2+(m_2-v_{12})^2)} + p_2 e^{-((m_1-v_{21})^2+(m_2-v_{22})^2)}\Big)$$

where p_0 is the probability of occurrence of no defect (normal operating condition), p_1 is the probability of occurrence of a single defect, p_2 is the probability of simultaneous occurrence of the two defects, m_1 and m_2 are the DOCs associated with the occurrence of two most significant defects.

The entropy measure, H, is given by:

$$H = -m_1\log(m_1) - (1-m_1)\log(1-m_1) - m_2\log(m_2) - (1-m_2)\log(1-m_2)$$

If the probability of normal system operation (i.e., when $m_1 = 0$ and $m_2 = 0$) is maximum, then the Reliability Index also takes its maximum at that value. If the system indicates the occurrence of multiple defects, the RI takes a relatively low value. It can also be seen that the overall RI surface is concave, since the belief of occurrence and non-occurrence of a defect is low at the center of the surface.

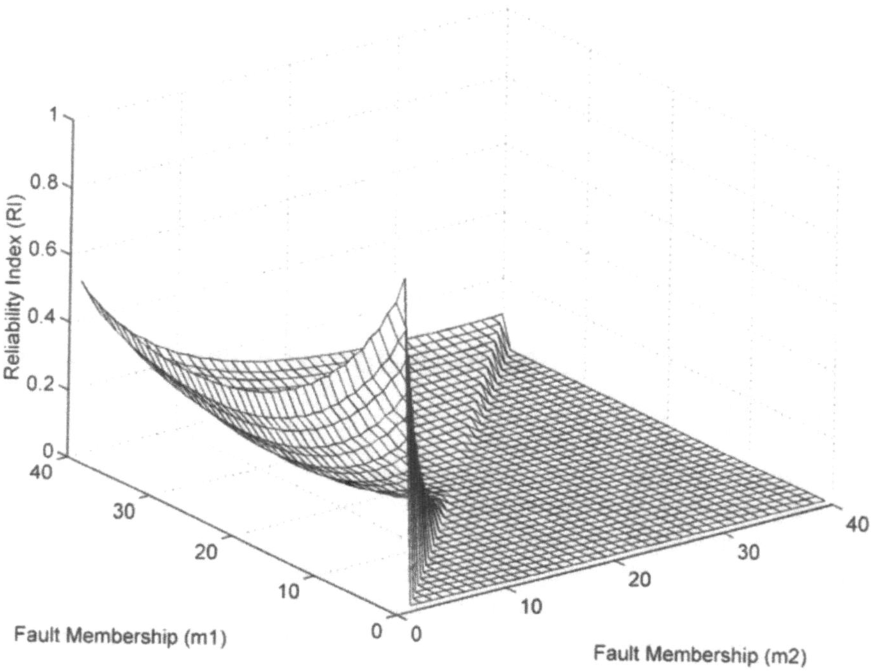

Figure 10.3. Reliability surface of two variables.

10.8 Detectability and Identifiability Measures

Identifiability and detectability are measures of *robustness* of the identification/detection scheme, which aim at minimizing the sensitivity of the identification/detection performance to modelling errors, uncertainties and system noise. *Detectability* is the extent to which the diagnostic scheme can detect the presence of a particular defect; it relates to the smallest defect signature that can be detected and the percentage of false alarms. *Identifiability* goes one step further in distinguishing between various defect modes, once the presence of a defect has been established. Identifiability targets questions like the source, location, type and consequence of defect.

10.9 Learning

An important attribute of intelligent systems is their ability to restructure or update the knowledge base, based on information from past experience. This is termed as learning. The FWA uses both on-line and off-line learning to increase the knowledge about the system and improve the detection and identification processes. On-line learning attempts to nullify the effects of factors such as variation in lighting and illumination, sensor imperfections, lens aberration, etc. by adjusting the FWA parameters. This adjustment is carried out in two stages:

i. during preprocessing of the input signal;
ii. at the time of fuzzification.

The off-line learning process provides the FWA with the ability to generate its knowledge base from sample defects or faults. This is an optimisation process that selects the best values for the given set of defects/faults and stores the corresponding features in the rulebase.

10.10 Practical Implementation of Fuzzy Wavelet Analysis

The FWA inspection system has been applied to a number of applications including inspection of woven textile fabrics, fault detection and identification in a jet engine, grading of lumber and quality monitoring of carpet tiles. Each application has unique challenges and requirements. Simulation studies include testing of the concepts and algorithms in MATLAB® and C language and implementation on a DSP platform. The implementation of the FWA on textile fabrics is discussed here.

The basic hardware consists of an array of six cameras which scan a moving fabric, while the images are processed and the results are sent to the controller for necessary action. Implementation of the FWA software requires a computationally intensive platform capable of performing in an expedient and efficient manner, the pre-processing and inferencing stages of the analysis. A general purpose DSP is very suitable for this type of application. Implementation of the FWA algorithms for the textile fabric inspection system requires dedicated control along with the signal processing tasks. The control sequence is programmed on a Motorola 68306 microprocessor operating at 20 MHz. This processor has its own external RAM for data storage and Flash memory for program and parameter storage. The use of flash memory allows the control sequence and operating parameters to be downloaded from an external device without removing the memory from the control board. The complete system is shown in Figure 10.4. The control module is connected to an array of detector modules which comprise mainly the CCD camera, TMS320C52 and memory units. The connection between the detector modules and the control module is accomplished via a 64-pin bus that allows for dual access to the memory on the detector module via a DSP chip and the 68306 in a multi-processor environment. The final results from the detector modules are stored in their respective memories and a flag is set to indicate to the controller that the DSP operation is complete. The controller then takes over the control of memory of each detector module one-by-one and retrieves the final results. The controller is also connected to the operator interface to which an additional hand-held terminal can be attached. The encoder for measuring the yardage of the fabric and network connections also feed to the controller.

A survey was conducted by collecting data from five major textile fabric producers in Georgia and South Carolina. The defects were rated as the most common defects and the most costly ones. The algorithm was tuned to detect and identify such defects as Broken Pick, Coarse End, Double End, End Out, Harness Drop, Knot, Oily Spot, Slub, Start Mark and Woven in Waste. The approach works at two hierarchical levels: the first one is the gross filter that detects the presence of the fault (or defect). This is mainly the detection process with primary classification of the defect into three categories: Line, Point, and Area. The image data is two-dimensional so it must be converted to a 1-D stream of data to make it compatible with the FWA algorithm. The image is scanned using fractal scanning 0,0,0 which is useful for scanning image areas while preserving the neighbourhood relationship. This scanning technique also helps in data reduction and hence processing time. If the gross filter detects a fault (or possible fault), the processing is handed over to the FWA, which classifies the fault into one of the anticipated faults or no fault, as the case may be.

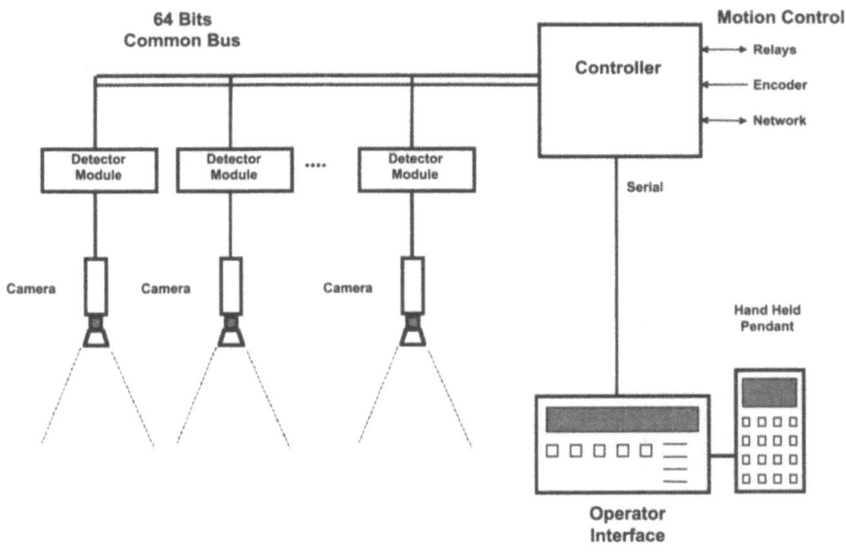

Figure 10.4. Hardware platform for implementation of fuzzy wavelet analysis and control.

The results of a space search for values of the scale coefficients varying from 0.01 to 0.82 for detectability and identifiability are shown in Figure 10.5. It can be seen that the detectability for all the defects drops around a scale of 0.02. This is the scale at which the texture of the fabric becomes dominant. The detectability is fairly uniform over a wide range of scales from 0.05 to 0.7. Identifiability on the other hand varies from one defect to another. It is more erratic near a scale of 0.02 and much more uniform in most other regions. The learning procedure was implemented using MATLAB® for five wavelet scales. The convergence of the objective function is shown in Figure 10.6. The values of the scales obtained by implementation of the optimization routines are {0.4144,0.1544,0.6739,0.7526, 0.2222}.

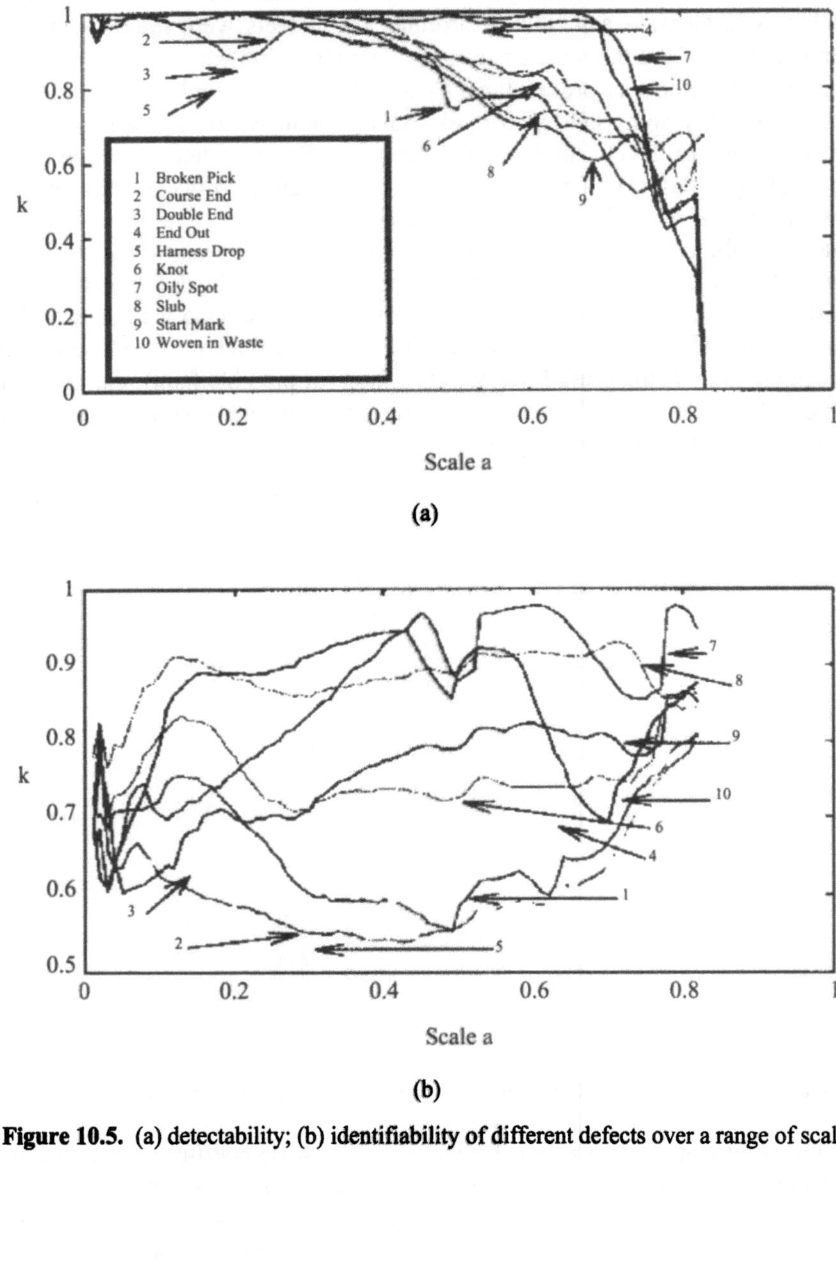

Figure 10.5. (a) detectability; (b) identifiability of different defects over a range of scales.

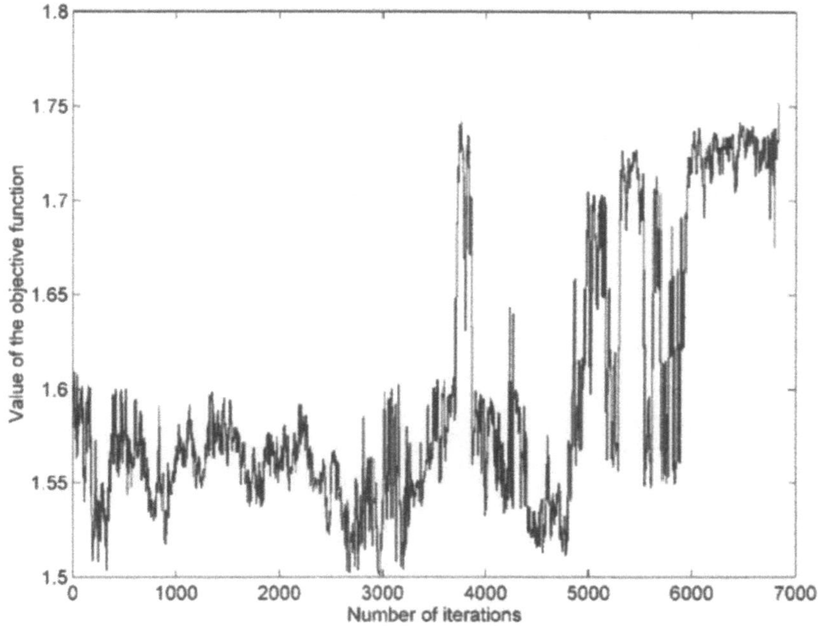

Figure 10.6. Convergence of objective function for the textile inspection problem.

The defects given here as examples are mispick and oil spot. These defects are shown in Figure 10.7 and Figure 10.8. The corresponding wavelet coefficients for scales of a=0.1544, 0.4144, 0.7526 are labelled c1,c2,c3 and are shown in Figure 10.9 and Figure 10.10. It can be seen that certain wavelet functions are more sensitive to some fault features as compared to others. The learning mechanism optimises this for the classification process. Since the textile structure is periodic in nature, it is also critical to avoid scales that respond to the inherent frequencies of the fabric.

Figure 10.7. Mispick defect in a textile fabric.

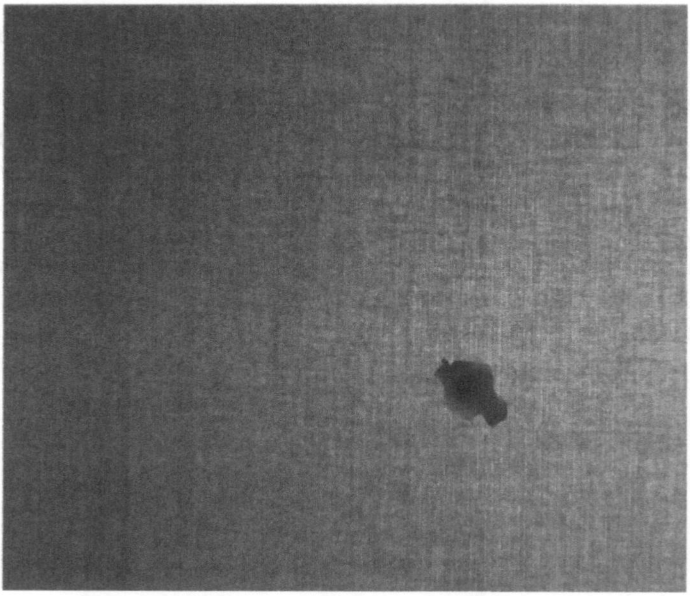

Figure 10.8. Oil spot defect in a textile fabric.

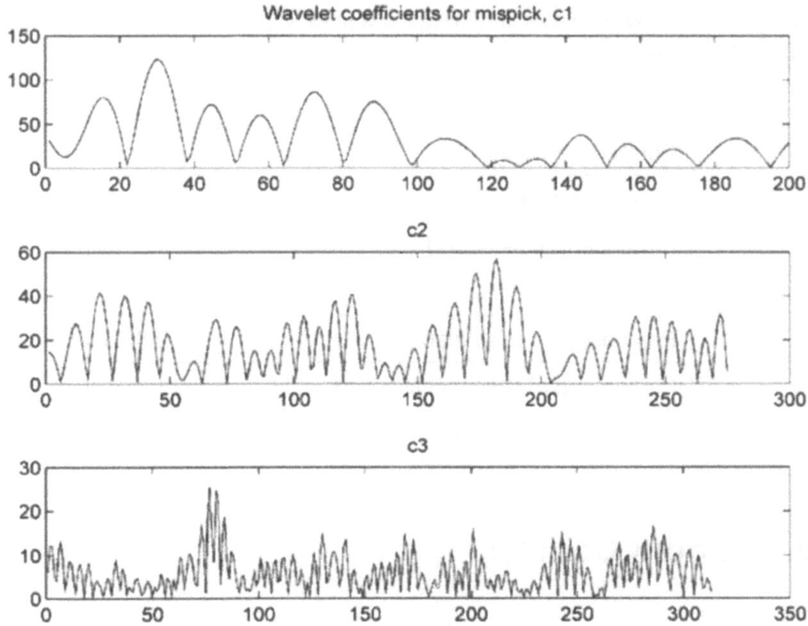

Figure 10.9. Wavelet coefficients for mispick.

An output summary of the experimental run is listed in Table 10.1.

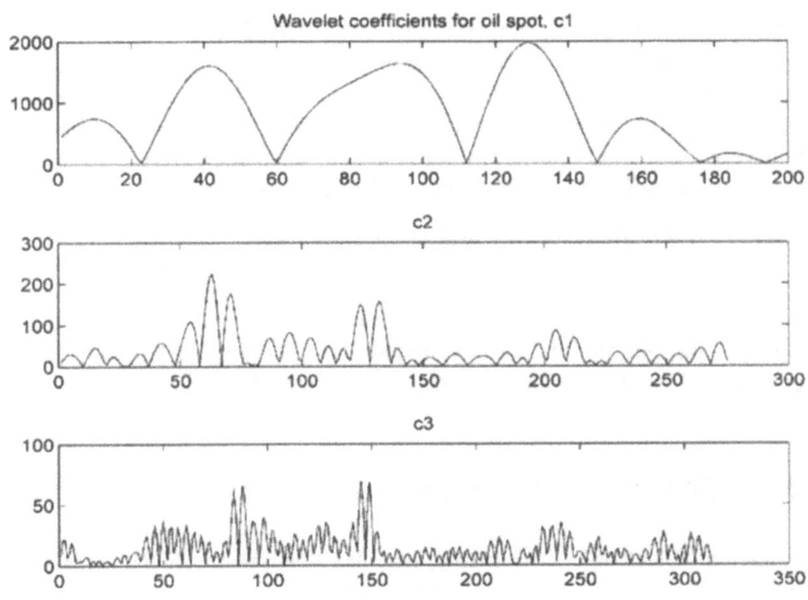

Figure 10.10. Wavelet coefficients for oil spot.

Table 10.1 False alarm (FA) indicates cases where the algorithm falsely declared a defect, while the cases in which, although a defect was detected, it was incorrectly classified, are indicated by false identification (FI).

Total number of defects identified	40
Correct identification	33
False alarms (FA)	5
False identifications (FI)	3
Total number of image frames analyzed	175
Percentage of false alarms	2.86%
Percentage of false identification	1.71%
Overall correct identification	95.43%

10.11 Loom Control

Once a reliable system of defect detection and identification has been implemented, there is a distinct possibility of the on-line inspection apparatus sending control information to the machine for performance adjustment purposes. A hierarchical control architecture is conceptualised which may perform a number of functions: defect information may be employed for statistical control purposes, condition-based machine maintenance, etc. A knowledge base consisting of production rules may represent the cause-effect templates, i.e., defect type (output of FWA) and possible machine parameter deviations that are the principal cause of the observed defects. A fuzzy inference engine attempts to match the incoming defect data with the stored templates. This fuzzy logic controller addresses effectively uncertainty in the process operation and performs robustly in the presence of disturbances. Typical rules in the fuzzy rule-base are of the form:

Rule i *If broken pich rate high, Then slow speed*
Rule i+1 *If broken pich rate very high, Then change package*

The proposed control architecture has not been implemented on an actual loom at this stage. Further studies are envisioned and a suitable test platform will be of such a control strategy employed to test and evaluate the performance.

10.12 Commercial Implementation

Appalachian Electronic Instruments, Inc. has licensed the fabric defect detection technology described in this paper from the Georgia Institute of Technology and has developed an inspection system for on-loom real-time monitoring of defects. A brief description of the apparatus and its intended application follows:

The textile fabric defect detection system using wavelet transforms employs inexpensive area-view camera technology and DSP technology combined in modular electronic units. Modern weaving machines are capable of inserting from 500 to 1000 picks/minute (weft yarns/minute) in the cross-machine direction. The production rate of the fabric in cm/minute would be a function of the fabric construction (i.e., yarns/cm). For example, a fabric with 20 yarns/cm on a machine weaving at 500 yarns/minute would produce fabric at a rate of 25 cm/minute. On a machine running at 1000 picks/minute it would be produced at 50 cm/minute. These are relatively slow production rates and if the field of view is 10 cm in the vertical dimension, the analysis time of an image would only have to occur in about 12 seconds. Typical analysis times are less than one second with even slow processors.

Figure 10.11 shows the "black box" installed underneath a weaving machine. It is located within about one metre of the point of fabric formation (called the "fell" of the cloth). Any running defect exceeding that length would be detected and the machine would be stopped for repair. An electronic module is used for each 28 cm of fabric width. Typical fabric widths would range from 160 to 300 cm. A microprocessor at one end of the box controls image capture and after receiving results of the analysis from each of the modules, combines the results and stores the results or sends them to a host machine.

An electronic encoder is used to indicate when each 15cm of cloth has been produced and the control microprocessor triggers the image capture. All modules capture and analyse images in parallel and report over a bus to the controller. The controller then analyses the data received from all modules and "pieces" together defects that continue from module to module. A full-width defect (filling) would thus have appeared as being defective in all modules. Actually, some weak defects may exceed defect thresholds in some modules and not others. An intelligent algorithm recognises the situation and determines continuity.

A defect oriented in the machine direction (warp) would show repeatedly in a single camera as the defect passes through. As in the case of the filling defect, a defect may not exceed the threshold in each successive image. In this case the algorithm in the controller must "piece" together these images and indicate continuity. It is common for yarn defects to fade in and out even to the visual inspector, but such a defect would be recognised as one.

Certain defects are randomly repetitive (with perhaps several metres between occurrences) and these would be much more difficult for the machine operator to recognise as problematic since the operator is responsible for 10 to 20 machines.

Troublesome defects are relatively rare in modern weaving production with major defects occurring on average only once or twice in 100 metres of fabric length. One might reasonably ask how to justify the cost of monitoring 100% of the fabric. Of course, the problem is that once a machine has a problem that causes a continuous or repetitive defect, it will continue to make off-quality fabric until the problem is corrected. This can prove very costly in many fabric styles. Waiting until the roll is complete (perhaps 1000 metres) then removed from the machine and manually inspected is unacceptable. There is a new weaving technology capable of producing at 3000 yarns/minute. As machine production speeds increase, the problem machine increases in importance.

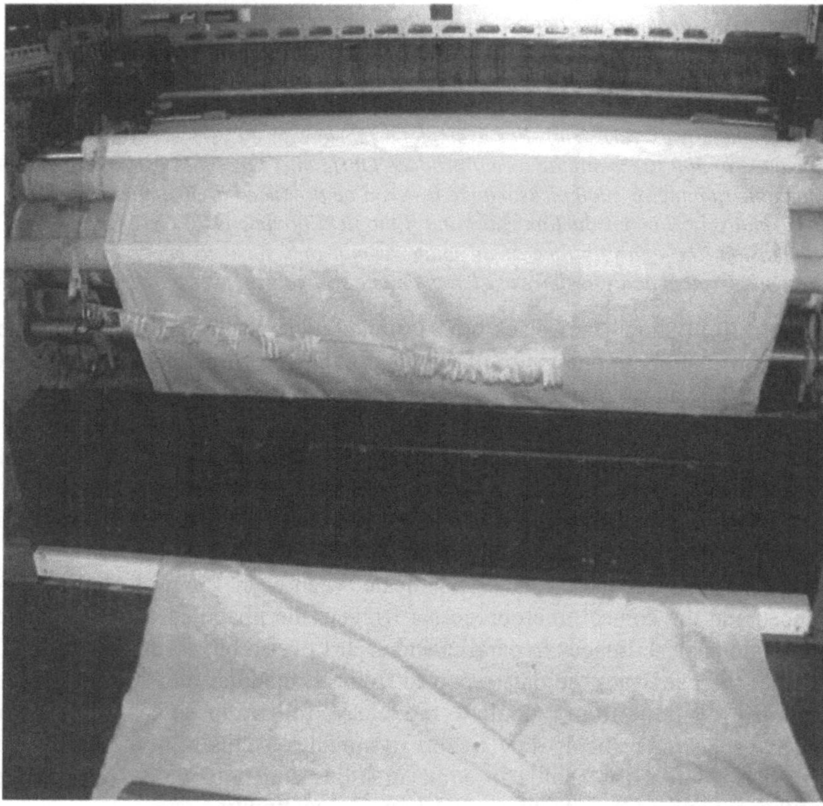

Figure 10.11. Detector "Black Box" on a weaving machine (Hightower).

10.13 Conclusions

The combination of fuzzy tools and wavelet theory has opened a new avenue for feature extraction, failure defect detection and identification of products and complex systems. The FWA algorithm gives a complete architecture for an intelligent approach and has been shown to perform much better than traditional signal processing techniques in many applications. This paper introduced a new fuzzy similarity measure that has a variable coefficient that controls the relative sensitivity of the inputs. This additional control provides greater versatility and robustness to the algorithm.

The definitions of detectability and identifiability measures have provided an intuitive notion of system performance, along with the capability of the system to improve its performance. Maximising these measures assists in reducing the fuzziness of the final output while increasing the level of confidence.

A US patent has been secured on the basic elements of the FWA technology and commercial versions of the inspection system are currently available.

Experimental testing and simulation results have shown the feasibility and viability of the technology.

10.14 Acknowledgment

We would like to acknowledge the support of the National Textile Center in the execution of this project.

10.15 References

[1] Juran, J. (1989) Quality Control Handbook, McGraw-Hill, 3rd edition.
[2] Pao, Y.H, (1989) *Adaptive Pattern Recognition and Neural Networks*, Addison-Wesley, New York.
[3] Pal, S.K., Majunder, D.D. (1977) Fuzzy sets and decision-making approaches in vowel and speaker recognition, *IEEE Transactions on Systems, Man and Cybernetics*, vol. 7, pp. 625–629.
[4] Binaghi, E., Orban, D., Rampani, A. (1989), Fuzzy logic based tools for classifiaction and reasoning with uncertainty, IEEE international workshop on tools for artificial intelligence. Architecture, languages and algorithms, pp. 572–577.
[5] Ishibuchi, H., Nozaki, K., Tanaka, H. (1993) Efficient fuzzy partition of pattern space for classification problems, *Fuzzy Sets and Systems*, Vol 59, pp. 295-304.
[6] Vachtsevanos, G., Kang, H., and Cheng, J. (1990) Managing ignorance and uncertainty in system fault detection and identification, Proceedings of the 5th IEEE International Symposium on Intelligent Control 1990, Philadelphia, Pennsylvania, pp. 558–563, 1990.
[7] Ikoma, N., Pedrycz, W., Hirota, K. (1993) Estimation of fuzzy relational matrix by using probabilistic descent method, *Fuzzy Sets and Systems*, vol. 57, 335–349.
[8] Kosko, B. (1991) Fuzzy associative memory systems, in *Fuzzy Expert Systems Theory*, CRC Press, Ann Arbor.
[9] Dumitrescu, D. (1994) Fuzzy training procedure II, *Fuzzy Sets and Systems*, vol. 67, 277–291.
[10] Loskiewicz-Buczak, A., Uhrig, R.E. (1993) Aggregation of evidence by fuzzy set operations for vibration monitorin, IEEE Conf. on Fuzzy Systems, Vol. 1, pp. 204–209.
[11] Kang, H., Vachtsevanos, G. (1992) Fuzzy hypercubes: a possibilistic inferencing paradigm, IEEE Conf. on Fuzzy Systems, vol. 1, pp. 553–560.
[12] Pedrycz, W. (1993) *Fuzzy Control and Fuzzy Systems*, John Wiley and Sons Inc. New York.
[13] Wang, L. (1994) *Adaptive Fuzzy Systems and Control: Design and Stability Analysis*, PTR Prentice Hall.
[14] Zadeh, L.A. (1965) Fuzzy sets, *Information and Control*, No. 8, 338–353.

[15] Kaiser, G. (1994) *A Friendly Guide to Wavelets*, Birkhauser, Boston MA.
[16] Daubechies, I. (1992) *Ten Lectures on Wavelets*, Capital City Press, Montpelier, Vermont.
[17] Terano, T., Asai, K., Sugeno, M. (1994) *Applied Fuzzy Systems*, AP Professional, Cambridge MA.
[18] Mufti, M., Vachtsevans, G. (1995) *Automated fault detection and identification using a fuzzy-wavelet analysis technique*, AUTOTESTCON Proc, IEEE cat n95CH35786, pp. 169–175.
[19] Phuvan, S., Oh, T.K., Carvis, N., Li, Y., Szu, H. (June 1992) Optoelectronic fractal scanning technique for wavelet transform and neural net pattern classifiers, IJCNN, International Joint Conference on Neural Networks, Vol. 1, pp. 40–6.
[20] Gartner, I., Zeevi, Y.Y. (1991) Generalized scanning and multiresolution image compression, Data Compression Conference, pp. 434.

Chapter 11

Editorial Introduction

Even seemingly simple everyday tasks can present formidable computational problems. Achieving optimal packing for large numbers of rectangular objects of randomly varying size (e.g., customer orders for carpets cut from a broad-loom roll) is very difficult. Packing random blob-like shapes, for example templates for leather shoe components on an animal hide, is even worse! Established optimisation techniques, for example, those based on dynamic programming, do not offer much help. This particular packing problem is known to be NP-complete, which as most mathematicians believe, implies that no solution, except exhaustive search, can guarantee finding an optimal solution. In such a situation, heuristics are our only resort but they cannot ensure that we will always find the best possible packing configuration. In such a situation, *ad hoc* measures of packing efficiency have to be used. We must not apologise for this but it may necessitate some hard work to convince would-be customers that a machine that we are hoping to sell cannot reasonably be expected to find the best possible solution. A rule-based system, exploiting application knowledge, can enable a visually guided robot to obtain valuable improvements in packing efficiency. This is achieved by reducing the volume of the search space. We can, for example, constrain the orientation of the shapes when the material being cut has a well-defined grain, or its characteristics vary from one place to another. Neither classical Artificial Intelligence nor image processing alone can solve the packing problem; both are needed. Producing an integrated system in which two different computational methodologies work harmoniously together is clearly essential but may be far from easy to achieve in practice.

Chapter 11

Automated Cutting of Natural Products: A Practical Packing Strategy

P.F. Whelan

"No program can deal flexibly with components of arbitrary shapes or with unanticipated failures that a human being would easily detect. It seems wiser, then, to abandon the goal of human flexibility and seek engineering solutions consistent with the limited capacities of present-generation robots. Better methods of standardization can obviate the need for human flexibility, and they have the advantage of working"

Dreyfus and Dreyfus 1986 [1]

11.1 Introduction

This chapter is concerned with the issues involved in the automated packing of two-dimensional objects into a two-dimensional bounded region, where the size and shape of both the objects and the region in which they are to be packed are not known prior to packing. The packing of such regions is directly related to the stock cutting task, one in which we try to maximise the amount of items that can be extracted for a given material (e.g., how are leather templates arranged on an animal hide so as to cut the hide efficiently?). Problems relating to the automated packing and nesting of irregular shapes are not only of theoretical importance, but have considerable industrial interest.

The automated cutting and packing strategy outlined in this chapter consists of two main components. The first provides a means of manipulating the shape and scene image at a geometric level. The second component consists of a rule-based geometric reasoning unit capable of deciding the ordering and orientation of the shapes to be packed. The heuristic component must be capable of dealing with the system issues arising from a specific application demand. This task can be simplified by maximising the use of the information available from the product, process and the environment for a specific industrial application. The use of heuristic methods increases the generality of the packing system, thus making the development of procedures for new applications less cumbersome. One of the key features of this system is that it works towards an efficient solution, accepting that we cannot guarantee reaching an optimal solution. Therefore a mechanism for quantifying the packing systems performance will be necessary. This will enable a quantitative comparison of packing procedures [2,3].

11.2 The Packing/Cutting Problem

Early research into determining optimal packing/nesting configurations can be traced back to Johannes Kepler in 1611 when he tried to determine if the most efficient method of packing identical spheres was an arrangement now known as face-centred cubic lattice. This consists of placing a bottom layer of spheres in a bounded region. Each successive layer is then arranged so that the spheres occupy the gaps of the layer below [4,5] (i.e., the arrangement greengrocers commonly use to stack oranges). Although the stacking of items such as oranges in this way seems intuitive, researchers are still unable to prove that this stacking configuration is the most efficient (Hinrichsen, Feder and Jossang [6] concentrated on a simplified version of this problem and developed strategies for the random packing of discs in two dimensions).

Research has shown that an optimal algorithmic solution for even the simplest, well-defined packing problem, such as pallet packing, is unlikely [7]. However, the aim of the system described in this chapter is more ambitious and complex. Its objective is to allow the flexible packing of random two-dimensional shapes into previously undefined scenes (the term *scene* is used when referring to a region of space into which we are required to place an arbitrary *shape*). The NP-complete nature of the simpler packing problem has a major bearing on the line of research taken. Any attempt at developing an optimal solution to the packing of random shapes, even if it did exist[1], would be difficult. It would also be difficult to constrain the problem, especially considering that it must deal with unpredictable shapes. Hence, the aim has been to produce an *efficient* packing strategy that is flexible enough for industrial use. To achieve this objective, the systems approach to the packing problem is essential.

Other approaches to the packing problem include single pattern techniques such as dynamic programming, and multi-pattern strategies such as linear programming [8,9]. Unfortunately these techniques do not meet all the requirements of a flexible packing system. In the former case, the aim is the generation of an optimal solution, and as such the approach would not seem to hold promise. The second technique, while useful for one-dimensional packing applications, is difficult to implement and use in two dimensions. Alternatively, a heuristic approach to packing can be adopted. This line of research would seem to be the way forward if the system is to remain flexible and have the ability to cope with random shapes.

Another key element in the development of automated packing systems, concerns the method of shape description. The majority of current systems rely on correlation methods and low-level features such as curve [10,11] and/or critical

---

[1] It is impossible to guarantee that an optimal procedure for the more general packing problem can be found, especially when you consider the NP-complete nature of the simpler pallet-packing task.

point matching techniques [12]. Other techniques enclose the shape of interest within a bounding rectangle [13], polygon [14] or convex hull, prior to paving the region to be packed with these predefined shapes.

11.2.1 The One-dimensional Packing Problem

Initial work on the development of automated packing systems concentrated on one-dimensional packing [15]. This can be best explained by considering a group of holes of similar size and cross-section, and a selection of boxes of varying length, but which have the same cross-section as the holes. The one-dimensional packing problem consists of placing all the boxes into the holes without any protruding. This area of research is more commonly referred to as *bin packing* [16], and it is one of the "celebrated problems of computer science" according to Hofri [17]. Sample one-dimensional packing applications include process scheduling (industrial and computer level), timetabling and the efficient packing of advertisements into a time slot between programmes. Chandra, Hirschberg and Wong [18] relate the problems involved in the design of distributed computer systems, such as processor allocation and file distribution, to bin packing.

11.2.2 The Two-dimensional Packing Problem

One-dimensional packing problems can be extended to two and three dimensions. For example, the objective of the simplest two-dimensional task is the packing of a number of flat rectangular pieces into a rectangular scene. In the case of two-dimensional depletion, the task is the division of a large rectangle into smaller ones. Applications of two-dimensional stock cutting include the cutting up of materials, such as sheet metal [19], fabrics, wooden planks, glass and rolls of paper, into smaller pieces. The aim of such systems is to minimise the amount of waste material, referred to as trim loss, produced by the cutting process. One of the main applications for these techniques is in the area of cloth and leather cutting [20,21]. The minimisation of waste material is especially important in the leather industry, for example shoe making [22], since the waste leather material cannot be recycled.

11.2.3 The Three-dimensional Packing Problem

In the three-dimensional case the objective is to pack rectangular blocks into a large empty space such as a rectangular container. This is usually referred to as the *knapsack problem*[2] due to its original formulation as a problem concerned with packing as many items as possible into a knapsack before a hike. The three-dimensional depletion task consists of segmenting a rectangular box into a number of smaller rectangular boxes.

---

[2] Also referred to as the *flyaway kit problem* by some authors (15).

11.3 Review of Current Research

The survey by Dowsland and Dowsland [9] is one of the more complete reviews of the application of operational research techniques to the solution of two and three-dimensional packing problems such as pallet packing and container loading. As well as modelling and solving problems, the authors review a number of algorithmic and heuristic approaches. The emphasis of this work is practical solutions to real issues. Sweeney and Paternoster's [8] review of the stock cutting and packing problem contains over 400 categorised application oriented references including books, dissertations and working papers. The authors have also grouped the publications according to the three main solution methodologies, these are summarised below:

- *sequential assignment heuristics:* packing of patterns based on a set of assignment rules. The majority of heuristic approaches consist of determining what order and orientation the pieces should be packed in;
- *single-pattern generating procedures:* such as dynamic programming based algorithms, which try to reuse a single *optimal* packing configuration. For example in the two-dimensional rectangular packing problem, the solution is built up by considering partial solutions within smaller containing rectangles [9];
- *multi-pattern generating procedures:* such as linear programming based approaches, which consider the interaction between patterns. This approach requires the solutions to be rounded and are, therefore, also heuristic in nature [15]. The packing task can also be formulated as a binary integer-programming problem in which a single variable represents each possible shape position. A major concern with this approach is the production of a physical design from the values of the variables in the integer programming solution [9].

Dychoff [23] develops a consistent and systematic approach for integrating various kinds of cutting and packing tasks to try to unify the various concepts found in the operational research literature. By doing so the author attempts to find appropriate methods for each relevant problem type and conversely to identify problem types that can be solved by a certain method. A practical review of two and three-dimensional packing issues and solution methods can be found in Dowsland's [7] paper. The majority of the applications outlined in this review are based on two-dimensional packing techniques. Many of the three-dimensional problems are tackled by applying two-dimensional techniques on a layer-by-layer basis. Most published work in the area of three-dimensional packing is limited due to its complexity, and the applications that are discussed tend to be concerned with the loading of shipping containers. The paper also summarises some of the practical requirements in pallet loading; these include the stability of the loading stack, the load bearing ability of the items in the stack, ease of stacking and the air circulation requirements of certain products in a stack.

Dowsland [7] reviews some of the heuristic approaches used for packing a given set of identical rectangular items into a containing rectangle. A summary of the techniques used in the packing of non-identical rectangles is also included. This extensive review covers the key areas in automated packing, such as optimality versus efficiency and the measurement of a packing systems performance. The basic conclusion of the author is that although some very high packing densities have been reported in the literature, as yet there is no generic heuristic approach that can be applied to the two-dimensional packing task. Solutions reported tend to be very application specific.

11.3.1 Packing of Regular Shapes

The main emphasis of the early research into packing issues tended to concentrate on the well-constrained problem of packing regular shapes. This task usually consists of packing two-dimensional regular shapes into a well-defined scene, such as a rectangle [6,15,16,24,25,26,27]. The main industrial applications are in the area of pallet packing [24,25] and container loading [28]. Other applications include efficient VLSI design and automated warehousing [29,30]. Hall, Shell and Slutzky's [29] work combines automated packing techniques developed in the field of operational research, with systems engineering and artificial intelligence approaches to packing. It outlines the issues associated with the arrival of packages at the packing station and relates this to the single server queuing problem, which is commonly discussed in the operational research literature. The authors also discuss a number of systems issues, such as the importance of the product information. A practical example would be the packing of foodstuffs and toxic products. In this case the packing strategy has not only to consider the efficiency of the packing procedure, it must also consider the product type. The foodstuffs and the toxic products should be packed in different boxes. These boxes should be well separated on the pallet to prevent contamination of the food. The authors also highlight the importance of how the pallet data is represented and how to determine the correct placement location for the robot. Other related areas of research discussed include, bin-picking, automated storage and retrieval, automated kitting of parts for assembly, automated warehousing, and line balancing.

Bischoff [31] discusses the methodologies of the pallet and container packing problem. The main emphasis of this paper is a discussion of the techniques used in the interactive tuning of packing algorithms. The author points out that the pallet packing stability criterion is application dependent. If the pallet load is wrapped or strapped down, then this issue becomes less important. This is a significant point as there is often a conflict between stability constraint and need to minimise waste space on the pallet. The concept of 'cargo fragility' in container loading, and its relationship to the stability requirements, is also discussed.

11.3.2 Packing of Irregular Shapes

More recently researchers have begun to concentrate on the issues involved in the packing of irregular shapes. Batchelor [13] outlines a technique for the packing of irregular shapes based on the use of the minimum area-bounding rectangle. In this approach each shape is enclosed by its minimum fit bounding rectangle, and these rectangles are packed using the techniques developed for the packing of regular shapes. Qu and Sanders [32] discuss a heuristic nesting algorithm for irregular parts and the factors affecting trim loss. The application discussed is the cutting of a bill-of-materials from rectangular stock sheets. The authors take a systems approach to the problem and produce some good results. These are discussed in the context of performance measurements, which they have developed. While the authors review the published work in this area, they make the important point that although a number of techniques have been developed to enable the flexible packing of irregular shapes, very few of these have been published due to commercial confidentiality.

Qu and Sanders [32] describe irregular shapes in terms of a set of non-overlapping rectangles. The authors state that each of the parts in their study can be represented by no more than five non-overlapping orthogonal rectangles. The system places each part in an orientation such that a. its length > height and b. the largest complimentary (void) area is in the upper-right corner. The parts are then sorted by non-increasing part height. The shapes are packed into a rectangular scene in a raster fashion, building up layers of intermeshed packed shapes. The major disadvantages with this approach are a. the use of rectangles to approximate the shape to be packed and b. the assumption that good packing patterns will be orthogonal. Dori and Ben-Bassat [14] and Chazelle [33] were the first to investigate the nesting of shapes within a polygon rather than a rectangle. The authors discuss the optimal packing of two-dimensional polygons with a view to minimising waste. The algorithm is only applicable to the nesting of congruent convex figures. The problem involves cutting a number of similar but irregular pieces from a steel board, this is referred to as the *template-layout problem*. The authors decompose the task into two sub-problems. The first consists of the optimal (minimal waste) circumscription of the original irregular shape by the most appropriate convex polygon. The remaining problem consists of circumscribing the convex polygon by another polygon that can pave the plane, that is, cover the plane by replications of the same figure without gaps or overlap. This is referred to as the *paver polygon*. Limitations of this approach include the fact that it is only applicable to congruent convex figures and the assumption that the packing plane is infinite; hence waste in the margin is not considered. Another limitation of this approach is that it can only be applied to convex components with straight sides.

Koroupi and Loftus [34] address the issues raised by Dori and Ben-Bassat [14], by enclosing the component within a polygon so that the area added is minimal. The identical components, whether regular or irregular, are then nested using paving techniques. Martin and Stephenson's [35] paper deals with the packing of two and three-dimensional objects into a well-defined scene. In this paper the authors tackle the task of automated nesting from a computer-aided design perspective. That is, given an arbitrary polygon and a rectangular box, will the polygon fit in the box, and if so how should the polygon be translated and rotated to implement this fitting. Prasad and Somasundaram [19] outline a heuristic-based computer-aided system that will allow the nesting of irregular-shaped sheet-metal blanks. This paper also contains a comprehensive list of the practical constraints one must consider in developing a packing system for sheet metal stamping operations. Constraints such as bridge width, blank separation, grain orientation, and the minimisation of scrap. They also highlight the need to align the pressure centre of the blank to be cut out with the axis of the press ram to reduce wear in the guideways of the press. Design requirements, such as maximising the strength of the part when subsequent bending is involved, are also considered.

Chow [36] discusses the optimal packing of templates of uniform orientation under limited conditions. This paper is useful as it discusses the edge effect issues in packing, which in general tend to be neglected. The author also outlines some of the concerns associated with manual packing. Kothari and Klinkhachorn [37] present a two-dimensional packing strategy capable of achieving dense packing of convex polygon shapes. The techniques described have been applied to stock cutting in the hardwood manufacturing industry. This consists of efficiently cutting wooden pieces from a hardwood board so that the pieces are free of defects and aligned in the direction of the grain. This last constraint is needed for strength and aesthetic reasons. Albano and Sapuppo [38] outline a procedure that will produce an optimal arrangement of irregular pieces. Manual and semi-automatic approaches to this nesting task are also discussed. The techniques described show how the optimal allocation of a set of irregular pieces can be transformed into the problem of finding an optimal path through a space of problem states from the initial state to the goal state. The search approach developed makes certain assumptions about the task; (a) the pieces are irregular polygons without holes and (b) the scene is rectangular. The main application discussed is that of cloth layout and leather cutting.

Vincent [39] discusses the application of morphological techniques to the *tailor suit* or space allocation problem. This addresses the problem of translating two shape pieces, A and B, such that both are included in a larger shape piece X without overlapping. Although this can be shown mathematically for two pieces [39], there is no general solution to this problem involving simple morphological techniques only. While the author does not make the link between this technique and the automated packing and nesting in an industrial system, the use of such a powerful technique to manipulate shapes does seem to point the way forward.

11.4 System Implementation

The packing scheme consists of two major components. The first is referred to as the *geometric packer*, and is based upon the principles of mathematical morphology. This component takes an arbitrary shape in a given orientation and puts the shape into place, *in that orientation*. A key element in the success of this approach is that it removes the limitations imposed by having to recognise and describe the object under analysis in order to pack it, thus increasing the systems flexibility. The second component is referred to as the *heuristic packer*, and is concerned with the ordering and alignment of shapes prior to the application of the *geometric packer*. This component also deals with other general considerations, such as the conflicts in problem constraints and the measurement of packing performance. In addition, it deals with practical considerations, such as the effects of the robot gripper on the packing strategy, packing in the presence of defective regions, anisotropy ("grain" in the material being handled) and pattern matching considerations.

By using heuristics in the packing strategy, it is hoped to produce an efficient, but not necessarily optimal solution. However, the main problem with such an approach is that there is a tendency to generate a set of overly complex rules, incorporating a variety of paradoxes and logical conflicts. It is necessary, therefore, to keep all the logic decisions as simple as possible. Another key aspect of applying heuristics to any complex problem is knowing when the solution is "good enough" so that the process can be terminated and a result produced [40]. To this end, a mechanism for the measurement of the packing systems performance must also be included in the overall system design.

One of the key features in such a system is that it should work towards an efficient solution, accepting that we cannot guarantee reaching an optimal solution. Therefore a mechanism for quantifying the packing systems performance will be necessary. This will enable a quantitative comparison of packing procedures. Burdea and Wolfson [11] suggest that the integration of such a heuristic approach with a packing verification procedure should ensure convergence to an efficient solution. The general packing strategy outlined in this chapter is illustrated in Figure 11.1 (a more detailed discussion of the system implementation can be found in [2,3]). Together the geometric and heuristic elements form a flexible strategy that allows the packing of arbitrary, two-dimensional shapes in a previously undefined scene. The aim of the design was to produce a flexible system capable of dealing with the majority of packing/cutting problems, and as such, the system was not designed around a specific application. As Burdea and Wolfson [11] point out, no single strategy, however efficient, will succeed in dealing with all shapes equally well. Therefore, when faced with a specific application, the system can be tuned to that task.

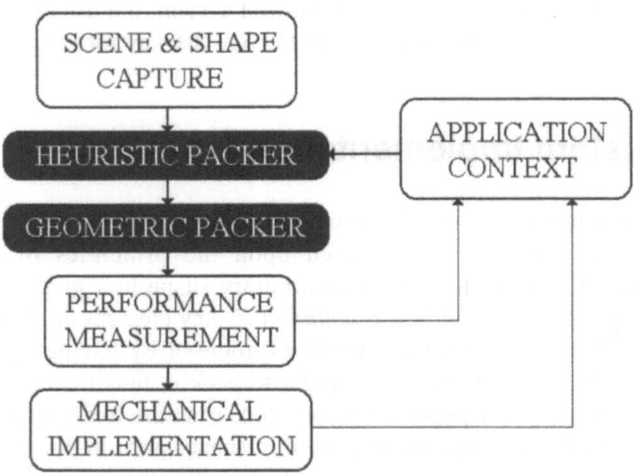

Figure 11.1. General packing strategy.

11.4.1 Geometric Packer: Implementation

The concept of enclosing an arbitrary shape within its bounding rectangle, convex hull or polygon approximation (known as paver polygons) is common to many of the irregular packing techniques discussed previously. Since these techniques involve the packing of the paver polygons rather than the original shape, only approximate packing solutions can be generated due to the loss in original shape information. Other strategies, such as contour matching, describe the shapes in terms of their critical points or chain codes. Although such approaches tend to be precise, they are also computationally expensive, especially for complex shapes. For this reason these techniques are rarely implemented on complex shapes without some degree of shape approximation. It would be advantageous to avoid using such estimates of the arbitrary shape. Therefore an approach that deals *directly* with shapes would be of great benefit in the development of a flexible packing strategy. Such an approach can be found in the set-theoretic concepts of mathematical morphology [41,42], which is concerned with the extraction, or imposition of shape structure. One of the key features of the application of morphological operations to automated packing is that the shape to be packed and its scene do not have to be formally described to enable their manipulation by the packing system.

The function of the geometric packer is to take any arbitrary shape in a given orientation and to put that shape into place in the scene, efficiently in that orientation. Providing the shape(s) to be packed, and the scene to be examined, can be captured and stored as binary images then morphological techniques can be applied to these images. These techniques allow the packing of a structuring element into a given image scene. In the case of the automated packing system, the shape to be packed will be represented by a morphological structuring element, while the scene will be represented by an image set on which this structuring element will act.

In the sample geometric packing problem illustrated in Figure 11.2a, the image scene is a rectangular bounded region (in which a *star* shape has already been packed) and is denoted by the image set A. The *star* shape to be packed is applied to the image set A and is denoted by the structuring element B. The image scene A is eroded by the structuring element B, to produce the erosion residue image C. Every white pixel in this residue represents a valid packing location. The erosion residue image is then scanned, in a raster fashion, for the location of the first (white) pixel. This location is denoted by *(fitx, fity)*. Experimentation has shown that further erosion of the residue image C by a standard 3×3 square structuring, prior to searching for the first packing location, enables control of the spacing between the packed shapes. That is, the number of pixel stripping operations on the erosion residue is related to the spacing between the packed shapes. This relationship can also be shown mathematically [43]. The translation of the shape to be packed, B, to the location *(fitx, fity)* effectively places B at the co-ordinate of the first possible packing location of B in the scene A. This image is denoted by $B_{(fitx, fity)}$. The resultant image is subtracted from the original image set A to produce a new value for the image set A, therefore effectively packing B into the scene

(Figure 11.2(b)). This procedure can then be reapplied to the image set A until an attempt to pack all the input shapes has been made (Figure 11.2(c)).

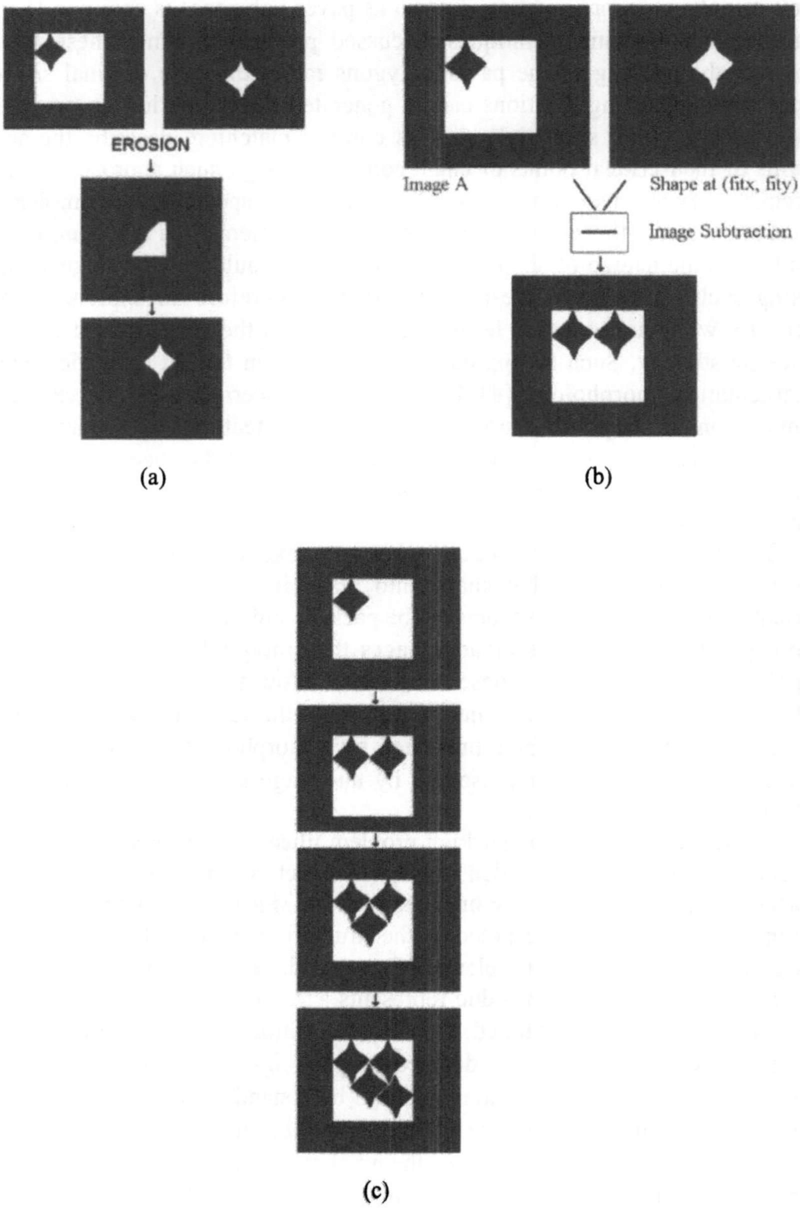

Figure 11.2. Geometric packing. This illustrates the steps involved in packing a star shape in a rectangular scene: a. Image A is eroded by star shape to produce an erosion residue; b. the star shape is then relocated to *(fitx, fity)*; c. this process is repeated until a terminating condition is reached.

11.4.2 Heuristic Packer: Implementation

As outlined earlier, the *heuristic packer* is concerned with the ordering and alignment of shapes prior to the application of the *geometric packer*. The *heuristic packer* operates on two classes of shapes: blobs (shapes with a high degree of curvature and/or significant concavities) and simple polygons. Details of these procedures can be found in [2]. It is necessary to consider both these general shape classes separately, since no single scheme exists for all shapes, and while the *geometric packer* is independent of the shape class and application context, the *heuristic packer* is not. The heuristic component also deals with other general considerations, such as the conflict in problem constraints and the measurement of packing performance. In addition, it deals with a number of practical issues, such as the effects of the robot gripper on the packing strategy [44], packing in the presence of defective regions [2,45], anisotropy and pattern matching considerations.

Blob Packing

This section outlines some of the heuristics that have been devised to deal with two-dimensional binary images of random shape and size, prior to the application of the geometric packer. The approach outlined was designed specifically for off-line packing but the techniques developed could equally well be applied to an on-line packing application.

All the shapes to be packed are presented simultaneously to the vision system. The shapes are then ranked according to their bay sizes; the shape with the largest bay is the first to be applied to the geometric packer. Once the shape ordering has been decided, it is necessary to orientate each shape so that an efficient local packing strategy can be implemented. Four orientation rules are used to align the shape to be packed in the scene.

The order in which the shapes are placed by the packer is determined by the *sort_by_bay* predicate defined below. If the area of the largest bay is significant compared to the area of the current shape, then the shape is sorted by its largest bay size (largest first). Otherwise the shapes are sorted by their size (largest first). The *bay_rot* predicate rotates a shape such that the largest bay is aligned with the scene's angle of least moment of inertia. This predicate also ensures that the biggest bay is facing into the scene (that is facing to the right and upwards). The operation of this predicate is summarised below:

- if *object_Y_coordinate* > *bay_Y_coordinate* then rotate shape by 180°
- if *object_Y_coordinate* = *bay_Y_coordinate* and *object_X_coordinate* *bay_X_coordinate* then rotate shape by 180°
- if *object_Y_coordinate* = *bay_Y_coordinate* and *object_X_coordinate* *bay_X_coordinate* then no action required as in correct orientation
- if *object_Y_coordinate* < *bay_Y_coordinate* then no action required as in correct orientation

Figure 11.3(a) shows the result of packing hand tools into a rectangular tray. These shapes were initially presented directly to the geometric packer, without the aid of the heuristic packer. This has the effect of packing each tool at whatever orientation it was in when it was presented to the vision system. Figure 11.3(b) shows the resultant packing configuration when the heuristic packer precedes the geometric packer; each shape is aligned and ordered, before it is applied to the geometric packer. Figure 11.3(c) shows the packing of the tools into a "random" blob region. The full packing strategy was used again here, as in Figure 11.3b. Figure 11.3(d) and (e) illustrate the packing of some general items, such as scissors, keys and pens, into a rectangular tray and an irregular scene using this approach.

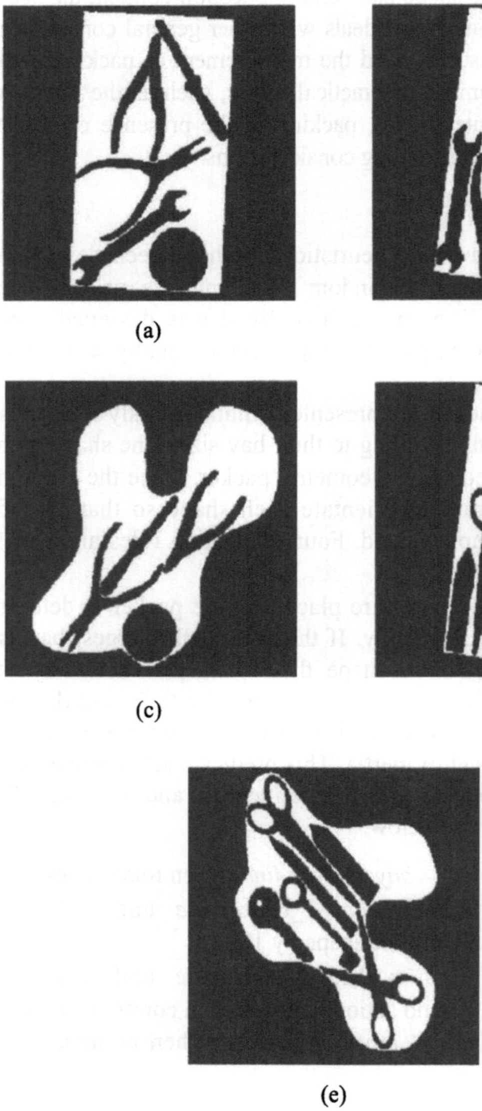

(e)

Figure 11.3. Automated packing implementation: (a) tools packed in their current orientation; (b) tools reorientated for better efficiency. (c) tools packed in an irregular scene. Packing general items into: (d) a rectangular tray and (e) an irregular scene.

Polygon Packing

The previous approach is not efficient when packing shapes which do not contain bays of significant area. Hence, a different packing procedure is used to pack simple polygons, which do not possess large bays. As before, this procedure was designed to work within an off-line packing system but could also be applied to on-line packing applications. Unlike the previous approach, however, this second procedure has the ability to determine the local packing efficiency for each shape and will reorientate it, if necessary, to ensure a more efficient configuration. (This local efficiency check could also be applied to the blob packing strategy.) In the second sample application, we chose to pack non-uniform box shapes (squares and rectangles) into a square scene (Figure 11.4(a)). Once all the shapes have been presented to the packing system, they are ordered according to size, with the largest shape being packed first. The shapes must then be orientated, prior to the application of the geometric packer.

(a) (b)

Figure 11.4. Automated packing of simple polygons: a. non-uniform boxes in a square tray; b. block polygons.

In the initial versions of this packing procedure, each shape was aligned in such a way that its axis of least moment of inertia was matched to that of the scene under investigation. However, this method proved unreliable for packing squares, because quantisation effects produce a digital image with a jagged edge. (An image resolution of 256 × 256 pixels was used.) Furthermore, a square has no well-defined axis of minimum second moment. This can cause errors in the calculation of the moment of inertia. The problem was overcome by aligning the longest straight edge of the shape to be packed with the longest straight edge of the scene. The edge angles for the shape and scene were found by applying an edge detection operator, followed by the Hough transform. The latter was used, because it is tolerant of local variations in edge straightness. Once the peaks in the Hough transform image were enhanced and separated from the background, the largest peak was found. This peak corresponds, of course, to the longest straight edge within the image under investigation, whether it is the shape or the scene. Since the position of the peak in Hough space defines the radial and the angular position of the longest straight edge, aligning the shape and the scene is straightforward.

Once a polygonal shape has been packed, a local packing efficiency check is carried out. This ensures that the number of unpacked regions within the scene is kept to a minimum. The shape to be packed is rotated through a number of predefined angular positions. After each rotation, the number of unpacked regions in the scene is checked. If a single unpacked region is found, then a local optimum has been reached. In this case, the local packing efficiency routine is terminated and the next shape is examined. Otherwise, the local packing efficiency check is continued, ensuring that, when a shape is packed, a minimum number of unpacked regions exist. This reduces the chance of producing large voids in the packed scene, and improves its overall efficiency of packing.

The packing order is determined by the sizes of the shapes to be packed (largest first). The rotation of the shapes by the packer is based on the angle of the largest face (longest straight side of the polygon) of the unpacked region. The predicate *shape_face_angles* finds the largest face angle and stores it in the face angle database. This database also contains a selection of rotational variations for the current shape. The face angles are sorted such that the angle of the largest face appears at the top of the database. The other entries are modified (by a fixed angle rotation factor) versions of this value. The *blob count* refers to the number of "free space blobs", that is the number of blocks of free space available to the packer. The polygon packer operates according to the following rules:

Rule 1 *If* blob count *is 1 then the best fit has occurred, so exit and view the next shape.*

Rule 2 *If* blob count *is 0 then read the new angle from face angles database and retry.*

Rule 3 *If* blob count < local optimum *then update blob count and update the local optimal storage buffer before trying the next angle in the face angles database.*

Rule 4 *If* blob count ≥ local optimum *then try the next angle in database.*

11.5 Performance Measures

To ensure that we have confidence in the global efficiency of any packing strategy, there must be some way of measuring its performance. Traditionally, packing performance has been measured by a single number, called the *packing density* [46]. This is the ratio of the total area of all the packed shapes to that of the total area of the scene. This is referred to as the *worst-case analysis* packing measure.

A number of other performance measures have been developed in the field of operational research, particularly for comparing different heuristics for packing rectangular bins with odd-sized boxes [7]. These performance metrics fall into two main categories: *probabilistic* and *statistical* analysis [47]. While these performance measurements can be quite useful in well-constrained packing problems, they are of little use in dealing with the packing of arbitrary shapes. Since it is unlikely that real data will fall neatly into a uniform, or any other easily analysable distribution. The performance measures used in our strategy are based on the traditional *worst-case analysis*. After a packing procedure has been applied to a given scene, the result is assessed by a number of performance parameters [2].

- *packing density* is the ratio of the total area of all the shapes packed, to the area of their (collective) convex hull after packing (minus the area of the scene defects). This measure has a maximum value of 1 (Figure 11.5);
- the *performance index* is a modified version of the *packing density* in which a weighting factor is applied. This is referred to as the *count ratio* and is defined as the ratio of the total number of shapes packed, to the number of shapes initially presented to the scene. The *performance index* is equal to the product of the *packing density* and the *count ratio*. The *performance index* also has a maximum value of 1. This measure accounts for any shapes that remain unpacked when the procedure terminates (Figure 11.6).

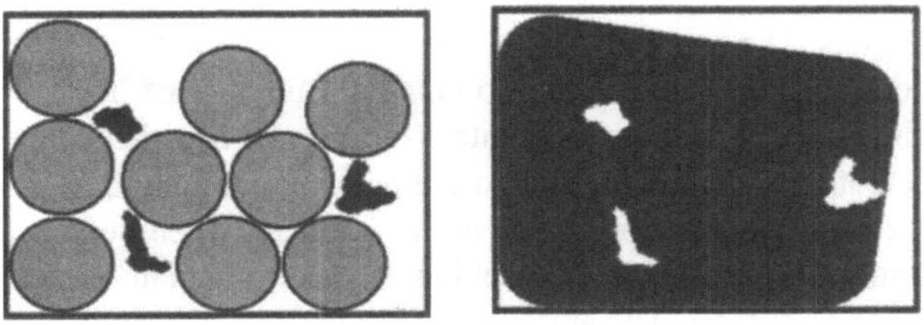

(a) (b)

Figure 11.5. *Packing density* calculation: (a) the *approximated optimal* packing area is calculated by summing the area of the individual packed shapes (indicated by the shaded discs). The black blob regions indicate scene defects; (b) the *actual* packing area is denoted by the area of the convex hull of all the shapes packed in the scene (minus the area of any scene defects). This area is indicated by the shaded region in (b).

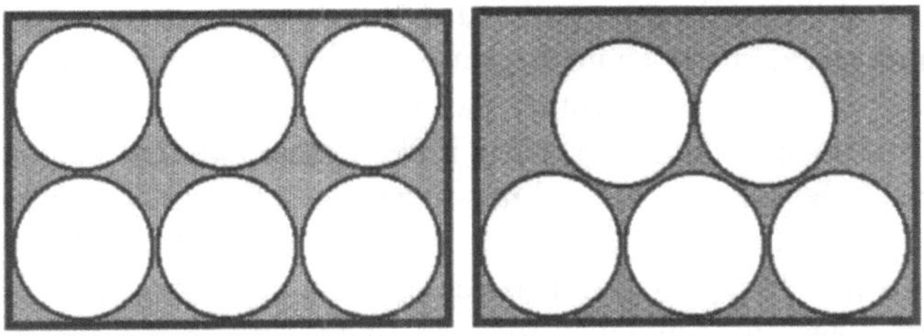

(a) (b)

Figure 11.6. Quantitative comparison of packing configurations using, *packing density*, *performance index* and *count ratio* values: (a) in this packing configuration all six shapes presented to the rectangular scene were packed, giving a *count ratio* of 1. Therefore the *performance index* equals the *packing density* (calculated as 0.77); (b) in the second configuration only five of the six shapes presented were packed, resulting in a *count ratio* of 0.83. Although the *packing density* for this configuration is better, at 0.82, the *performance index* is only 0.68, due to the fact that not all the shapes were packed.

11.6 System Issues

In general, the design of the packing system can be greatly simplified the more application constraints that can be incorporated into the heuristic packer. The design of packing is made easier by the fact that many natural materials have a pronounced grain. For example, in certain applications only two orientations of a given shoe component may be permissible, a fact which can greatly enhance the speed of the packing procedure. Again, the heuristic packer can easily take this type of application constraint into account. Alternatively, some practical considerations can increase the complexity of the packing procedure.

11.6.1 Packing Scenes with Defective Regions

Any practical automated packing system for use in such industries as leather or timber processing must be able to pack objects into a scene that may contain defective regions. The heuristic packer can readily accommodate defective regions; by simply defining the initial scene to contain a number of holes. Figure 11.7(a) illustrates the effect of packing tools into a rectangular tray that contains four small blob-like defects. By comparing this to the packing configuration shown in Figure 11.3(b), it is clear that the packing is not as compact when defects are taken into account. Figure 11.7(b) shows the packing of jacket template pieces on to a piece of fabric, prior to cutting. The small blob-like regions indicate the defective areas in the fabric. These defective regions are not to be included in the jacket pieces to be cut. These results illustrate the flexibility of the packing strategies adopted, when applied to the automated cutting of natural materials.

11.7 Packing of Templates on Leather Hides

The purpose of the application outlined in this in this section, is to automatically arrange, and place, shape templates on to an arbitrary shaped non-homogenous leather hide in an efficient manner (so as to minimize the leather waste). The importance of good packing procedures in the leather industry is obvious, since the raw material is both expensive and non-recyclable[3]. Presently, as much as 40% of the hide is wasted [48], and the European Union funded ALCUT project aimed to reduce this waste by up to 8%. This represents a significant saving for any firm dealing with a large hide turnover, for example the fashion industry in which a new range of footwear is introduced to the market about twice a year.

---

[3] In general, any waste material produced in the cutting of leather hides is sold to external companies who deal in small leather goods.

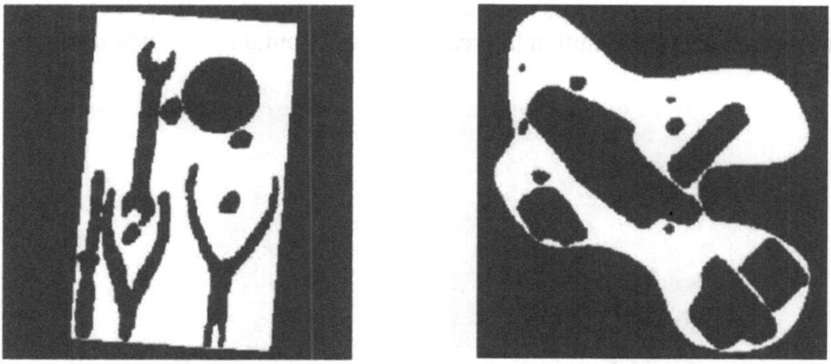

(a) (b)

Figure 11.7. Packing items into defective regions: (a) packing tools into a defective tray; (b) cutting jacket template pieces from a fabric segment, which contains defective regions.

In the current generation of leather cutting systems the hides pass underneath a bank of linescan cameras, which generates a two-dimensional image of the hide. The region of the hide to be cut is then determined by the placing of shape templates on the scanned hide (generally, the shape templates are entered into the CAD system via digitising tablets or external databases). The placing of the shape templates is done interactively by trained operators at CAD workstations. Once the position of the shape templates have been finalised, the corresponding regions are automatically cut from the hide [20]. The size and shape of these templates are application dependent. The specific application addressed in this discussion concerns the cutting of shape pieces for use in the upholstery of high quality leather car seats.

The CAD operators have only a short time period in which they can place the shape templates on the hide. This is the time between when the hide is imaged by the vision system and when the hide has progressed underneath the cutting station (in leather upholstery the operators will have to deal with 35–40 different shapes over a global surface of 55 to 60 ft^2 [49]). To this end, semi-automated interactive CAD systems have been developed to aid the operator in maximising (a) the speed at which the shape templates are placed on the hide, and (b) the number of shape templates to be cut from a given hide. The interactive functions allow operations such as bumping, sliding, automatic repetition of shapes, quality matching, grouping and area filling. The main packing strategy used by these operators involves the packing of the larger template pieces on the outer edges of the hide, and progressively moving in towards the centre with the remaining shapes.

The approach outlined in this chapter has been used to automate the template layout process (i.e., the automatic layout of the hide template shapes in an *efficient* manner). Initial investigations have concentrated on the application of the unmodified packing strategies outlined previously. The application of the *polygon* packing procedure to the automatic placement of the template pieces on a leather hide, produces an efficient packing configuration (Figure 11.8). This highlights the flexibility of the approach taken. (A medium-resolution CCD array camera was used in the prototype system to capture the hide and shape template images. This resulted in quantisation errors on rotation of the shape template pieces. A full-scale system requires high-resolution linescan cameras to build up the hide image.)

(a)

(b)

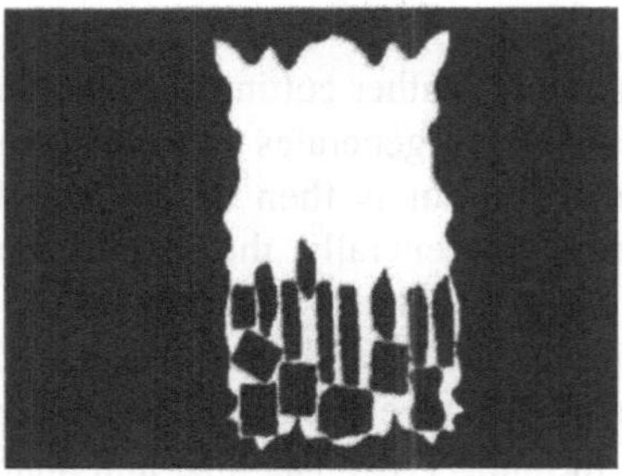

(c)

Figure 11.8. Automatic placement of car seat template pieces: a. car seat template pieces; b. leather hide image; c. the resultant packed image using the *polygon* packing procedure. The *polygon* packing strategy is implemented since there are no significant bay regions in the shapes to be packed.

11.7.1 Packing of Templates on Defective Hides

Any practical automated packing system for use in such industries as leather, textile or timber processing must be able to pack "objects" into a scene that may contain defective regions. In a natural hide there are a number of regions that cannot be used (The ideas outlined can also be applied to the cutting of synthetic leather, these are produced as more regular shapes.) These consist mainly of the spine and corner regions, although defects can occur anywhere within the hide. Currently, the hide is marked using chalk or removable inks, to indicate the stress directions and to aid in defect and quality recognition during the hide scanning process. The automated packer can readily accommodate defects like these; by initially defining the scene to contain a number of holes. Figure 11.9 illustrates the packing of leather templates onto such a hide. The black blob-like regions, illustrated in Figure 11.9(a), indicate the defective areas of the hide. These defective regions are not to be included in the leather pieces to be cut.

While the packing systems ability to automatically image the shape templates is of little benefit in this application (since the shapes are generally stored in databases and do not change frequently) the automated packing procedures do enable a more adaptive packing strategy to be implemented. The procedures ability to deal with random shapes and previously undefined defective regions, without the need for software modifications or human interaction, is a significant benefit of this approach.

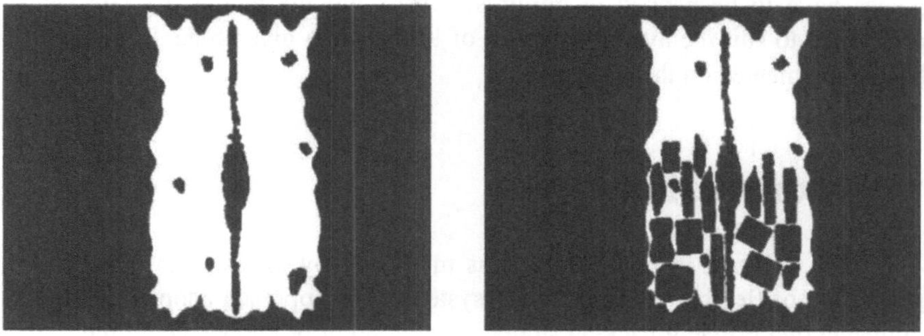

(a) (b)

Figure 11.9. Automatic placement of car seat template pieces illustrated in Figure 11.8(a) onto a defective leather hide: (a) the black blob-like regions indicate the hide defects; (b) the resultant defective hide after the shape templates have been packed.

11.7.2 Additional Points on Packing in the Leather Industry

The design of packing systems for the automated cutting of leather hides is made easier by the fact that leather, like fabric, wood, marble and many other natural materials has a pronounced grain or stress direction. This means that quite often only two orientations of a given leather component are permissible, a fact which can greatly enhance the speed of the packing procedure. Again, the *heuristic packer* can easily take this type of application constraint into account. Packing leather component templates onto a hide is not quite as simple as suggested, because the leather is not uniform in its thickness and suppleness. When making shoes, for example, the components which will make up the soft leather uppers are cut from the stomach region of the hide, while the tougher, more rigid sole is take from the back. Adding heuristic rules to assist packing under these constraints should not be difficult, although this has not yet been attempted.

Further complications arise from the fact that natural leather hides contain a number of quality levels (or grades). Each region of a shape, depending on its importance and visibility, must satisfy a *quality matching criteria*. The hide is subdivided into several areas of constant average quality. The shapes are also given a well-defined quality, therefore each single shape, or part of a shape, can only be positioned on a portion of the hide with the *same or higher quality level* [20]. In the cutting of shape templates for the manufacture of high-quality leather furniture, there may be up to 40 grades of leather, whereas in the application discussed above (high quality car seats) there are only five grades. One objective of such a layout system is to keep high-quality parts of the hide for those components of the object which are the most visible and to try to utilise lower quality regions for non-visible parts. For example, in cutting the leather component of a car seat armrest, some of the leather will not be exposed to the driver, and as such it can be of a lower leather grade. This also influences the speed of the cutting operation, since lower quality parts are cut less precisely and at a higher speed. Therefore, not only do the leather pieces have to be packed to minimise waste, but the template grades must be positioned to suit the available grades of leather on a given hide. This has not yet been implemented on the system.

11.8 Conclusion

The work outlined in this chapter was motivated by the need to produce a new generation of flexible packing/cutting systems. The approach adopted is capable of implementing efficient packing strategies, with no prior knowledge of the shapes to be packed or the scenes into which the shapes were to be placed. Automated packing systems have a wide range of possible industrial applications, including flexible assembly and automated cutting systems.

The strengths of the adopted approach become more evident when the systems issues of a specific application are considered. The packing system outlined has the ability to deal with a range of such issues. These include the ability to pack shapes into defective regions. This is not a trivial task for a human operator. Other issues that must be considered include the ability of the automated packing procedure to control the spacing between packed items in a consistent manner. This is a task that manual operators would find difficult, especially for irregular shapes about which they had no prior information.

11.9 References

[1] Dreyfus H.L. and Dreyfus S.E. (1986) *Mind over Machine*, The Free Press.

[2] Whelan P.F. and Batchelor B.G. (1993) Flexible packing of arbitrary two-dimensional shapes, *Optical Engineering*, 32(12), 3278–3287.

[3] Whelan P.F. and Batchelor B.G. (1996) Automated packing systems - a systems engineering approach, *IEEE Trans. on Systems, Man and Cybernetics – Part A: Systems and Humans*, 26(05), 533–544.

[4] Stewart I. (1991) How to succeed in stacking, *New Scientist*, July, 29–32.

[5] Stewart I. (1992) Has the sphere packing problem been solved?, *New Scientist*, May.

[6] Hinrichsen E.L., Feder J., Jossang T. (1990) Random packing of disks in two dimensions, *Physical Review A*, vol. 41, no. 8, 4199–4209.

[7] Dowsland W.B. (1985) Two and three dimensional packing problems and solution methods, *New Zealand Operational Research*, vol. 13, no. 1, 1–18.

[8] Sweeney P.E., Paternoster E.R. (1992) Cutting and packing problems: A categorized, application-orientated research bibliography, *Journal of the Operational Research Society*, vol. 43, no. 7, 691–706.

[9] Dowsland K.A., Dowsland W.B. (1992) Packing problems, *European Journal of Operational Research*, vol. 56, 2–14.

[10] Wolfson H., Schonberg E., Kalvin A., Lamdan Y. (1988) Solving jigsaw puzzles by computer, *Annals of Operations Research*, vol. 12, 51–64.

[11] Burdea G.C. and Wolfson H.J. (1989) Solving jigsaw puzzles by a robot, *Proc. IEEE Trans. on Robotics and Automation*, vol. 5, no. 6, 752–764.

[12] Oh S.R., Lee J.H., Kim K.J., Bien Z. (1985) An intelligent robot system with jigsaw-puzzle matching capability, Proc. 15'th Int. Symp. On Industrial Robots (Tokyo), pp. 103–112.

[13] Batchelor B.G. (1991) *Intelligent Image Processing in Prolog*, Springer-Verlag, London.

[14] Dori D., Ben-Bassat M. (1984) Efficient nesting of congruent convex figures, *Communications of the ACM*, vol. 27, no. 3, 228–235.

[15] Brown A.R. (1971) *Optimum Packing and Depletion: The Computer in Space- and Resource-Usage Problems*, American Elsevier Publishing.

[16] Ong H.L., Magazine M.J., Wee T.S. (1984) Probabilistic analysis of bin packing Heuristics, *Operations Research*, vol. 32, no. 5, pp. 983–998.

[17] Hofri M. (1982) Bin packing: an analysis of the next-fit algorithm, Technical Report #242, Israel Institute of Technology.

[18] Chandra A.K., Hirschberg D.S., Wong C.K. (1978) Bin packing with geometric constraints in computer network design, *Operations Research,* 26(5), 760–772.

[19] Prasad Y.K.D.V, Somasundaram S. (1991) CASNS - A heuristic algorithm for the nesting of irregular-shaped sheet-metal blanks, *Computer-Aided Engineering Journal,* 8(2), 69–73.

[20] Dulio S. (1990) Application of automation technology to leather cutting, Proceedings of the 23rd International Symposium on Automotive Technology and Automation, pp. 83–96.

[21] Chetverikov D., Lerch A. (1992) Prototype machine vision system for segmentation of hide images, *International Journal of Imaging Systems and Technology,* 4(1), 46–50.

[22] Cuninghame-Green R. (1989) Geometry, shoemaking and the Milk Tray problem, *New Scientist,* (12 Aug), 50–53.

[23] Dyckhoff H. (1990) A typology of cutting and packing problems, *European Journal of Operational Research*, vol. 44, pp. 145–159.

[24] Carpenter H., Dowsland W.B. (1985) Practical considerations of the pallet-loading problem, *J. Operational. Res. Soc.*, vol. 36, no. 6, pp. 489–497.

[25] Chen C.S., Sarin S., Ram B. (1991) The pallet packing problem for non-uniform box sizes, *International Journal of Production Research*, vol. 29, no. 10, 1963–1968.

[26] Chuang F.R.K., Garey M.R., Johnson D.S. (1982) On packing two-dimensional bins, *SIAM J. Alg. Disc. Meth.*, vol. 3, no. 1, 66–76.

[27] Baker B.S., Coffman E.G., Rivest R.L. (1980) Orthogonal packing in two dimensions, *SIAM J. Comput.*, vol. 9, no. 4, 846–855.

[28] Bischoff E.E., Marriott M.D. (1990) A comparative evaluation of heuristics for container loading, *European Journal of Operational Research,* vol. 44, 267–276.

[29] Hall E., Shell R., Slutzky G. (1990) Intelligent packing and material handling, Proc. SPIE Intelligent Robots and Computer Vision IX: Algorithms and Techniques, vol. 1381, 162–170.

[30] Wilson R.C. (1965) A packaging problem, *Management Science*, vol. 12, no. 4, B135–B145.

[31] Bischoff E.E. (1989) Interactive approaches to packing problems, 10th ICPR, pp. 55–61.

[32] Qu W., Sanders J.L. (1987) A nesting algorithm for irregular parts and factors affecting trim losses, *Int. J. Prod. Res.*, vol. 25, no. 3, 381–397.

[33] Chazelle B. (1983) The polygon containment problem, *Advances in Computing Research*, vol. 1, 1–33.

[34] Koroupi F., Loftus M. (1991) Accommodating diverse shapes within hexagonal pavers, *Int. J. Prod. Res.*, vol. 29, no. 8, 1507–1519.

[35] Martin R.R., Stephenson P.C. (1988) Putting objects into boxes, *Computer-Aided Design*, vol. 20, no. 9, 506–514.

[36] Chow W.W. (1980) The packing of a template on a flat surface, *Trans. of the ASME – Journal of Mechanical Design*, vol. 102, 490–496.

[37] Kothari R., Klinkhachorn P. (1989) *Packing of convex polygons in a rectangularly bounded, non-homogeneous space*, IEEE Proc. of 21st Southeastern Symp. on System Theory, pp. 200–203.

[38] Albano A., Sapuppo G. (1980) Optimal allocation of two-dimensional irregular shapes using heuristic search methods, *IEEE Transactions on Systems, Man, and Cybernetics*, vol. 10, no. 5, pp. 242–248.

[39] Vincent L. (1991) Morphological transformations of binary images with arbitrary structuring elements, *Signal Processing*, vol. 22, pp. 3–23.

[40] Silver E.A., Vidal R.V.V., De Werra D. (1980) A tutorial on heuristic methods, *European Journal of Operational Research* 5, 153–162.

[41] Dougherty E.R. (1992) *An Introduction to Morphological Image Processing*, Tutorial Text TT9, SPIE Press.

[42] Whelan P.F., Molloy D. (2000) *Machine Vision Algorithms in Java: Techniques and Implementation*, Springer, London.

[43] Haralick R.M., Shapiro L.G. (1992) Mathematical morphology, Chapter 5 of *Computer and Robot Vision: Volume 1*, Addison-Wesley.

[44] Whelan P.F., Batchelor B.G. (1992) Development of a vision system for the flexible packing of random shapes, in *Machine Vision Applications, Architectures, and Systems Integration, Proc. SPIE*, vol. 1823, 223–232.

[45] Whelan P.F., Batchelor B.G. (1993) Automated packing systems: Review of industrial implementations, *Machine Vision Applications, Architectures, and Systems Integration II, Proc. SPIE*, vol. 2064, 358–369.

[46] Fowler R., Paterson M., Tanimoto S. (1981) Optimal packing and covering in the plane are NP complete, *Inf. Proc. Letters*, vol. 12, no. 3, 133–137.

[47] Garey M.R., Johnson D.S. (1979) Computers and Intractability – *A Guide to the Theory of NP-Completeness*, W.H. Freeman and Co.

[48] EUREKA (1989) *Robotics and Production Automation*, European Community.

[49] Dulio S. (1992), Private communication.

Chapter 12

Editorial Introduction

The identification and measurement of free-swimming fish in their natural environment represents a particularly difficult challenge for designers of Machine Vision systems, since the images are inevitably highly variable. The water can be clear or turbid, while sun light, diacaustic and shadows can all create large intensity changes, even within a single image. In a sea cage there is some scope for controlling the viewing environment, although the options open to us may be limited by the need to avoid disturbing the natural behaviour of the fish. For example, some species of fish avoid bright lights, while others are attracted to them. Over a longer time scale, bright lights will stimulate the growth of certain species of marine plants and sessile animals nearby. Eventually, this will alter the feeding habits of the fish. Even within a controlled viewing environment, the variations in image quality can be very large indeed; it is even possible that the contrast between the fish and background can become inverted. Hence, intelligent image interpretation is necessary. We must exploit application knowledge in whatever way we can, to ensure that the problem is simplified as far as is possible, without compromising the overall effectiveness of the system. For example, we have to limit the camera–fish viewing range and ignore all fish outside a certain size range. Although fish theoretically have six degrees of freedom in their movement, in practice fish that feed in open water, or on tall plants, control both roll and pitch angles. However, yaw angle and the three positional parameters are totally uncontrolled. Moreover, apparent body shape changes as fish swim by flexing their posterior ends. Fish do not always present themselves for viewing in isolation from other fish. The result is that fish outlines may overlap. Attempts to limit the effects of social behaviour, by forcing the fish to swim individually through narrow channels, may be counter-productive. Finally, we must prevent, or cure, fouling of the optical surfaces in a submarine vision system due to microbe growth. Vision systems can use viewing techniques that are not found in nature. Using stereo image pairs derived from cameras that are displaced vertically, rather than horizontally, may seem to represent an insignificant variation from nature. However, this change can make a big difference in practice, since it exploits the well-defined edge contours presented by the dorsal and belly surfaces. It is often appropriate in applications such as this to study the task first in carefully controlled conditions and then gradually remove the constraint, to approach more natural viewing conditions. In this way, we can see which factors create the greatest difficulties. A proper experimental study inevitably takes a long time and costs a lot of money, so we must be careful to justify it on commercial, social, ethical and (whole world) environmental grounds first. Once the best possible images have been obtained, their analysis proceeds using quite sophisticated techniques, based first on straightforward motion analysis and then on N-tuple feature recognition, self-adaptive learning and model fitting. This is a good example illustrating the general maxim that advanced (i.e., most "intelligent") analysis methods are needed to solve problems where there is a high level of variability.

Chapter 12

Model-based Stereo Imaging for Estimating the Biomass of Live Fish

R.D. Tillett, J.A. Lines, D. Chan, N.J.B. McFarlane and L.G. Ross

12.1 Introduction

It is important to salmon farmers to be able to monitor the size of their fish. This chapter describes a new approach to mass estimation using stereo cameras, automatic image analysis, and improved morphometric analysis. The approach requires the capture and analysis of underwater stereo pairs of images as the fish swim freely within the sea cage. All the components of the system have been designed with the complex and variable nature of the sea cage environment in mind, although most of the testing of the image analysis system has so far taken place under more controlled tank conditions. The components of the system have been tested separately and, to a limited extent, together. This chapter provides an overview of the system and a more detailed presentation of the image processing algorithms developed.

12.2 Typical Sea-cage Images

Two examples of stereo pairs of images of salmon in a sea cage are given in Figure 12.1. The left-hand images were collected in relatively clear water with the sea surface illuminated by direct sunlight. The upper image is a slightly downward looking view from the upper camera of the stereo pair. The lower image was captured simultaneously with the upper by a horizontal looking camera about 0.5 m below the upper. The pair of images on the right were collected by the same equipment in slightly more turbid water when the water surface was in shadow.

Images captured in sea cages typically show low contrast between the fish and the background. The fish being examined may be either lighter or darker than the background and this may vary from point to point on a single fish. The surface features of the fish and the fins are also frequently obscure. Strong highlights may appear on the fish owing to specular reflection of the sunlight. Highlights are also common in the background due to reflection of the sunlight off fish or debris.

Where the water surface is visible, its brightness is likely to vary widely across the image due to refraction of the sunlight through the surface. The water surface curvature can focus the light creating diacaustics, which result in uneven and rapidly changing illumination. Owing to the high densities of fish in a sea cage it is also likely that some images of fish will overlap.

Figure 12.1. Two pairs of stereo images of salmon in a sea cage.

In the work described below we have developed techniques to identify and measure individual fish when they appear as approximate silhouettes, as shown in the right-hand images in Figure 12.1.

12.3 Stereo Image Collection

12.3.1 Stereo Cameras

Stereo pair video images are collected using a vertically oriented pair of video cameras with a base line of about 0.5 m and optical axes converging 1.5 m in front of the cameras as shown in Figure 12.2. Vertical orientation allows the relatively well defined top and bottom edges of the fish to be used to identify the distance of the fish from the cameras. Images of salmon are collected when they are between 1 and 2 metres from the cameras. This represents a compromise between aiming to capture images from a zone close to the cameras, which the fish might avoid, and capturing images from a greater distance where imaging might be prevented by poor water clarity and by fish between the camera and target. The chosen separation of the cameras represents a compromise between the requirement for precision in the depth estimates and the requirement that the two images are similar enough to enable corresponding points on the object to be identified in both images. Lens length was chosen to give a field of view width approximately twice the length of the fish. The system has been used with fish of length between 0.4 and 0.7 m. Fish significantly smaller or larger than this might require a stereo system of different dimensions.

The cameras, which have been used, are shuttered, raster scan, $^1/_2$ inch monochrome CCD cameras with 4.5 mm lenses. In order to minimise the effects of fish motion only one of the two interlaced fields from each image is used, giving an image resolution of 384×288. The two cameras use a single external sync to ensure they collect simultaneous images. Since video recorders are unable to store pairs of simultaneous images, direct image collection by computer is used.

Figure 12.2. Stereo pair of underwater cameras.

12.3.2 Calibration

The interpretation of stereo images requires that corresponding points in each image of the pair be identified. Triangulation is then used to map a pair of matched images onto a real-world 3D co-ordinate system. Because the geometrical relationship between the two cameras is fixed, a point in one image maps onto some point on a line in the second image. These lines, known as epipolar lines, are identified in the calibration of the cameras. Calibration of the stereo system taking account of the camera geometry and lens distortion is essential before 3D measurements can be made. This is achieved by identifying common points on a 3D grid of known geometry on each image of a stereo pair. The grid is shown in Figure 12.3. The calibration target was a chequered board of black and white alternate squares (100×100 mm) which produced good contrast in underwater conditions. The well-defined "V" shape of the calibration target *al.*lows a non-coplanar calibration to be carried out.

Figure 12.3. The calibration grid.

A pair of stereo images was grabbed and the corner of each visible square (calibration points) was selected manually with sub-pixel accuracy with the aid of corner extraction software which were designed and implemented specifically for this calibration purpose. Well-established methods for calibration, such as that presented by Tsai [1], are available in the literature. The calibration software used for estimating the intrinsic and extrinsic camera parameters in this work were developed by R. Willson [2] at Carnegie-Mellon University, who has implemented the software according to the Tsai [1] camera calibration methods. The software is publicly available on the Internet [3].

12.3.3 Accuracy Achieved

The stereo system was calibrated by using the test points extracted from an underwater image of the calibration grid (70 calibration test points were used for the stereo camera calibration process). Images were also collected of the grid placed further and nearer to the cameras. The calibration could then be used to recover the horizontal and vertical edges of each square in each image. The true length was known to be 100mm. Table 12.1 shows the mean and standard deviation of the lengths measured using the stereo system and so shows the accuracy achieved.

Table 12.1 Accuracy achieved from the calibrated stereo system.

Calibration target position relative to the cameras	**Measured horizontal lengths (mm)**		**Measured vertical lengths (mm)**	
	Mean	Standard deviation	Mean	Standard deviation
Far (2 m)	99.62	2.44	100.03	1.08
Mid-range (1.5 m)	99.86	1.43	99.97	0.40
Close (1 m)	100.31	0.9	100.52	0.30

12.3.4 Tank-based Trials

In order to develop and test the image processing algorithms, images were captured in a tank stocked with salmon of known mass and dimensions. These salmon were marked to enable them to be individually identified from the video images. The tank was 3.5 m in diameter with a water depth of 0.9 m and is shown in Figure 12.4. A black back-board was placed 1.8 m from the cameras and a similar base-board was placed on the floor of the tank. These boards provided a more uniform background to the images than the fibreglass tank wall, and also encouraged the fish to position themselves in front of the cameras. Shading cloth covering the tank reduced and diffused the natural illumination. The entire sequence of tasks from collection of the images to prediction of fish mass has been followed through using this facility. These images are simpler to analyse than many images collected in sea cages because of the uniform lighting, the background and the high water clarity. However, the outline-based techniques described here should also be applicable to the silhouette type images available in the sea cage. Typical images from the tank are shown in Figure 12.5.

Figure 12.4. Overhead view of the tank used for fish image collection.

Figure 12.5. A stereo pair of images of salmon in the tank.

12.4 Locating Fish with a Trainable Classifier

The trainable classifier is used to select suitable images and to estimate the location of the fish in the images prior to attempting to identify its outline. Simple image segmentation techniques such as thresholding and edge detection do not perform well due to the low levels of contrast typical of these images. The method developed for detecting objects that are likely to be fish uses a binary pattern classifier. This is outlined below and has been reported in more detail by Chan *et al.* [4,5,6].

When an image of a swimming fish is subtracted from a second image slightly separated in time, a characteristic crescent shape consistently appears at the head of the fish. This is created by the change in image content from background to fish as the fish moves in front of the cameras. This pattern is robust enough to be enhanced by thresholding the processed image so reducing it to a simple binary representation which can then be recognised by a binary pattern classifier (Figure 12.6). Recognition of the crescent appears to be relatively independent of the size of the fish image, since a small part of a large crescent and a large part of a smaller crescent can match equally well. The corresponding feature that is created at the tail of the fish is much more variable in form than the crescent and so is of less use for recognition.

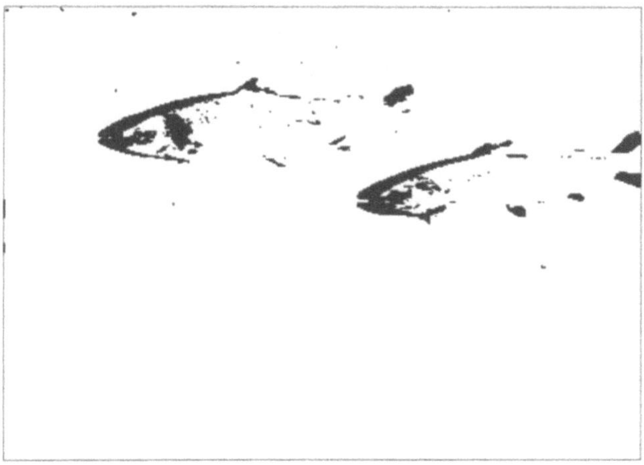

Figure 12.6. The distinctive crescent shapes due to the fish head in the difference images from a sequence.

Recognition of the crescent is achieved using an n-tuple binary pattern classifier. This established pattern recognition system is both simple and fast. The version used for this development is a simulation of the WISARD adaptive image classifier developed by Aleksander and Stonham [7] and Stonham [8]. The pixels in the image area to be examined are divided into groups of eight randomly selected pixels, known as n-tuples. The pattern to be recognised is then learnt by examination of a small number of sample or training images. A typical set of training images is shown in Figure 12.7. For each training image, the binary values of the pixels in each n-tuple define the value assigned to the n-tuple. For each image, each n-tuple can therefore take one of 256 possible values. The observed values for each n-tuple for the whole set of training images are calculated and stored. These values characterise the pattern.

Figure 12.7. The training images used to train the classifier to detect fish position.

During the recognition process the test image area is broken into the pre-determined n-tuples, and their values are calculated. These values are compared with those that occurred in the set of training images. The proportion of n-tuples in the test image area that have a value which matches that of the corresponding n-tuple, in any one of the set of training images, is used as a measure of the probability that the test image pattern belongs to the same class as the training image patterns.

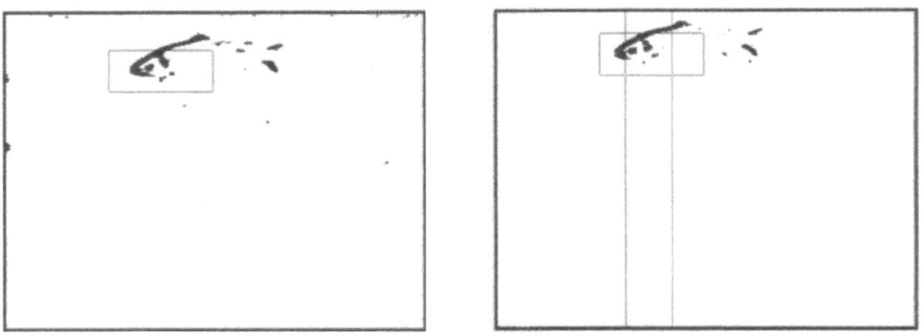

Figure 12.8. The result of using the classifier on a stereo pair.

This pattern recognition process is performed on all possible areas of one of the images of the stereo pair and the area with the highest probability score is identified. A search is then performed in the second image for a corresponding object. In this image it is, however, only necessary to search for the highest scoring area close to the epipolar line associated with the highest scoring area of the first image. If the scores of these two areas are sufficiently high and if they are similar in value, then the 3D position of the area identified is calculated and considered to be a candidate position for a fish head. A successful result is shown in Figure 12.8. In practice since fish are likely to be swimming either way, two searches will be needed using mirror images of the characteristic crescent. The position and direction of the candidate fish identified in this way is used to initiate the process of searching for the outline of the fish.

12.5 Taking Measurements Using a Model-based Approach

The edges of the fish in sea cage or tank images can be difficult to identify due to the poor contrast of the fish with the background, the uneven illumination, and partial occlusion by other fish. A model-based approach to image segmentation is therefore used since this enables weak fish-shaped edges to be selected in preference to other, stronger, edges and allows areas where no edge is visible to be interpreted correctly. The method selected is the Point Distribution Model (PDM) [9]. In this technique, the computer holds a shape template, which in our work comprises the relative locations of a number of points on the fish shape together with the principal modes of variation of this shape. This shape is then fitted by working through a series of iterations where the strength and proximity of local edges are used to identify candidate fish edges which are then tested against the template. This work is described below, and in more detail by McFarlane and Tillett [10] and by Tillett *et al.* [11].

The shape template comprises 26 landmark points on the boundary of the fish. Some of the landmarks correspond to physical features of the fish, such as the tip of the tail or the junction of a fin with the body, while others are added to fill in the gaps along featureless parts of the shape. This template is built from data derived from a small set of training images. For each fish in this training set, the landmarks are placed by hand in both images of the stereo pair and the 3D positions of the landmarks are calculated. These fish shapes are then normalised to the same scale, rotation and translation, and the main modes of variation of the normalised shape are calculated using principal components analysis. Examination of these modes of variation suggests that the most significant mode is due to the swimming motion of the fish. The first mode of variation is illustrated in Figure 12.9.

The PDM is fitted to the candidate fish by minimising an energy-like function. This function comprises three components representing the energy required to deform and move the model to a particular position, shape and orientation, the image energy (edge strength) pulling the model to specific edge points and a residual energy due to the distance between the model and the edge points. Each component is configured as of a probability distribution, to avoid the need for empirical weightings of the individual energy terms.

The energy required to deform and move the model is calculated assuming that there is a Gaussian distribution of values of scale, translation, rotation and shape variation in the training set.

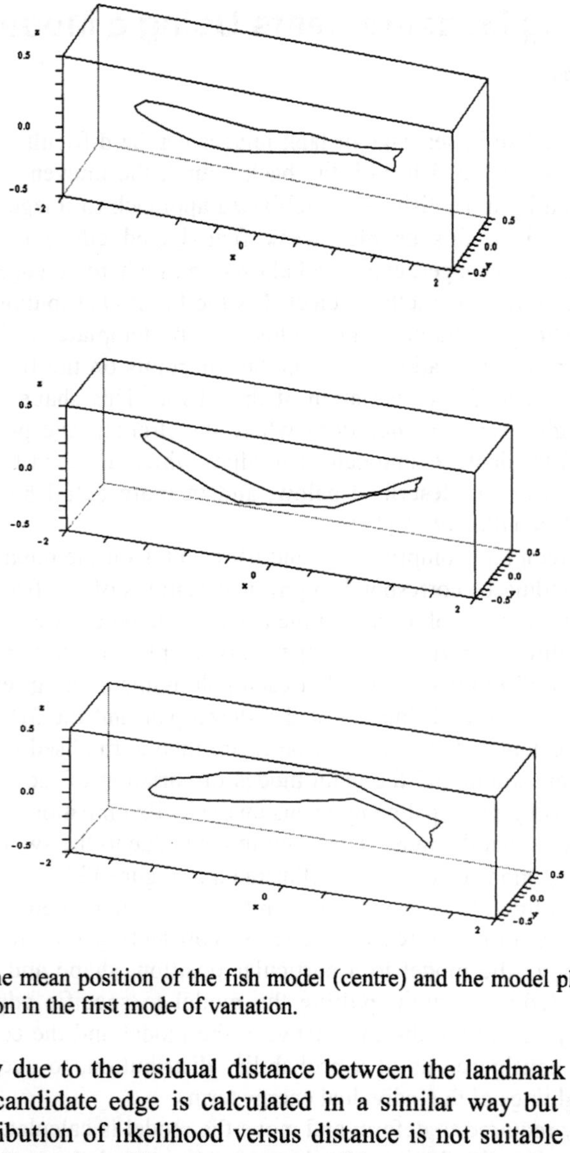

Figure 12.9. The mean position of the fish model (centre) and the model plus or minus one standard deviation in the first mode of variation.

The energy due to the residual distance between the landmark points and the point on the candidate edge is calculated in a similar way but in this case, a Gaussian distribution of likelihood versus distance is not suitable because of the presence of outlier points. When incorrect candidate edges have been selected, they can occur at significant distances from the corresponding landmark points. To avoid a small number of such errors exerting too much influence on the overall fit of the PDM, a function corresponding to the Cauchy distribution is used since this does not increase as fast as a Gaussian distribution when distance errors become large. The image energy is calculated from the sum of the absolute values of the gray level gradient at the edges and the deviation of the angle of orientation of the edges from the angle of the edge of the model. The absolute grey-levels and knowledge as to whether the fish are lighter or darker than the background are not used, because of the danger of learning features which are only applicable to specific lighting conditions and camera angles.

The PDM is fitted to the fish using a two-step iteration. In the first step, the PDM is held fixed, while a set of candidate edges is identified for each landmark point. In each image, a search is performed along the normals to the landmarks, identifying edges. At this point, the edge candidates for each landmark consist of a list of 2D image co-ordinates for each of the images in the stereo pair. These are combined into a list of 3D positions. The most likely edge to be associated with each point is then identified as that with the minimum sum of image energy and residual distance energy. In the second step, these candidate edges are held fixed while the PDM is fitted to them by minimising the sum of the model energy and the residual energy. These two steps are iterated several times to allow the model to converge to a stable position. Figure 12.10 shows the model in the start position, part way through fitting, and in the final fit to the fish.

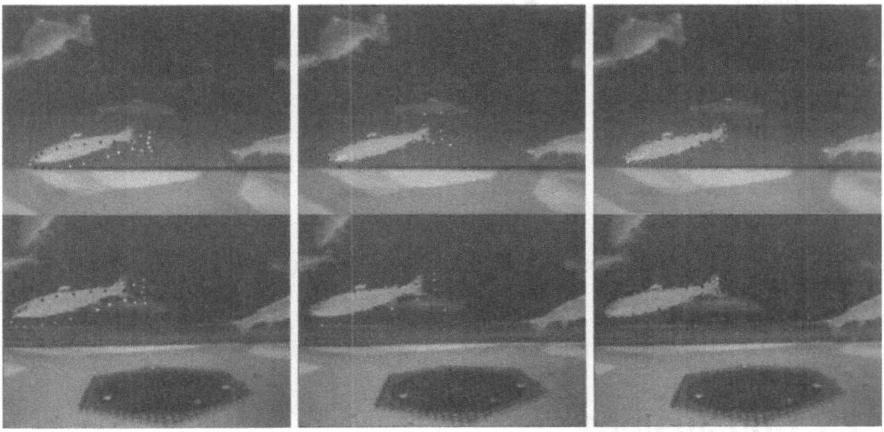

(a) (b) (c)

Figure 12.10. The Point Distribution Model fitting to a stereo pair: (a.)model starting position; (b) part way through the model-fitting; (c) the final fit to the fish.

12.6 Estimating Fish Mass

The mass of individual fish can be estimated from a combination of length measurements from the fish [12,13,14,15]. The positions of the landmark points in a fitted Point Distribution Model were used to estimate the lengths of individual fish in the stereo images. These lengths were then used to predict the fish's mass. The fish were also anaesthetised and weighed and measured out of the water. Hand selection of points on the images was also used to estimate lengths for each fish. Figure 12.11 shows the results of predicting the fish mass from these different measurements. It can be seen that the calliper measurements (taken out of the water) give the best mass estimates. The fitted Point Distribution Model tends to give an underestimate of the mass of each fish, probably due to the model fitting inside the outline of the fish in some places.

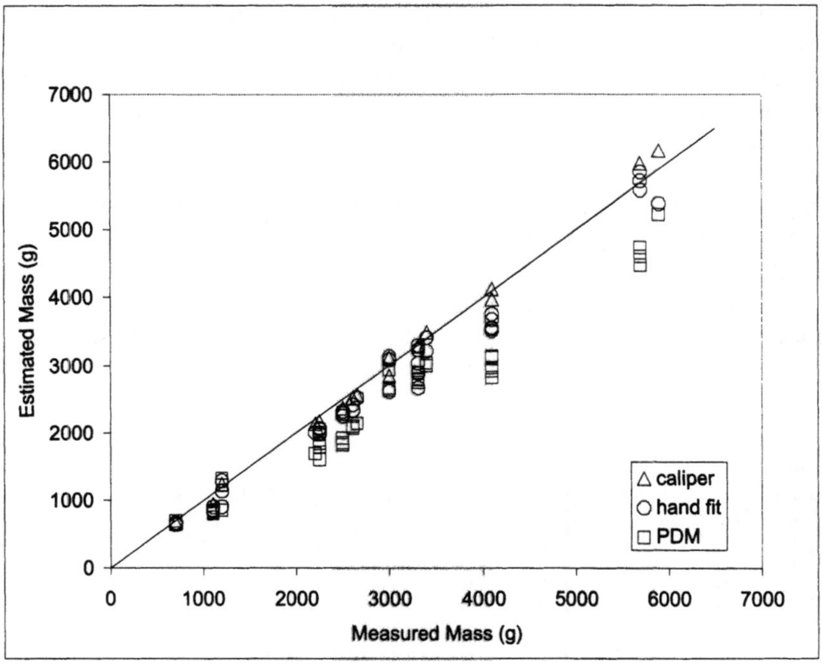

Figure 12.11. The estimates of fish mass from length measurements from calliper measurements, hand selected landmark points in images, and automatically fitted Point Distribution Models.

12.7 Conclusions

This chapter describes the image analysis algorithms and approach used to measure the mass of freely swimming farmed fish. A pair of underwater cameras is used to collect sequences of images of fish as they swim past. The cameras are calibrated to allow the recovery of real-world measurements. The difference images, formed by subtracting images taken at different times in the sequence, show a distinctive crescent shape around the fish head. A trainable classifier has been used to detect the crescent shape within the image. A Point Distribution Model has been used to capture the typical shape of the fish. A novel method for fitting this model to the image has been presented. This model-based approach allows the fish outline to be recovered even in the presence of confusing features elsewhere in the image. A mass estimation model has been developed which allows the mass of individual fish to be estimated from length measurements of the fish. The components have been developed separately and tested on images collected in a tank. The results are promising and further work is planned to test the application of these techniques to the more variable images found in a sea cage.

12.8 References

[1] Tsai, R.Y. (1986) An efficient and accurate camera calibration technique for 3D vision, Proc. IEEE Conf. On Computer Vision and Pattern Recognition, Miami Beach, USA, 22–26 June 1986.

[2] Wilson R. (1995) Carnegie-Mellon University, http://www.cs.cmu.edu/~rgw/.

[3] http://www.cs.cmu.edu/afs/cs.cmu.edu/user/rgw/www/TsaiCode.html.

[4] Chan D., McFarlane N., Hockaday S., Tillet R.D. and Ross L.G. (1998) Image Processing for underwater measurement of salmon biomass, Proc. IEE Colloquium on Image Processing in Underwater Applications, 25 March 1998. IEE Colloquium (Digest), No217 12/1–12/6.

[5] Chan D, Hockaday S., Tillet R.D. and Ross L.G. (1999) A trainable n-tuple pattern classifier and its application for monitoring fish underware, Proc. IEE Conf. on Image Processing and its Applications, 13–15 July 1999, Conference publication No. 465: 255–259. Institution of Electrical Engineers, London. ISBN 0 85296 717 9.

[6] Chan D., Hockaday S., Tillet R.D. and Ross L.G. (2000) Automatic initiation of a model fitting algorithm using an n-tuple Classifier for monitoring fish underwater, Fourth Asian Conference on Computer Vision (ACCV 2000) Taipei, Taiwan, 8–11 Jan 2000. IEEE (Asia).

[7] Aleksander I., Stonham T.J. (1979) Guide to pattern recognition using random-access memories, *Computers and Digital Techniques* 2(1), 29–40.

[8] Stonham I.J. (1986) *Practical Face Recognition and Verification with WISARD: Aspects of Face Processing,* Nartinus Nighoff.

[9] Cootes T.F. *et al.* (1992). Training methods of shape from sets of examples, Proc 3rd British Machine Vision Conf. Leeds, UK, 22–24 Sept, 1992, pp. 9–18.

[10] McFarlane N. and Tillett R.D. (1997). Fitting 3D point distribution models of fish to stereo images, Proc. British Machine Vision Conference, BMVC 1997, Vol 1: pp. 330–339, University of Essex, UK, 8–11 September 1997. BMVA Publishing. ISBN 0952 189887.

[11] Tillett R.D., McFarlane N. and Lines J.A. (2000) Estimating dimensions of free-swimming fish using 3D point distribution models, *Computer Vision and Image Understanding,* 79, 123–141.

[12] Beddow T.A. and Ross L.G. (1996)Predicting biomass of Atlantic salmon from morphometric lateral measurements, *Journal of Fish Biology*, 49, 469–482.

[13] Beddow, T.A., Ross, L.G. and Marchant, J.A. (1996). Predicting salmon biomass remotely using a digital technique, *Acquaculture*, 146, 189–203.

[14] Hockaday S., Ross L.G. and Beddow T.A. (1997) Using stereo pairs to measure mass in strains of Atlantic salmon (Salmo salar L), Paper presented at Sensors and their applications VIII, Section A Environmental and Biomedical Sensors, 7–10 Sept. Institute of Physics Publishing, Bristol, UK.

[15] Hockaday S., Ross L.G. and Beddow T.A. (2000) A comparison of models built to estimate the mass of different strains of Atlantic salmon, Salmo salar L, using morphometric techniques, Submitted to *Aquaculture*.

Chapter 13

Editorial Introduction

Using a vision system to examine livestock provides the potential for significant commercial benefit, enabling farmers to maximise production efficiency. It can, for example, assist in husbandry, by providing information about an individual animal's pattern of growth and physical development, thereby enabling its feeding regime to be optimised automatically. In such an application, Machine Vision merely provides a sensing tool, feeding data about an animal's size, shape and weight into an expert system that performs the appropriate management control function. Good engineering is vitally important, since a piggery provides a hostile operating environment for a Machine Vision system. There is a lot of material that can contaminate delicate optical and electronic equipment, which must be enclosed within a rugged protective housing. Since pigs are essentially non-cooperative, they must be coaxed into a suitable position for viewing, by luring them with food or water. An electronic ear tag provides information about the pig's identity, thereby saving on computing effort for image processing. "Intelligent" image analysis procedures are needed to accommodate the high degree of variability of the pigs and their living quarters, and to cope with unpredictable events, such as illness and fighting for access to the feeding station. The image processing is necessarily heuristic in nature and therefore must be tested rigorously before a robust and reliable commercial unit can be designed. Careful attention to the broad systems-level issues is essential in applications like this. Such imaginative use of Machine Vision must always be tempered by careful consideration of all aspects of the application. There is no suggestion in this chapter that a machine can currently match an experienced pig farmer's ability to detect injuries and ill health in the animals. Hence, systems like the one described here must be seen as *aids* not *replacements* for human beings. In the past, overselling of Machine Vision systems, for certain industrial applications, has seriously damaged the reputation of the technology. It must not happen again.

Chapter 13

A System for Estimating the Size and Shape of Live Pigs

J.A. Marchant and C.P. Schofield

13.1 Introduction

In modern pig production, the large number of animals raised in a unit precludes a stockman from giving animals individual attention. The ability to observe pig size, shape, and growth rate is vital to the proper management of stock. Using machine vision for animal monitoring could provide the necessary management information for maximising production efficiency and also for monitoring health by detecting abnormalities in growth rates. The potential for computer controlled systems in livestock monitoring has been reviewed by Frost *et al.* [1], and various authors [2,3,4,5,6,] have proposed the use of machine vision for monitoring development and growth.

An obvious characteristic of pigs is that they are variable in shape and size, they move, and they are not always co-operative in presenting themselves to an imaging system in the most appropriate manner. Also, although lighting can be controlled to a certain extent, economic constraints and the requirement that maintenance be simple mean that only rudimentary lighting schemes are practicable. These factors make the automatic inspection of growing pigs (like most objects in agriculture) a significant challenge.

This chapter describes research at the BBSRC Silsoe Research Institute and further development by Osborne (Europe) Ltd that applies machine vision to measure dimensions and areas from images of pigs as they use a feeder or drinker in their pen. These dimensions, and in particular the plan view area, can be related directly to the weight of the pig, enabling its growth to be monitored day by day. This information is of immediate use to pig breeding companies who wish to monitor the performance of different strains of pig for breeding development purposes, and a developing market with larger producers is taking the system on board to monitor and improve their production programme.

As machine vision can estimate changes in the shape, weight and condition of pigs from their images, it follows that the data produced could be used as an input to control the supply of feed to the pig. The quality and quantity of feed required would be predicted by the system, based on the pig's actual daily performance, thus the potential for improvements in pig production is significant and far reaching [7].

As well as advising on the required feed for the pigs to grow efficiently and at a desired rate, the system could be developed to indicate when best to send a particular pig or pen of pigs to market. The information on growth rates and a suitable pig growth model could provide control information for the production of individual pigs or the pen in question, and be applied to predict the amount of lean gained by the pigs each day until a desired target is reached.

13.2 Description of the System

Figure 13.1. Feed station with associated vision system.

The system is designed to be installed in a pen containing a small group of pigs (say 15–20). In normal commercial practice the animals will be free to move around the pen. The first problem to be solved is how to get the animals to present themselves to an imaging system in a reasonably consistent pose. The solution is to mount a camera above a feeding station. Such a station (FIRE, Osborne (Europe) Ltd) is shown in Figure 13.1. The station incorporates an identification system triggered by a passive transponder fitted in an ear tag worn by each animal. This is normally used to dispense feed and to measure the weight consumed on each visit so recording individual feed intake. We use this facility to trigger image capture and to associate our measurements with each pig seen by the system. The monochrome CCD camera is fitted above the feeder in a protective box. Lighting is provided by tungsten filament lamps to provide good illumination over the whole plan view of the pig. Particular emphasis is placed on providing good edge illumination. Image analysis is done with a PC, each image being analysed in a few seconds (at the time of writing). Little attempt has been made to speed up processing further as this rate is adequate to deal with the rate of data acquisition currently envisaged (say 16 pens per PC, 30 images per pig per day).

Animals will visit of their own accord and will generally be standing singly with their head at the feed trough end. This may sound a fairly well-controlled situation but problems can and do occur. Even if the background starts off being a contrasting colour to the pigs, it soon becomes contaminated whereupon the contrast cannot be relied on. Occasionally pigs can fight for food and more than one animal can be in the station at any time, sometimes one on top of another. Pigs can lie down or even go to sleep in the station.

To deal with the uncertain nature of the imaging, the philosophy in this work is to collect many more images than would be necessary to monitor the animals in ideal circumstances. Where possible, algorithms are designed to degrade disgracefully, the opposite to the normal ideal, so that erroneous processing can be detected easily. Intermediate results have to pass checks at stages through the processing and any doubtful performance leads to discarding the image. If an image passes all checks, the derived measurements are then analysed by robust statistical methods that are tolerant to outliers in the data.

13.3 Calibration

13.3.1 General Method

Calibration can be regarded as converting measurements in pixel units into quantities that describe the shape and size of the animal. We consider three aspects of calibration: i. removing lens distortion; ii. determining the magnification; iii. dealing with the curvature of the animal's surface.

When calibrating in a piggery, where non-specialist staff must sometimes be used, only simple calibration methods are practicable. We fix the camera in a known location in the stall so that the distance to the ground and the camera attitude are pre-determined. We use a target of known dimensions and position it horizontally in the stall at a fixed distance above the ground. A typical camera view of the target is shown in Figure 13.2. Our image analysis methods (see below) are designed to divide the pig plan view into three parts, shoulder (minus head), abdomen, and rump. The boundaries between the parts are based on the sudden curvature changes in the outlines. The same method is used to process the calibration target and the result can be seen in Figure 13.2. As the dimensions of the target are known, the required calibration parameters can be derived as follows.

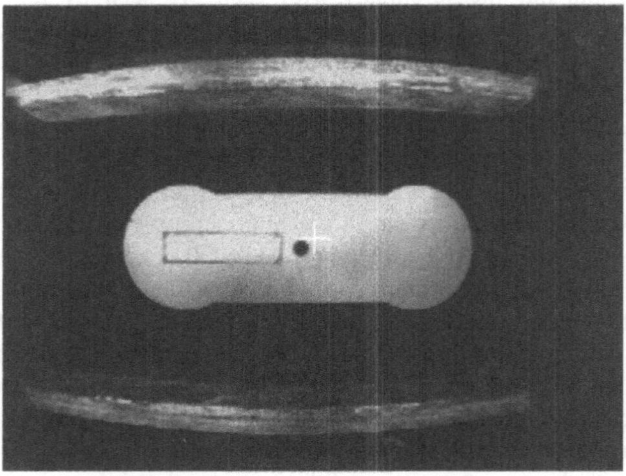

Figure 13.2. Camera view of calibration target.

13.3.2 Lens Distortion

Van der Stuyft *et al.* [8] characterised lens distortion by placing a square grid pattern under their camera and measuring the grid point positions in the image. From this they derived a mapping to warp the image back so that the grid pattern appeared undistorted in the image. This mapping was then used to correct subsequent animal images. Minagawa and Ichikawa [4] mention lens distortion but it is not clear from their paper how they compensated for it. We use an approach popular in machine vision for robotics where the distortion is characterised as a function that is circularly symmetric about the centre of the image [9]. In Tsai's work the distortion function was characterised by an even polynomial function of distance from the optical centre using two coefficients, those of the square and the fourth power of the distance. However, we have found that this over-parameterises the problem. It is possible to derive pairs of coefficients which very nearly account for the calibration data but which vary greatly. Thus only slight differences in data give very variable coefficients. We have found much more robust and stable performance using only one coefficient, than for the square of the distance.

We denote the shoulder area, abdomen area, and rump area by $A1$, $A2$, and $A3$ respectively. For a trial value of lens distortion parameter, we calculate $(A1+A3)/A2$ from the image. This is independent of the magnification (which might be different in the x and y directions), which is not yet known. We then adjust the lens distortion parameter and undistort the image, iterating until $(A1+A3)/A2$ equals the known value from the template.

13.3.3 Magnification

Next we operate on the final undistorted image to calculate the magnification. Here we make use of the pinhole camera model. This model is very commonly used in machine vision to characterise camera parameters [9]. Figure 13.3 shows the features of the model.

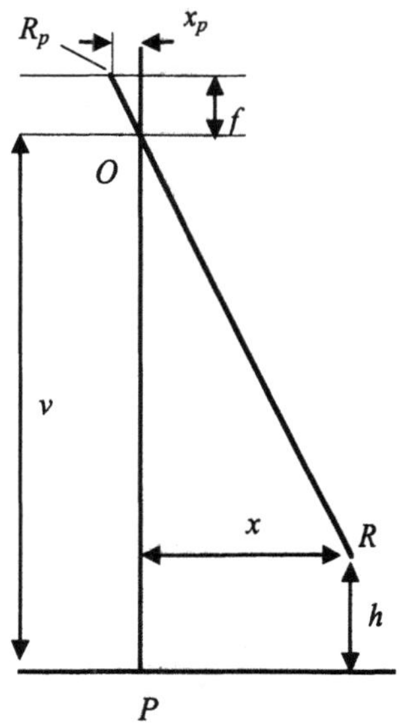

Figure 13.3. Camera geometry and pinhole model.

The camera is pointing with its optical axis vertically downwards at a height v above the ground. P is the projection of the axis on to the ground. Any ray from a point in space (e.g., point R, height h distance from the axis x) is assumed to pass through a pinhole, O, on the optical axis. This ray strikes a plane, distance f behind the pinhole, on which the image is formed. The ray strikes at point R_p , distance x_p from the axis.

From Figure 13.3

$$x = x_p(v-h)/f \tag{13.1}$$

If we assume that a ray along the optical axis strikes the focal plane to give a pixel at the centre of the frame grabber area then it can be shown [9] that

$$x = x_i(v-h)/F_x \tag{13.2}$$

where x_i is the pixel offset in the image from the centre on the frame grabber area and F_x is a constant that includes f and other non adjustable camera parameters. The assumption that the frame grabber is aligned to the optical axis is, in fact, only necessary for lens distortion correction because all other measurements depend on pixel differences rather than absolute locations. If this did cause a problem the offset between camera and frame grabber could be established in a separate calibration exercise.

For a point distance y out of the plane of the paper in Figure 13.3, a similar relationship holds

$$y = y_i(v - h) / F_y \tag{13.3}$$

Note that F_x can be different from F_y, most obviously when image pixels are rectangular rather than square.

Next we operate on the undistorted image to calculate the ratio of F_x and F_y by comparing the ratio *L7*/*L5* in the image to the known template values. *L5* is the width across the rump and *L7* is the template length. This operation depends only on the ratio of the magnifications, not on their absolute values. We then correct the image using the F_x,F_y ratio and calculate F_y by comparing (*A1*+*A2*+*A3*) to the known value from the template. F_x follows from F_y and the F_x,F_y ratio.

13.3.4 Curvature of the Animal's Surface

As an animal is a complex and changeable three-dimensional object, it can be difficult to define exactly which physical dimensions to measure and so some subjectivity and variability is unavoidable. In this work we make measurements from the boundary of the plan view of the animal. However, because the surfaces of the animal are curved, the physical points that makes up the apparent boundary change as the camera is moved. In order to make measurements that depend on the animal and not on the camera position we must base our measurements on the plan view that would have been obtained if the camera could "see" along vertical lines instead of the diverging lines from the camera optical centre. To do this we must assume a model of animal shape. Van der Stuyft *et al.* [8] represented the curvature of the animal's flanks by a circular cross section that was everywhere normal to the apparent boundary. They assumed that the radius was constant and the circle touched the floor. We use a similar approach but allow a height above the ground, h (our animals are standing), and a variable radius, r (they also grow). In earlier work we related h and r to the animal's age using published tables [10]). It can be shown [11] that errors in r cause little problem but errors in h can be more significant. Consequently we are now investigating a laser-based method of height measurement, while calculating r from h using the published tables.

Figure 13.4 represents our model for animal flank shape compensation. Consider a cylindrical cross-section orientated so that its axis is at an angle γ. It is of radius r, height h, and is offset from the camera axis. Point C (which appears on the image plane at point F, offsets of x_i and y_i) contributes to the apparent boundary whereas point D is the one we wish to measure as this is independent of camera position.

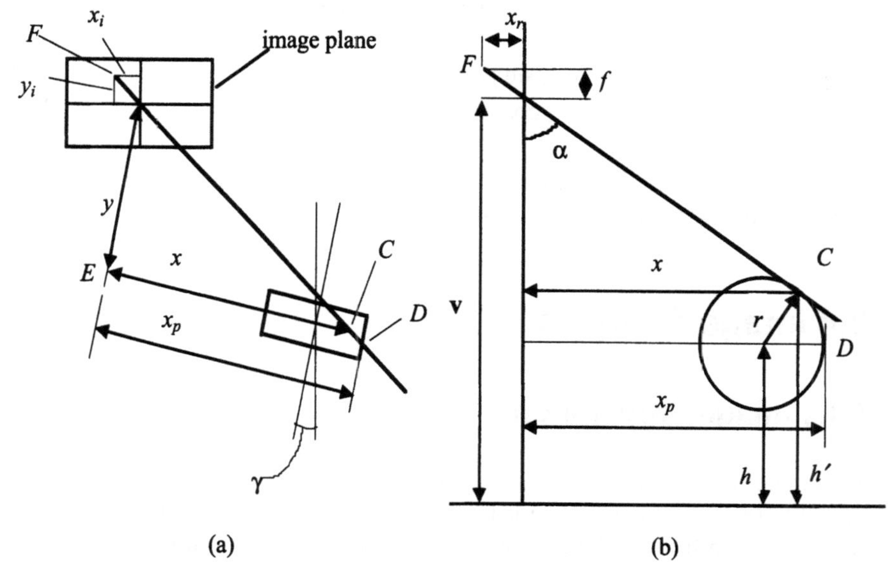

Figure 13.4. Model for animal height and radius calculations: (a) plan view; (b) view along cylinder axis.

We first transform co-ordinates in the image plane to a rotated image plane where the x axis is parallel to the line EC. Thus

$$x_r = x_i \cos\gamma + y_i \sin\gamma \tag{13.4}$$
$$y_r = -x_i \sin\gamma + y_i \cos\gamma \tag{13.5}$$

where x_r and y_r are the offsets in the rotated plane. From Figure 13.4(b)

$$x/(v-h') = \tan\alpha \tag{13.6}$$

Changing the variable in Equation 13.2 above and combining with Equation 13.6

$$\tan\alpha = x_r / F_x \tag{13.7}$$

From Figure 13.4(b)

$$x_p = x + r - r/(\cos\alpha) \tag{13.8}$$

We now wish to correct point F in the image to a corresponding point that would have come from a ray from point D. The offsets in the rotated image plane to the corrected point are therefore

$$x_{rcorr} = x_p F_x /(v-h) = x_r + rF_x(1 - 1/\cos\alpha)/(v-h) \tag{13.9}$$
$$y_{rcorr} = y_p F_y /(v-h) = y_r \tag{13.10}$$

Transforming back into the unrotated image plane gives

$$x_{ircorr} = x_{rcorr} \cos\gamma - y_{rcorr} \sin\gamma \tag{13.11}$$

$$y_{ircorr} = x_{rcorr} \sin\gamma + y_{rcorr} \cos\gamma \tag{13.12}$$

which are the corrected values of x_i and y_i , i.e., the values which would be obtained if point *D* were imaged instead of point *C*.

13.4 Image Analysis

13.4.1 Image Preparation

An example of the image processing is shown in Figure 13.5. The initial image, Figure 13.5(a) is first corrected for lens distortion (Section 13.3.2). Then the area of feeder wall in front of the animal's head is set to black so that it can be ignored in subsequent processing. A small area dimension 9 × 9 pixels is established centred on a point where we expect to find an animal. This point is midway between the points where the side walls intersect the front wall and the floor in the image and we term it the "hot spot". Note that these operations are possible because the wall positions and hence their projections into the image are known beforehand. Figure 13.5(b) shows the corrected image, the blanked wall area, and the hot spot.

(a) (b) (c)

Figure 13.5. Stages of image analysis: (a) original image; (b) lens distortion removed, front wall of feeder blanked, hot spot; (c) division of body.

13.4.2 Initial and Improved Boundary

The next series of operations follows Schofield and Marchant [2] whereby advantage is taken of the fact that the animal's body appears brighter towards the middle. This enables an approximate boundary to be established by thresholding at a relatively high level. We establish the threshold by sampling the average grey level in the hot spot with an animal present and setting the threshold to 0.8 times this value. An approximate outline is encoded with a Freeman chain code [12]. The boundary is then improved by searching outwards from each chain code point for a high value of image gradient. The new points are joined by minimising a weighted sum of distance between points and direction change [2]. This process is repeated once more to give the improved boundary, after which the tail is removed by eroding the area within the boundary. The erosion operation is equivalent to rolling a circle around the inside of the boundary and tracing out a new boundary which is the boundary of the area swept by the circle. Thus a region joined to the main body by an isthmus narrower than the circle diameter will be cut off. The diameter of the circle must be greater than the tail width but less than the diameter of other significant parts of the boundary. Note that the erosion operation will not affect concave parts of the boundary such as the junctions between the rump and abdomen that we wish to use for division of the body. A suitable value of eroding circle radius is five pixels. The original outline (less tail) is then recovered by dilating the outline by a similar amount.

13.4.3 Division into Rump and Abdomen

In the following, where the width of an object is used, it is measured with respect to some characteristic angle of the object such as its principal axis. Imagine the jaws of a vernier calliper parallel to the principal axis, which are closed together so they eventually touch the object. The width is the distance apart of the jaws.

First, the principal axis of the improved boundary is calculated. An animal width is then established using the calliper analogy above at the angle of the principal axis. Junctions between the rump, abdomen, and shoulder segments of the body are characterised by kinks (positions of high curvature) in the boundary. Initial estimates of the kink positions are made at distances along the principal axis equal to the width (for the rump/abdomen kinks) and twice the width (abdomen/shoulder kinks) and at right angles to the principal axis so that they are just inside the boundary. Location of the kinks requires an analysis of the curvature of the boundary in these areas. For this analysis the boundary must be smoothed to avoid problems caused by representing the image as discrete pixels. Also, because the kinks are sometimes rather subtle, we use a method of boundary refinement (a snake) capable of sub-pixel representation. The snake method was introduced by Kass *et al.* [13] and follows a boundary by simulating the mechanics of an elastic string. Normally a complete loop of elastic is used. Elastic forces keep the elements of the string together and produce a controllably smooth boundary, while image forces, from the grey level and the grey-level gradient, attract the string to the object edges. In this work we have modified the original concept and use a string of a finite length where the end points can move only in the image x direction [14]. Each of four snakes is initialised with its middle on the estimated kink point and its ends either side of the estimated point similarly just inside the improved boundary. After the snakes have converged the kink points are found as the points of maximum curvature. The improved boundary is divided into rump and abdomen components by joining the appropriate pairs of kinks.

13.4.4 Shoulder Estimation

When done by a human analyst, outlining the shoulder component is rather subjective. One would probably follow the shoulder boundaries until reaching the neck and then interpolate across the neck with a smooth curve. In our setup the shoulder is not always completely visible. We therefore attempt to estimate the shoulder extent with the data available, filling in the rest assisted by prior knowledge of pig shape.

A snake is initialised to stretch between the abdomen/shoulder kinks. An internal pressure is simulated [14,15] to expand the snake into a circular segment extending from the abdomen/shoulder junction forwards. Sufficient pressure is applied so that the area of the segment, in the absence of image forces, would be approximately the same area as the rump. The required pressure can be calculated automatically from the known area of the rump and the elasticity of the elastic string. The forces are then allowed to balance with the image forces until the snake comes to a rest position. The final result is that the snake clings to the shoulder boundaries but makes a smooth interpolation across the neck. Where the shoulder is partially obscured by the front wall of the feeder the algorithm makes as much use of the visible boundary as it can and estimates the rest with a smooth curve. The final division into rump, abdomen, and shoulder is shown in Figure 13.5(c).

13.4.5 Quality Control Checks

It is extremely difficult to design an image analysis algorithm to work on all images presented to it, especially when analysing natural objects. Our approach is to grab a large number of images and to reject the result if the analysis fails. This approach is only feasible if it is possible to detect failures. Fortunately, many of our failure modes produce results that are catastrophically wrong and are easily detected. The major checks are as follows:

a. Grey level at hotspot too low. Indicates no pig present.
b. Error in boundary improvement. Usually because the boundary tracker has not returned to the start point within a reasonable length of boundary. This commonly happens when the pig body merges with the sidewall in the image.
c. Improved boundary has wrong shape. The ratios of length to width and area to (perimeter squared) must be within certain bounds.
d. Improved boundary in wrong position. The x position of the centroid of the boundary must be approximately halfway between the sidewalls. This also happens when the pig body merges with the sidewall.
e. Error in initialising snakes for kinks. Part of a snake has been placed outside the improved boundary. This typically occurs if the animal has too much of its head in the feeder (while still passing check c above) whereupon the abdomen/shoulder kink point is placed too low in the image.
f. Kink not significant. The curvature analysis in the kink region has failed to find a point of sufficiently high curvature.

g. Initial shoulder radius too small. The diameter of the initial circular shoulder segment is less than the distance between the abdomen/shoulder kinks.

13.5 Experiments and Results

13.5.1 Experimental Arrangement

Image collection was carried out in a piggery with a pen measuring approximately 6m by 7m. The pen was equipped with a feed station supplied by Osborne (Europe) Ltd. The system identified each feeding pig by its electronic ear tag, and recorded the time. Sufficient natural lighting was available through daylight hours, and service lighting was available for husbandry operations. Initially, two low-level light sources were installed to allow images of acceptable quality to be collected day and night. One light source was positioned to the rear and above the feed station race, to illuminate the back of any pig using the feeder. The second light source was mounted inside the top of the feeder to illuminate the head and shoulder region of the feeding pig. These gave good illumination of the pigs inside the feed race until they reached an average weight of approximately 40kg. At this stage, the lighting pattern was improved by replacing the feeder and rear lamps with a single 60W lamp, positioned at the top rear of the feeder race and directed diagonally towards the feeder opening. This illuminated the top and rump area of the pig but not the race floor.

Ten female and ten male pigs were used for the experiment. They were taken from an outdoor breeding herd, directly upon weaning at between six and seven weeks old, and weighing on average 16kg. They were from Landrace X Duroc females, by a predominantly Large White boar.

A monochrome CCD camera was mounted on the top rail of the race, positioned centrally and looking vertically down at the race floor. In the early stages of the work, the camera lens was fixed at a height of 1.50m above the floor giving a field of view at ground level about 1.0 × 1.3m. After a time the height was changed to 1.42m and a different lens used giving a field of view about 1.2 × 1.6m. The rails of the feed station race were modified to permit the camera an unimpeded view of the floor area.

The output from the Osborne feeder and the CCD camera were connected to a PC fitted with a monochrome video frame grabbing board. The PC was loaded with the software provided by Osborne to interrogate the feed station and log and present results. About 1250 images were captured each day. In addition, the animals were weighed at weekly intervals using a conventional farm weigh crate.

13.5.2 Initial Filtering of the Data

Previous work has shown that the plan area of the animal without the head can be related to weight [2,16,17] and so we have presented the sum of the shoulder, abdomen, and rump apparent areas (denoted *A4*) plotted against animal age in Figure 13.6.

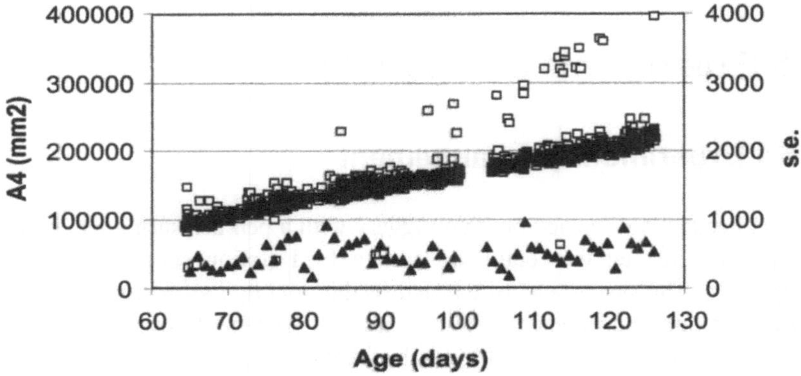

Figure 13.6. A4 area measurement and standard errors vs. age, raw data (□); median filtered data (■); s.e. (⊕). Note that a large number of data points appear superimposed on the central band.

This gives a visual impression of the spread of some typical data and its development over time. Due to an operational error there is a period of missing data from days 102–104. Notable are the outliers in the data where the image analysis has given a result that is clearly wrong but has passed the quality control checks. A feature of these outliers is that they generally occur singly or, more rarely, in groups of two. They can be removed therefore with a median filter with a window of length 5, e.g., a reading is grouped with the two preceding and two subsequent ones, the central reading is then replaced with the third in the rank order. This results in the rejection of up to two consecutive outliers with no effect on genuine readings as shown in Figure 13.6.

13.5.3 Image Analysis Repeatability

In this section we assess how much confidence can be placed in any set of image measurements. As there are a large number of readings (average of 38 per day for this pig) they can be grouped and averaged. After removing the outliers, we collect the data into daily sets and assume that the underlying trend of increasing weight is insignificant over one day. The one-day time period is a balance between smoothing out diurnal variations (e.g., lighting and animal metabolism) and resolving short-term changes in animal shape.

We can examine the repeatability of the image analysis by estimating the standard deviation of each daily average. In other words, if the standard error (s.e.) of the each day's readings is small then the image analysis procedure is adequately repeatable given the number of readings that can be taken during one day. More readings could be used to compensate for a less repeatable procedure and vice versa. Figure 13.6 includes a plot of the daily s.e. The scales have been chosen such that if the s.e. value is lower on the graph than the measurement, then the s.e. is less than 1% of the measurement. It can be seen that the standard errors are generally less than 0.5% for the early period. As the animal grows the error becomes progressively less in percentage terms.

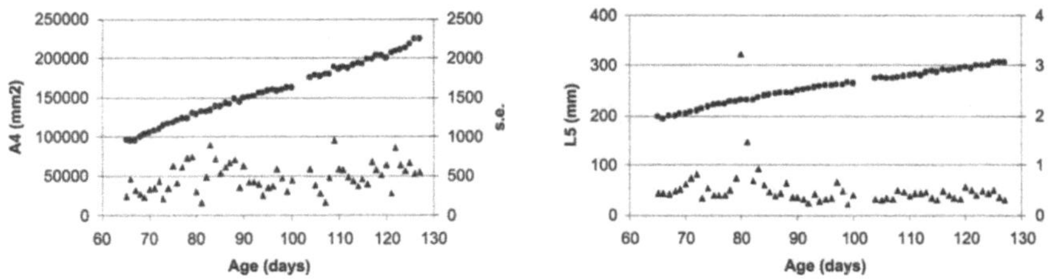

Figure 13.7. Daily averages (•) and standard errors (⊕) plotted against age: (a) *A4* area; (b) *L5* dimension.

Figure 13.7 shows plots of the daily averages and the standard errors for two typical dimensions measured, *A4*, the sum of shoulder, abdomen, and rump, and *L5*, the rump width. Generally the errors for area measurements are below 0.5% while for the linear measurements they are below 0.25%. Notable exceptions are the regions around the ages 80 and 109 days. At these times comparatively few images were collected and successfully analysed due to lighting problems, which were subsequently corrected. These periods gave relatively large standard errors. We conclude that, provided enough images are analysed (e.g., our average figure of 38 per day for each pig), the analysis is sufficiently reliable to move on to the next stage of estimating physical attributes from images.

13.5.4 Relationship with Weight

An important variable that users wish to estimate is animal weight. This can also be measured using manual methods and so forms a convenient quantity for testing the system. Previous work [2,16,17] has found a linear relationship between weight and *A4* area. This relationship is coincidental but enables simple statistical techniques to be used in when we estimate weight from area. The problem still remains however of finding the parameters of the fit.

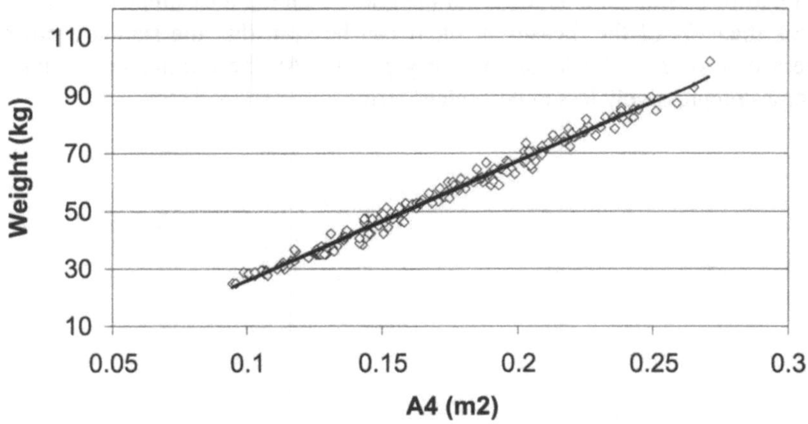

Figure 13.8. Relationship between area *A4* and manual weight for each pig (◊) with fitted line, $W = -15.6 + 411.3\,A4$.

Figure 13.8 shows the manually recorded weights plotted against the average of the *A4* measurement for the day of the weighing for all the pigs in the trial. A linear fit gives

$$W = -15.6 + 411.3\,A4 \tag{13.13}$$

where *W* is the weight in kg and *A4* is the area in m^2. We have carried out extensive statistical tests on our results [11]which show that the slope of fitted curves for the data for individual pigs is effectively constant (standard error 3.5 kg/m^2). However, there is some variation in the intercept for each pig. In fact if we use Equation 13.13 to predict the weight from *A4* for an individual pig (say pig no. 1493) we get figures that are consistently low by between 3 and 4kg. If we keep the slope constant and fit a line for the data of pig 1493 only, the intercept becomes –12.4 and predictions give errors of below 1kg. However, these fits only give parameters after trials have been done and may not be valid for the practical situation where a new batch of pigs needs to be monitored.

In practice we foresee that the area measurements will be used to predict pig weight on, say, a daily basis, along with less frequent manual weighings to calibrate for variations in shape between individual pigs. Carrying this to an extreme it should be possible to weigh manually at an early growth stage and calculate the intercept value for each pig. As the slope is considered to be the same for each pig (411.3), a pig-specific curve can then be used to calculate weight from *A4* measurements for the rest of the growth period. As an example, pig 1493 at 75 days old weighed 36 kg, and the intercept was calculated as –12.7. Using this to predict the weight at future ages gave the results in Figure 13.9 where the errors are no greater than 1.1kg.

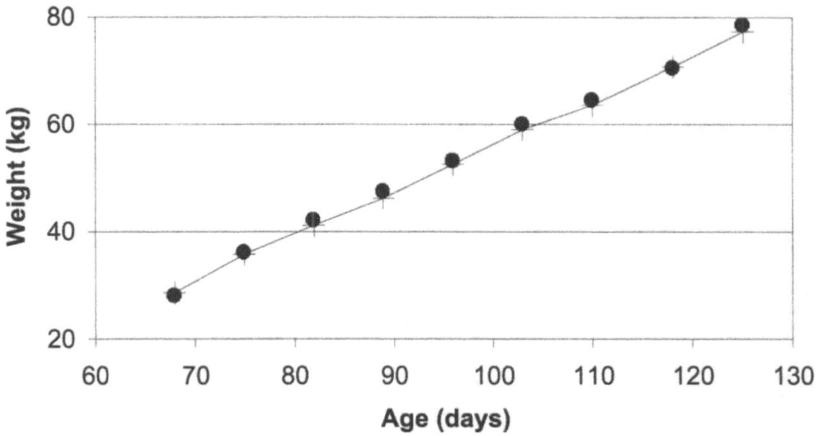

Figure 13.9. Example results for pig 1493 showing change with age of manual weight (•) and predicted weight (+) using intercept = –12.7 for *A4*.

13.6 Commercial Development

Close links with commercial partners have been maintained all through the development process. This has ensured that the system provides what the industry wants, and that the technology is both commercially viable and is made available to the customer as quickly as possible. Osborne (Europe) Ltd has collaborated with us from the testing of early prototypes, and now offers a commercial system for sale on the open market. This complements their existing range of pig monitoring and feeding equipment.

Development and programming of the novel imaging software has remained the responsibility of SRI. This software has been supplied to Osborne as a set of compiled programmes, which include packages for calibration of images and for processing these images to measure linear dimensions and areas as output.

Osborne has integrated the compiled programmes into a commercially viable package. In developing this, Osborne combined calibration, process monitoring, data analysis and data reduction as well as presentation and record keeping processes within a computational framework, which enables communication with the user via a Graphical User Interface (GUI). An example of the station set-up process-monitoring screen is presented in Figure 13.10. As well as preparing the GUI, Osborne developed facilities for integrating the imaging software with existing hardware, which can provide animal identification, feed intake and eating times. When this is added to the information from the imaging software, it provides a comprehensive, near real-time record of the growth and feeding behaviour of each pig being monitored.

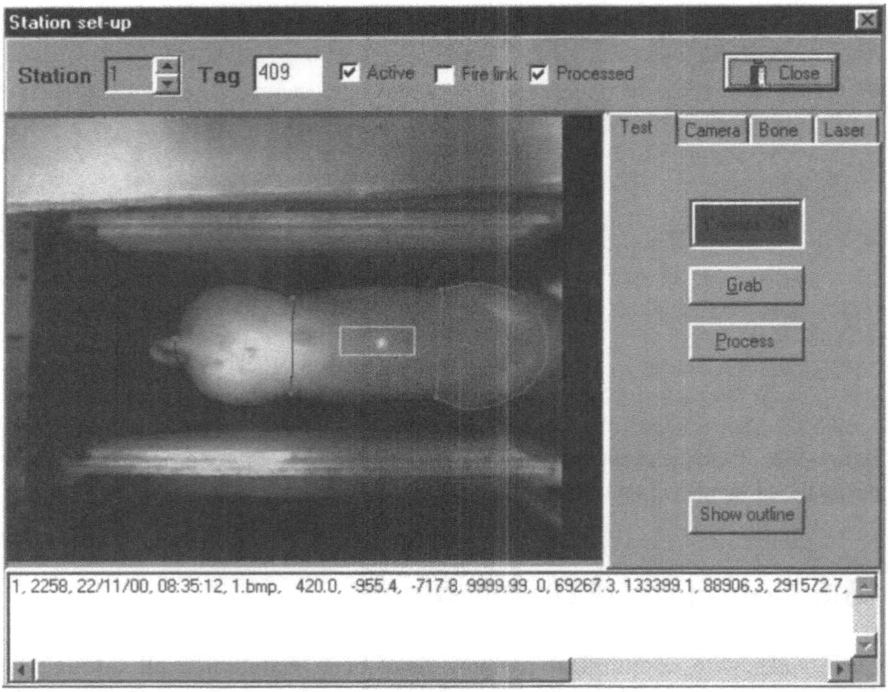

Figure 13.10. Example of a window from the Osborne graphical user interface.

13.7 Conclusions

We have presented work on a system for estimating the size and shape of growing pigs. The system uses machine vision to make various area and linear dimension measurements from a plan view of each animal. In order to obtain views of the animals in a reasonably constrained attitude and posture, the camera was sited over a feeding station that also identified each animal.

The machine vision system consists of a number of algorithms designed to obtain a clear outline of the animal's plan view and to detect features in the outline that are used to divide the body into specific areas. The system includes a number of quality control checks to minimise the amount of erroneous data entering the next phase. Outlier rejection techniques are then used to further improve the quality of the data.

We have described the method of removing lens distortion and determining the magnification. This, along with a method of measuring animal height and compensating for the curvature of the animals' flanks, allows us to estimate dimensions in real units. Previous work, supported by data presented here, leads us to conclude that weight can be related to a particular area measurement in a linear way. The slope of the relationship is constant for pigs of a particular strain but the intercept varies slightly from pig to pig. For accuracy of weight estimation we propose that each animal is weighed manually once, early on in its growth period, in order to establish the intercept.

The system has been developed commercially by Osborne (Europe) Ltd who have provided facilities for integrating the imaging software with existing hardware which can provide animal identification, feed intake and eating times. This package is controlled with a graphical user interface for ease of use. When this is added to the information from the imaging software, it provides a comprehensive, near real-time record of the growth and feeding behaviour of each pig being monitored.

13.8 References

[1] Frost AR, Schofield CP, Beaulah SA, Mottram TT, Lines JA, Wathes CM. A review of livestock monitoring and the need for integrated systems, *Computers and Electronics in Agriculture* 17: 139–157.

[2] Schofield CP, Marchant JA (1990) Image analysis for estimating the weight of live animals, Proceedings of Optics in Agriculture conference, SPIE, Boston MA, 7–8 Nov., pp. 209–219.

[3] Van der Stuyft E, Schofield CP, Randall JM, Wambacq P, Goedseels V (1991) Development and application of computer vision systems for use in livestock production, *Computers and Electronics in Agriculture* 6: 243–265.

[4] Minagawa H, Ichikawa T (1992) Measurements of pigs' weights by an image analysis, Paper No. 927023 Summer Meeting, American Society of Agricultural Engineers, Charlotte, June 21–24.

[5] Schofield CP (1993) Image analysis for non-intrusive weight and activity monitoring, Proceedings of the 4th International Symposium on Livestock Environment, ASAE, Univ of Warwick, pp. 503–510.

[6] Brandl N, Jorgensen E (1996) Determination of live weight of pigs from dimensions measured using image analysis, *Computers and Electronics in Agriculture* 15: 57–72.

[7] Whittemore CT, Schofield CP (2000) A case for size and shape scaling for understanding nutrient use in breeding sows and growing pigs, *Livestock Production Science* 65: 203–208.

[8] Van der Stuyft E, Goedseels V, Geers R (1990) Digital restoration of distorted geometric features of pigs, Proceedings of Optics in Agriculture conference, SPIE, Boston MA, 7–8 Nov., pp. 189–200.

[9] Tsai RY (1986) An efficient and accurate camera calibration technique for 3D machine vision, Proceedings of Computer Vision and Pattern Recognition conference, IEEE Computer Society, Miami Beach, pp. 364–374.

[10] ASAE (1987) *Handbook, American Society of Agricultural Engineers*, St Joseph pp. 391–398.

[11] Marchant JA, Schofield CP, White RP (1999) Pig growth and conformation monitoring using image analysis, *Animal Science* 68: 141–150.

[12] Davies ER (1990) *Machine Vision: Theory, Algorithms, Practicalities*, Academic Press, London.

[13] Kass M. Witkin A, Terzopoulos D (1988) Snakes: Active contour models, *International Journal of Computer Vision* 1: 321–331.
[14] Marchant JA, Onyango CM, Tillett RD (1995) A compartmented snake for model-based boundary location. In: Gill CA Mardia KV (eds) *Current Issues in Statistical Shape Analysis*, Leeds University Press, Leeds, pp. 100–107.
[15] Cohen LD, Cohen I (1990) A finite element method applied to new active contour models and 3D reconstruction from cross sections, Proceedings of the 3rd International Conference on Computer Vision, IEEE Computer Society, Osaka, pp. 587–591.
[16] Schofield CP (1993) Evaluation of image analysis as a means of estimating the weight of pigs, *Journal of Agricultural Engineering Research* 47: 287–296.
[17] Schofield CP, Marchant JA (1996) Measuring the size and shape of pigs using image analysis, Paper No. 96G-035 Ag Eng 96, European Society of Agricultural Engineers, Madrid, Sept. 23–26.

Chapter 14

Editorial Introduction

The examination and grading of sheep pelts illustrates almost all of the general points made in the earlier chapters regarding the inspection of natural products. The pelts are flexible and vary in size, thickness, shape, texture and colour. There is a broad range of ill-defined defects, which cannot be defined objectively using rule-based recognition criteria. Moreover, the significance of a flaw depends upon both its type and location on the pelt, so defect identification is important. Acceptable natural variations in the colour and texture of the surface must be tolerated, while certain types of defects render part of the pelt unusable. In this type of situation, Pattern Recognition methods are useful. However, the ubiquitous problem of obtaining a representative and reliably labelled set of training and test images remains. The input to the defect classifier is obtained from a flying-spot UV laser scanner. This particular means of image acquisition provides three precisely registered images derived from the transmission, reflection and fluorescence properties of the pelt. These images can be processed individually or merged, as most appropriate. This allows the ready identification of certain types of defect. However, in other cases, it is not easy to design recognition criteria, which is the reason for using self-adaptive learning techniques. In view of the high computational load imposed by the image processing and analysis, a multi-processor system, based on high-speed DSP (Digital Signal Processing) hardware was devised. By taking a broad systems-oriented view of the operational requirements, it was realised that a great deal of benefit could be obtained by adding a machine-readable code uniquely identifying each pelt. Hardware to imprint this code had to be integrated with the pelt-transport mechanism, laser scanner and special-purpose image processing hardware. All of this equipment must be safe and easy to use. It must ruggedised so that it can withstand the hostile working environment and the high-power UV laser requires careful screening and safety cutout switches to protect workers. Speech recognition is employed in the user interface, enabling hands-free operation. Clearly, a multi-discipline team is needed to design a system of this complexity. This requires significant financial investment, which can only be justified if there is an obvious and long-lasting need for such a vision system. It is all too easy for a plant manager to continue using tried and trusted methods (e.g., human inspectors), rather than investing in novel, expensive and "unproven" vision hardware. Indeed, we must always expect a battle when trying to convince a sceptical manager of the effectiveness of a complex vision system. Demonstrating the commercial viability of a project like this can be just as demanding as the technical problems that have to be overcome. For this reason, good systems-level design is essential from the beginning of any machine vision project.

Chapter 14

Sheep Pelt Inspection

P. Hilton, W. Power, M. Hayes and C. Bowman

14.1 Introduction

New Zealand is the largest exporter of sheep pelts in the world and supplies about 40% of the world market. Before being exported, the raw pelts have the wool removed and are then preserved by a pickling process. Subsequently, they are further processed and tanned before being turned into a wide variety of leather goods around the world.

At present, the grading of these pickled pelts is manual, and thus subjective, resulting in imprecise classifications. Since there is a significant difference in value between pelts of the highest grade and lower grades, studies have indicated that more accurate and consistent grading would provide a better match to market requirements resulting in an increase in the export value of the pelts.

Another issue facing fellmongers is the difficulty of reliably tracing the origins of any pelt. Once a skin is separated from the animal, it is impossible to determine from which animal and farm it came. Consequently, farmers are paid purely on quantity of skins rather than quality—thereby removing an incentive for them to care for their stock in such a way to produce a premium product. A similar situation exists when various meat works provide skins to independent fellmongers. This provides a strong motivation for the industry to develop a technique for uniquely tagging each skin at source and tracking it through the pickling process.

This chapter describes a range of technologies, developed by the authors and their colleagues, aimed at improving the grading and identification of pickled sheep-pelts. Because of the unique characteristics of this variable natural product and the features and measurements required, the application has necessitated a systems approach whereby several complementary technologies have been developed and integrated into a prototype grading and identification machine.

The work started in 1991, with the concept being to provide enhanced pelt images for operator grading assistance, rather than fully automated grading which was considered far too difficult. Early in the project, laser-induced fluorescence was identified as a useful technique for highlighting certain types of pelt defect that are otherwise invisible or very difficult to see with the naked eye. Such defects can become apparent during later processing.

An experimental MKI system was built in 1993/94 to assess the capabilities of laser-induced fluorescence imaging in this application. Design parameters were then established for a second system, which could:

- spread the pelt;
- acquire in real time an image of each pelt for viewing by the grader to enhance the visibility of those faults which fluoresce;
- read a pelt batch number branded system on the raw pelt;
- measure the pelt area consistently.

The MKII system was built in 1995/96 and trialed in a fellmongery.

In parallel with the trial system development, a number of techniques useful for automated pelt grading and identification have been developed, some of which are described in this chapter. A robust and reliable means of branding each pelt for identification purposes has been developed as have a coding method and machine vision algorithms for reading the applied codes. We have investigated in some depth the fluorescence properties of pelts and defects and developed pattern recognition algorithms to assess the viability of automated detection and classification of the various pelt defects.

14.2 Pelt Defects

There are three main sources of pelt defects:

- *On-farm defects:*
  - *seed marks:* caused by seed sticking into the flesh during the animal's lifetime;
  - *scars:* caused by a variety of mechanisms, such as shearing, barbed-wire fences and dog bites;
  - *cockle:* an allergic rash caused by a tick on the sheep. This defect is often difficult to see with the naked eye at the pickled stage and only becomes apparent after subsequent processing;
  - *fly strike:* caused by fly maggots in the sheep's flesh. This appears as puckered scarring.
- *Meat works or butcher damage:*
  - *strain:* a splitting of the top layer of the skin from the bottom layer;
  - *shape defects:* damage caused by butchering and processing actions that remove sections of the pelt;
  - *grain:* similar to strain;
  - *knife scores:* butcher damage which causes scores in the fleshy side of the pelt. This can lead to tearing.
- *Fellmongery or processing faults:*
  - *holes:* which can be caused at various stages in the processing;
  - *poor fleshing:* caused by inadequate removal of carcass flesh from the skin;

There are other sources of defects including breed related faults, for example:

- *ribbing:* a breed-fault commonly found in Merinos and consisting of a series of pronounced folds in the skin;
- *pinholes:* very small holes (around 200 microns in diameter) in the upper skin layer caused by clumping of wool follicles.

Laser-scanned images of various defects are presented later in this chapter in Section 14.4. The diverse range of defect types of widely differing visual appearance poses significant challenges for any automated visual inspection system.

14.3 Pelt Grading System

The pelt grading system comprises a number of complimentary modules, which together facilitate the inspection and grading task. Figure 14.1 shows the system layout.

The system uses a conveyor to transport the pelts and requires two operators: one feeding the pelts onto the conveyor and the other, located at the other end of the conveyor, grading the pelts.

As each pelt passes along the conveyor, it is first spread flat by the spreader. A roller clamps the leading edge firmly while a helical roller spreads over the pelt, flattening it. This ensures that the pelt lies flat on the conveyor, which enables consistent measurement of pelt area and thickness. The spreader can handle sheep and lamb pelts of varying sizes and thickness and achieves consistent spreading at a throughput of 15 pelts per minute.

After spreading, the pelt substance or thickness is measured. This is accomplished using a substance wheel, which is mounted over the conveyor, immediately after the spreader. The wheel rolls over the pelt as it travels down the conveyor. The measurement wheel is pivoted and has a linear displacement transducer, which records changes in height with pelt thickness. There is a downward force applied to the wheel, partially squashing the pelt. The resistance of the pelt to this squashing is a better measure of the amount of "body" in the pelt than thickness alone. Hence, the term substance is used by the industry rather than simple thickness.

The laser-imaging unit is in a separate bridge that can be rolled into place and bolted to the conveyor. The laser imager acquires three images of the pelt: reflection, transmission and fluorescence. The conveyor uses a translucent belt that enables laser light to reach a detector rail mounted underneath the belt. This provides for the acquisition of the transmission image. The area of the pelt is calculated using the transmission image, which is also used to locate and decode a unique identification pattern previously branded on each pelt. The fluorescence image is displayed to the grader to provide him with pictorial information on otherwise invisible defects.

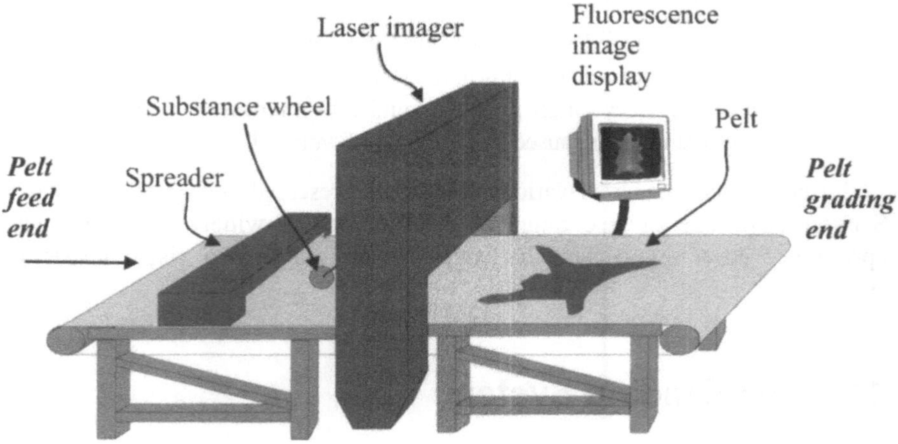

Figure 14.1. Pelt grading system.

As the pelt emerges from the laser imager, the grader visually inspects the pelt and makes a grading decision based on its appearance and the displayed fluorescence image. This grade decision is then entered on an operator console.

Pelt substance, area, grade and ID code are logged in a file, which is date and time stamped. The system has two Pentium computers and three digital signal processors to acquire and display the image data, calculate the pelt area, decode the identification symbol, and collate and log pelt characteristics such as grade, area and substance.

The environment in fellmongeries is extremely hostile to electronics and photonic equipment with hydrogen sulphide and sulphuric acid present. To withstand this environment, the electronic systems are in an enclosure pressurised with a continuously filtered and scrubbed air supply. The conveyor is constructed of fibreglass and stainless steel.

14.3.1 Laser Imaging

The imaging and automated detection of such a diversity of defect types on a variable natural product presents a significant challenge for a machine vision system. As with all machine vision applications, the choice of imaging technique has a profound effect on the success of subsequent data processing. In this application, conventional camera imaging techniques cannot highlight all of the defects of interest.

In this application, there are two main advantages for using a laser scanner, rather than a camera. Firstly, laser-induced fluorescence imaging has the ability to highlight defects not easily discernible using other imaging mechanisms. Secondly, a laser scanner can simultaneously acquire perfectly registered reflection, transmission and fluorescence images, which can be used either individually or in combination to detect and discriminate between different pelt defects.

14.3.2 Laser Imager

The laser imager combines advanced optical, electronic, and computing technologies.

A simplified cut-away diagram of the laser imager is shown in Figure 14.2. The laser and scanner are shown along with the relative positions of the three imaging rails.

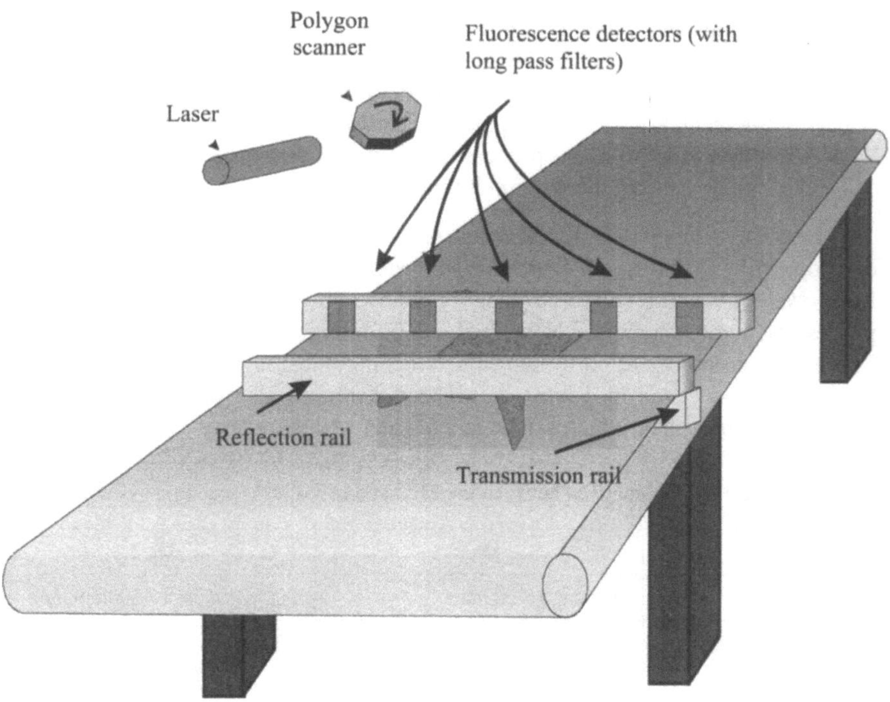

Figure 14.2. Laser scanner.

The rotating polygonal mirror produces a continuous sequence of laser-scan lines across the conveyor. As the pelt passes along the conveyor under the laser scanner, the three pelt images are acquired by simultaneously sampling the output of each of the three detection rails at multiple points across the laser scan. Each rail consists of multiple sensors (up to 60) connected in parallel. The rail output is the combined signal level from the light collected by each of the sensors. Normally, only two or three of the sensors have appreciable output at any one time depending on the location of the laser spot.

The reflection image is obtained using back-scattered light gathered from the reflection rail mounted above the conveyor. The transmission rail is mounted under the conveyor and collects light passing through both the pelt and the translucent belt. The fluorescence image is obtained using another rail consisting of five highly sensitive Photo Multiplier Tube (PMT) detectors mounted above the conveyor.

Optical filtering is used to block the laser light to these detectors while passing the fluorescence wavelengths.

Core technologies used in the scanner are:

- high power air-cooled argon ion laser;
- advanced acoustic-optic modulation devices and customised polygonal mirror scanning hardware to enable high scan rates (high-resolution imaging and high inspection throughput);
- sensitive miniaturised PMTs;
- advanced, high-speed digital down-converter electronics to enable digital filtering and demodulation to maximise signal to noise and system bandwidth;
- extensive use of field programmable logic arrays to enable rapid implementation and development of hardware design;
- multiple digital signal processors to control and process the large amount of image data acquired per pelt (6 Megabytes).

There have been two implementations of the system: a Mk I experimental development [1] and a Mk II prototype system. The main driving factor for the MKII system was the increased imaging bandwidth required to meet the real-time processing requirements of New Zealand fellmongeries of 15 pelts per minute. There were a number of performance and technological improvements between the two systems as listed in Table 14.1. The differences also serve to illustrate the alternative approaches, which can be taken when designing laser-scanning systems.

Table 14.1 Comparison between the MKI and MKII laser scanning systems.

	Mk I	**Mk II**
Imaging Area	1000mm × 1500mm	1300mm wide × 1500mm long
Pelt Spacing	2000mm	2000mm
Throughput rate	3 pelt.min^{-1}	15 pelts.min^{-1}
Scanning method	Galvanometer	Rotating mirror
Scan frequency	100Hz	500Hz, scan, active data rate
Laser spot size	1mm	3–4mm at 1/e^2 power point (with 1mm pixel resolution achieved through interpolation).
Laser modulator	Electro-optic	Acoustic-optic
Pixel bandwidth	70kHz per image	640 kHz for each image
Pixel size	1mm wide by 1mm long	1mm wide by 1mm long
Grey scale	6 bits R, 5 bits F and 6 bits T	8 bits each image
Image size in pixels	1024 wide × 1024 long	1300 wide × 1500 long

A high-power air-cooled argon ion gas laser is used, operating with multiple spectral lines from 514nm to 454nm. This provides sufficient optical power for the reflection and transmission images and excitation energy for the fluorescence image. Most of the detected fluorescence comes from excitation at 514nm (the longest wavelength) with most of the rest of the excited fluorescence light being filtered out with a long pass filter.

The optical assembly is mounted on a flat aluminium plate that is resistant to corrosion and rigid enough to avoid undesirable deflections of the beam. This criterion also applies to the whole laser enclosure relative to the pelt surface. The image spot must not deviate due to vibrations or flexing by more than 0.5mm at the pelt surface.

14.3.3 Pelt Images

Three perfectly registered images, representing reflection, transmission and fluorescent properties of the pelts, are acquired simultaneously by sampling the three different detector signals during the laser scan. Examples of the different types of images for the same pelt are shown in Figures 14.3 to 14.5.

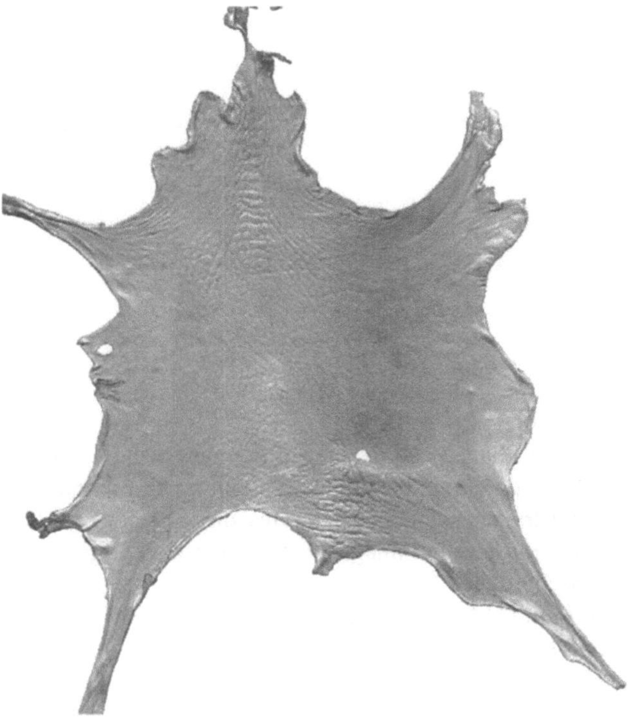

Figure 14.3. Reflection image.

The reflection image gives a view of the pelt similar to that seen by a grader under diffuse top lighting conditions. This is useful to get an overall impression of the quality of the pelt. Note the very bad scar damage and hole on the lower centre-right region in Figure 14.3.

Holes and variations in pelt thickness are better seen in the transmission image as shown in Figure 14.4.

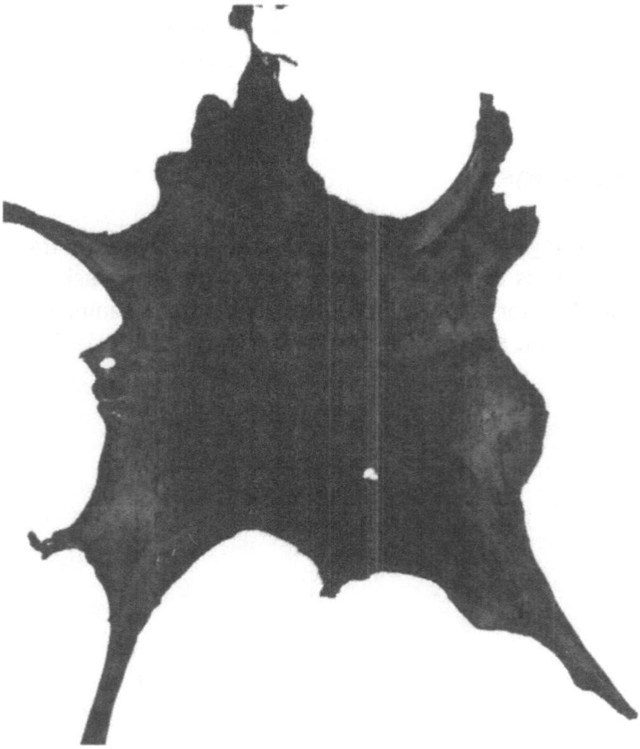

Figure 14.4. Transmission image.

The transmission image is equivalent to the view, which would be seen if the pelt is held up against a strong backlight. Without a pelt present, the level of laser light reaching the transmission rail saturates the detectors resulting in a white background to the transmission image. Hence there is very good contrast between the pelt and the background.

Note that in Figure 14.4, the hole can be clearly identified. A knife score can also be seen as a sweeping circular mark on the front right leg of the pelt (upper right-hand quadrant in the image).

The fluorescence image is shown in Figure 14.5. This image can highlight defects, such as cockle, that are very hard to see with the naked eye. In Figure 14.5, there is a large cockle spot on the lower right flank and some localised cockle on the neck of the pelt (near the top of the image). There is also severe fly strike around the butt region.

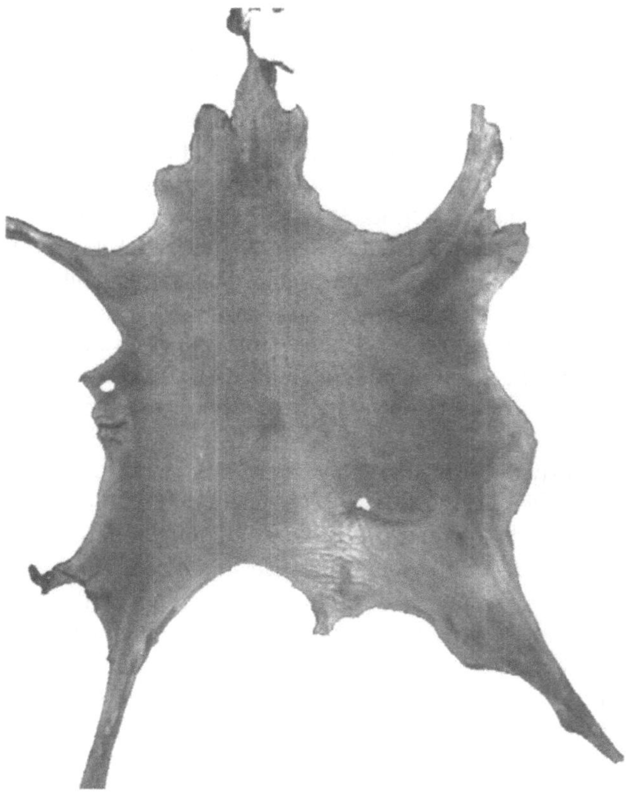

Figure 14.5. Fluorescence image.

The fact that the pixels are perfectly registered in each image is exploited in the multi-band pattern recognition algorithms described later.

14.3.4 Processing System Architecture

The data processing module comprises two PC computers running the Linux operating system. The PCs host three TMS320C30 Digital Signal Processing (DSP) cards, one for each image acquisition channel (the fluorescence DSP is installed in one of the host PCs; the reflection and transmission DSPs are in the other PC). The DSPs handle the time-critical functions of the system such as the image acquisition. Non-time-critical functions such as the image display, operator interface, substance monitoring, recognition and decoding of the pelt identification marks, monitoring of the laser operating conditions, and data logging are performed by the host computers.

The DSPs provide the interface with the system hardware and are configured identically except that one of them generates the actual laser modulation and polygonal scanner control signals. They are synchronised using start-of-scan and end-of-scan signals generated from photodiodes on either side of the conveyors and are triggered by a start-of-pelt signal from an opto-interrupter sensor. The outputs of the photodetector channels are demodulated using digital down-converters (DDCs) tuned to the laser modulation frequency. The quadrature base-band signals from the DDCs are transmitted to the DSPs, which compute the magnitude of each pixel sample using a lookup table. These are then assembled into image rows and transmitted to the host PC for further processing.

The DSPs run a multithreaded real-time operating system with a Unix-like applications programmer's interface to simplify porting of software developed on the host computers to the DSPs. Device drivers on the DSP control the polygonal scanner, the digital down-converters, and the shaft encoder monitoring the conveyor movement. Communication between the DSP and host is via a pair of device drivers that implement a set of virtual pipes using a shared dual port memory. These pipes, in conjunction with daemon processes running on the host, allow the DSP software to read or write any file or device mounted on the host file system.

The host software is written as a number of processes that communicate via the Unix socket interface using the TCP/IP protocols. The user interface program is based around a simple interpreter that parses a configuration script upon start-up. For each image acquisition channel specified in the configuration script, a server process is started to communicate with its corresponding DSP. These image server processes communicate with an X-windows server running on one of the hosts for the image display. Other processes are then started to perform the recognition of the pelt identification codes and to interface with the pelt substance measurement sub-system. Once the system is configured, the user interface program operates as a finite state machine to co-ordinate these processes and to log data. At the end of each day of operation, the data that has been logged is transferred from the factory to the office via a modem. This link also allows the system to be accessed remotely for software upgrades and system maintenance.

14.3.5 Trials

The system was trialed on-line in one of the major meat works in the South Island of New Zealand in 1997. Figure 14.6 shows the system ready to be transported to the fellmongery for trialing. During 11 months of batch operation, a large database of pelt images was acquired, the performance of the various systems was tested and a trial of pelt ID recording carried out. At this stage the system did not have automatic pelt brand reading software, so the trials used a code that was also human readable. The pelt was spread, substance measured, imaged and graded by a factory grader. The branded identification mark was read by another operator and spoken out aloud. Voice recognition software then interpreted the spoken code and transferred it to the pelt system computers to be added to the log file.

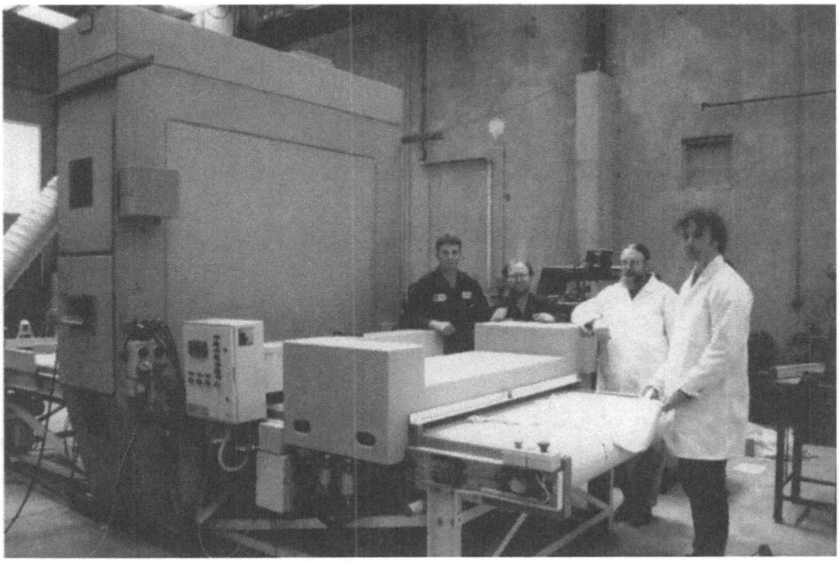

Figure 14.6. Pelt inspection system ready for trialing.

14.4 Automated Defect Recognition and Classification

14.4.1 Defect Appearance

Some of the defects are in the form of broadly circular spots, others have strong edge features, while the remainder have reasonably distinct intensity and textural properties. The following images show how some of the defects appear in the laser images.

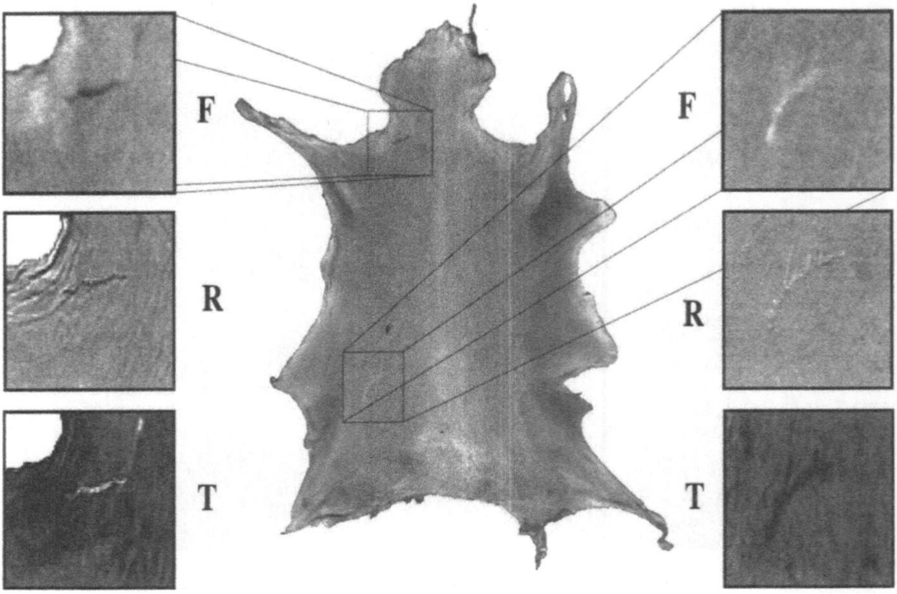

Figure 14.7. Strain (left) and scar (right) defects. F = Fluorescence image, R = Reflection image and T = Transmission image.

Figure 14.7 shows two different types of defect as they appear in the fluorescence, reflection and transmission images. The images on the left show a strain mark that appears as a long thin feature, dark in the fluorescence and reflection images but light in the transmission image. The strain defect is caused by the top layer of skin splitting during processing to expose the underlying layers. This results in a variation in thickness that can be readily seen in the transmission image. The defect appears dark in the other images, as the underlying layers are darker. Note the lack of fluorescence of the underlying layers.

The images on the right show a healed scar. Here, the skin tissue has become thicker during healing and so appears in the transmission image as a darker long thin feature. The scar is quite difficult to see in the reflection image, as the scar tissue is much the same colour as the surrounding skin. However, the fluorescence image shows a concentration of fluorescence within the scar tissue. This is a general property found in most defects resulting from a healing response in the animal, i.e., defects that cause skin damage before slaughter. This is believed to be due to some agent in the healing process producing additional fluorophores at the wound location.

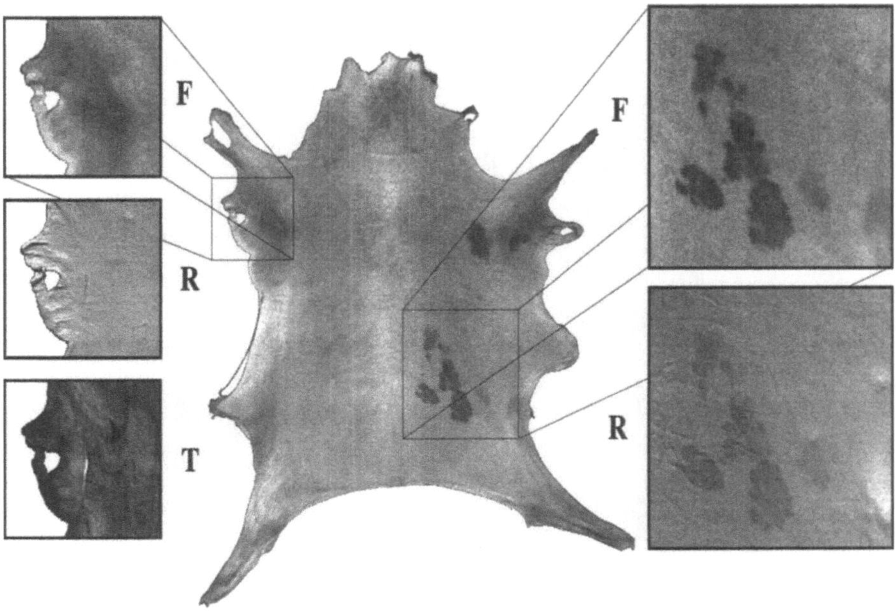

Figure 14.8. Knife score (left) and pigmentation (right) defects

The defects shown in Figure 14.8 are knife score on the left and pigmentation marks on the right. The knife score is a butchering fault caused by workers during carcass dressing operations on the processing chain. As it is primarily a variation in thickness, it is best seen in the transmission image. The pigmentation marks shown in the set of images on the right are caused at birth and not considered a defect for downgrading the pelt as the pigmentation marks are not visible after processing into leather. The pigmentation marks greatly reduce the pelt fluorescence so are best seen in that image. The transmission image contains little information on pigmentation and is not shown.

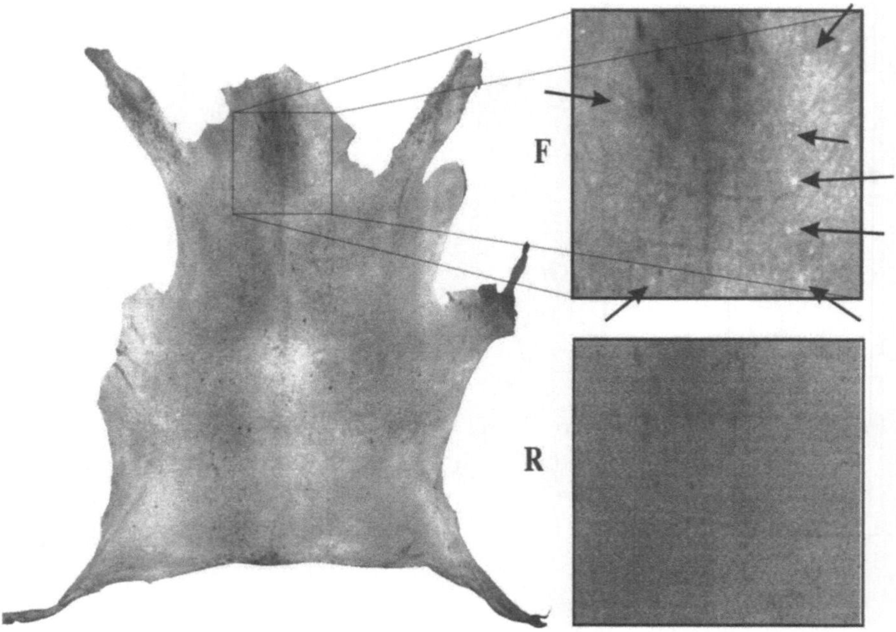

Figure 14.9. Cockle defects

The defects shown in Figure 14.9 are called 'cockle' which is one of the worst defects affecting pelt quality. Cockle is caused by an allergic reaction to a mite infestation during the lifetime of the animal and results in a rash on the skin, which concentrates the background fluorescence. This can be seen in the enlarged image on the top right that has several white spots corresponding to cockle. Note that in the corresponding section of the reflection image, the cockle is invisible. This is what the grader would normally see which explains why cockle is so hard for graders to identify.

As can be seen, many of the defects appear differently in the reflection, transmission and fluorescence images. Defects that produce a variation in pelt thickness are most readily apparent in the transmission image. These include: holes, dog bites, old scarring, knife scores, strain and poor fleshing. Defects resulting from recent damage to the skin are most easily detected in the fluorescence image. Other defects most apparent in the fluorescence image include cockle, ribbing and fly strike. Cockle is particularly important, as it is almost invisible to the naked eye at the pickled stage, but can have severe visual impact in the finished leather. Defect appearance in the reflection image is much as it would appear to the human eye or camera under diffuse top lighting.

The differing appearance of the defects in the various images can be used to distinguish between them. For example, strain and pigmentation marks both appear dark in the fluorescence and reflection images. However, strain, unlike pigmentation, results in a variation in thickness, which is apparent in the transmission image. Such differences are used as the basis for the defect detection and classification strategies described in a following section.

14.4.2 Supervised Learning

While there have been a number of studies of defect detection on processed leather images [4 – 9], we are not aware of any other work on wet pickled pelts.

Blob and edge detection techniques, applied to the most appropriate single pelt image, are adequate for some simple defects, but are inadequate for other more complex defects. Since each pelt is represented by three perfectly registered images, multi-band pattern recognition techniques become applicable. It is convenient to consider these three images as pseudo-colour RGB images. A basic arbitrary assignment of Fluorescence = R, Reflection = G and Transmission = B has been made. All the information in the images is then displayable using standard colour image viewers. Most of the defects can be recognised by their reasonably distinct intensity and textural properties in the three image channels. Thus, their detection is amenable to a supervised-learning approach [10,11,12].

The principle is as follows: A segment (e.g., a defect) of a given image can be outlined and classified by a human expert. Using a collection of such images (the training set), the learning program is trained to recognise similar instances in a set of images it has not seen before (the test set).

Before we outline how the program may be trained, it is important to note a number of pitfalls such an approach can have.

One is that that ground truth of any such system is only as accurate or consistent as the human expert or experts who train it. In the case of the New Zealand pelt industry, different regions of the country have different names for the same defect. Regional differences also occur, with defects having the same name but having significantly different colour and textural appearances. Similar differences were even traceable to individual graders in the same region. Hence, it was difficult for us to obtain consistent human grading information for training of the classifiers.

There is also the problem that the grading of a pelt is done by the human expert on the actual pelt in normal lighting conditions in a factory, whereas our system is required to classify the defects from the laser-scanned images. The human pelt graders are unused to classifying defects from images.

Perhaps the most fundamental problem for the pelt images is that which also applies to all natural images, namely limited scope of the training sets. We can never be sure that we have encompassed the full range of possible images, features and defects. It remains the challenge of a truly intelligent system to cope with a significantly atypical example. Thus, while our system does perform moderately well within the confines of the small training and testing sets so far available, its performance on a larger, less constrained sets implicit in commercial use has yet to demonstrated.

MBIS

MBIS (Multi Band Image Segmentation) is a software package developed by the authors for investigating and trialing a feature-vector supervised-learning approach to pattern recognition. It has also been used in other applications [13].

Training Phase. The expert examines each pelt image using a suitable image-editing program. Using the segmentation tools provided, he or she hand-segments all the classifiable features of the image and assigns each a class number from a list of possible features. The selected image areas (blobs) are copied to a grey-scale mask image, with each blob painted with its class number. The image mask is then saved.

When a set of images has been so annotated, it may then be processed in batch to produce a single large file of feature vectors. MBIS has a total of 22 feature-vector producing options. Five options are multi-band based and others relate to intensity, gradient or texture information in a given image plane. MBIS also provides some Markov Random Field (MRF) [12] options in a range of window sizes, currently spanning from 1×1 to 5×5 pixels, for each image plane. In the pure Markov options, all pixels in all planes of the image are placed equally into the feature vector and this is passed unaltered to the classifier. In the "mixed" Markov options, some or all of the pixels can first be transformed by a variety of algorithms based on emphasising intensity, edge, texture and, where appropriate, pseudo-colour information. Some of the second order textural options [15–18] are typically too computationally intensive to be a practical proposition, whereas the simple 3×3 or 5×5 Markov random field options are fast and typically perform well.

Classifiers Used. MBIS uses three classifiers: Bayesian (maximum likelihood), Decision Tree [14], and K Nearest Neighbours (KNN). For the large files typically generated, the latter requires too much processing time and we have concentrated mainly on the first two. In practice, there is little to separate the performance of these, but we tend to favour the maximum likelihood on the grounds of its firm statistical basis. We do not, however, use *a priori* biasing. Thus, frequency of occurrence of a class in an image is not used to push a split decision towards or away from it. This is because some defects tend to be infrequent, whereas it is important that they should still be identified.

Performance. The performance of the better performing of the 22 MBIS feature-vector options is summarised in Table 14.2.

The number of individual pixel samples used in the training and test sets was about 50,000, derived from various regions of about 100 pelt images. The test set was a randomly chosen set of 25,000 samples from the above data.

Generally, MBIS scored well in classifying the various defects. However, as can be seen, cockle, knife score and scar detection using MBIS gave mixed results. However, standard blob and edge detection techniques can be used to detect many such defects.

Table 14.2 Percentage score for the classifier to correctly identify a class in the test set.

Feature/Option No	1	7	10	11	12	20	21	22
Scar	92.3	94.5	96.4	62.7	88.9	94.5	92.5	92.1
Knife	85.3	90.8	94.1	63.5	82.4	90.1	86.8	88.6
Strain	99.7	99.6	99.8	98.2	99.3	99.6	99.5	99.6
Cockle	47.6	57.1	71.4	43.3	61.9	57.1	57.1	52.4
Good	99.8	99.9	99.9	99.3	99.8	99.9	99.8	99.8
Pigmentation	99.9	99.8	99.9	99.3	99.8	99.9	99.8	99.8
Fly Strike	94.9	95.5	96.6	72.0	89.3	95.2	94.3	94.4
Stain	97.8	98.8	98.7	90.5	96.3	98.8	98.0	98.3
Ribbing	96.7	97.8	98.4	83.4	94.0	97.5	96.3	96.5
Mottle	96.9	98.1	98.6	89.2	93.7	98.2	97.2	97.7
Background	100.0	100.0	100.0	100.0	100.0	100.0	100.0	100.0
Area cockle	96.5	98.6	100	95.0	98.6	99.3	96.5	97.9
Micotic Dermatitis	98.9	99.1	99.0	94.9	98.7	99.6	98.7	98.9
Patch	96.9	98.9	98.2	87.8	97.3	98.5	97.3	96.8

14.5 Pelt Identification

A reliable means of tagging each pelt with an identification mark allows the pelt to be tracked through the various processing stages and its origins to be established at any time. This provides a means by which farmers and meat processing works can be paid on the basis of final pelt quality instead of just by the number of pelts supplied. This in turn provides a better incentive for the farmers to care for the sheep and for the processors to carefully remove the skins, and leads to an overall increase in value of the product.

The pelt identification requirements necessitate at least 10,000 code combinations with a readability rate at the final grading stage of better than 90%. The ID mark needs to be applied preferably in the meat works with an application rate fast enough to keep up with the chain speed of around 15 animals per minute.

Previous approaches have used a separate plastic tag, which is mechanically attached to the pelt. Disadvantages of this approach include:

- relatively high cost of the tag;
- time consuming to apply the tag;
- tags become separated from the pelt during the harsh mechanical and chemical processing involved in pelt pickling;

- fat adheres to the surface of the tag making it difficult to read automatically.

14.5.1 Pelt Branding

For the above reasons, our colleagues and ourselves have investigated various approaches for physically marking the pelt itself with a unique identifying mark. Marking techniques investigated included radio frequency, electrical current, laser and mechanical. The method chosen and further developed uses an electric current to locally heat and 'cook' the pelt.

The brander developed uses a matrix of electrodes, each of which delivers a localised and controlled current to cook the flesh. The branding head is applied to the one side of the raw skin, at a suitable location, and the electrodes can be individually energised to generate a two-dimensional pattern of holes that constitute the identification mark. The cooked areas are removed by pelt processing, leaving a hole or pit in the pelt. The marks, which generally penetrate the skin, show up on the other side. Hence, the transmission image is best for detection of these holes at the inspection station.

In trials, this branding method has been shown to produce highly reliable and repeatable marks at the required application rate. The technique has a very low per-unit added cost, being quick to apply and using relatively small amounts of electricity. The mark also has good visibility for both human and machine readability at the pickled stage.

Distortion of the code occurs both upon branding and during pelt processing. Hence, the precise geometric structure of the code cannot be pre-determined. This provides a challenge for the reliable automated detection and reading of the code patterns and has implications for both the choice of code pattern used and the algorithms for decoding it.

There are restrictions on the location available for marking the pelt. Matters that need to be considered include: minimising pattern distortion, consistency of the medium, possible accidental removal during pelt trimming, and saturation of the transmission image due to thin areas of the pelt. One of the best locations for marking and reading is the butt region where the pelt is thick and reasonably stable. One disadvantage of this region is that there can be slippage between the skin layers which can cause occlusion of some of the ID holes.

An example of an ID mark branded pelt is shown in Figure 14.10. Note the zoomed portion of one of the branded ID marks, this shows the distortion of the normally rectangular matrix code pattern.

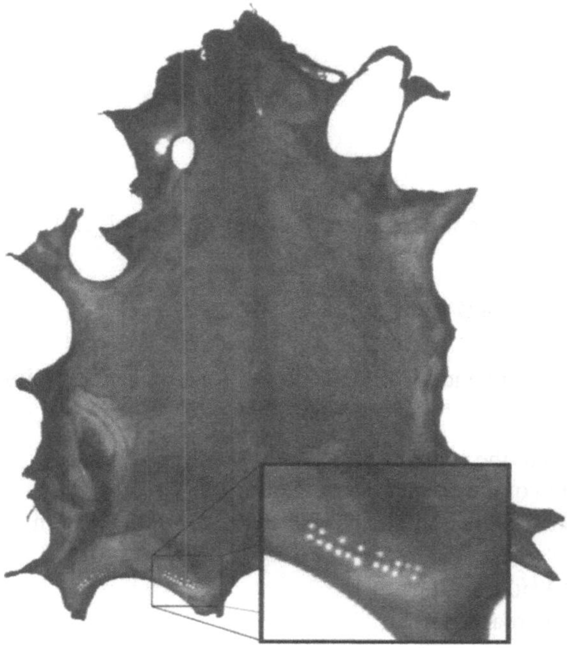

Figure 14.10. Pelt Identification Marks

14.5.2 The Code Structure

The code pattern currently being used copes with distortion by using a row of marks that are all 'ON', providing a local 1D sample of the pelt distortion within the marked area. This distortion is used to correct for the detected position of the other marks in the pattern.

Our colleagues and ourselves have developed and trialed various code structures and methods for automated reading. The most successful of these [19] is outlined here. The code consists of six numbers, each number represented by as many marks in a column. No number can be less than one (i.e., a column must have at least one mark) and no number may exceed five. Different groupings of columns have been tried, but our preferred grouping, for reasons that will become apparent, is that of three groups of two. Each pair of columns is separated from the next by a gap approximately twice that between its columns. A typical idealised example (free from noise and distortion) is given in Figure 14.11.

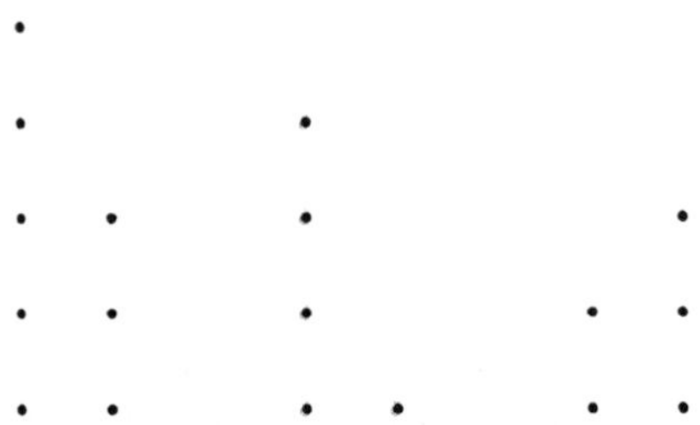

Figure 14.11. Example of hole pattern (reads 53 41 23).

14.5.3 Automated Code Reading

To automatically find and read this code from the transmission images, three main steps are involved:

- Location of the code pattern.
- Detection of the orientation of the pattern.
- Correction for distortion and final decoding of the pattern.

Location of the Hole Pattern

The following approach is used:

1. Mask the search area.
   The pelts are always presented in the same orientation. Hence, it is possible to set a simple rectangular area of interest in the bottom left of the image. Also, obviously the ID mark is always inside the pelt boundaries which, due to the high contrast of the transmission images, are easily obtained with automatic (entropy) thresholding. The reduced processing window greatly speeds up the search and also prevents those false positives which a wider search might have allowed.

2. High-pass filter the masked image and threshold.
   A kernel size is used which approximately matches the pattern hole sizes. This emphasises features of this size and greater, while suppressing smaller features. An automatic (entropy) threshold then produces a binary image.

3. Blob labelling and post-processing.
   The binary blobs are then labelled and listed. The list is purged of blobs whose size is outside the possible size range. A "binary convex" algorithm is then applied which separates approximately circular blobs which are lightly touching.

4. Cluster Analysis
   At this stage, the centroids of all remaining blobs are found. This final list of blobs includes those representing the code pattern but also includes blobs generated by holes that are not part of the code, as well as some resulting from noise or other thresholding artefacts. Cluster analysis is then performed. In this technique, a point is considered to be part of a cluster if the nearest member of that cluster is less than a certain distance away. Most

non-code points typically end up in clusters of only one or two in size. The largest cluster so detected is then deemed to be the code-containing one. The centroid of that cluster is then extracted, thus completing our first objective.

Detection of the Orientation of the Pattern

The key to this step is detection of the base line of the pattern and the first step is to isolate each of the three column-pairs in that pattern. The same cluster-finding algorithm is used as in the pattern-location stage, but with the cluster distance-gauge parameter now reduced to a value greater than the distance between points in any one column-pair but less than the distance between adjacent column-pairs. There is (in the absence of distortion) a 1:2 ratio between these two distances, which is a wide tolerance. This isolates and separately labels each of the column pairs.

For finding the orientation of the column pairs, and thus the whole pattern, a simple form of a cost function approach is used. This technique has, in general, no hard and fast rules, but the idea is to find the most probable fit to a set of constraints. Adapting it to this problem, we start by listing all the features the column-pair baselines have in the absence of noise and gross distortion:

- There will be no other point in the column-pair closer to a line drawn between the base-line pair than the distance between adjacent points in the pattern.
- All other points in the column-pair will be on only one side of the base-line pair.
- The distance between points that are candidate base-line pairs will be amongst the shortest between any two points.
- The base-line pair will always be present.

Having created, using the rules above, a short list of possible base-line pairs for each column pair, all possible combinations of actual base-line pairs are tested by taking one pair from each list and applying the following rules:

- The overall base line will have all other points on only one side of it (although some other lines may also meet this criterion).
- No other points in the overall pattern should be closer than a certain distance (nominally the inter-point spacing) to the overall base line.
- All three sets of base pairs must be approximately co-linear.

A scoring technique is then used which assigns a score corresponding to the likelihood of each pair of points being part of the baseline.

The algorithm starts at the column-pair level. For any given column-pair containing n points, there are n(n-1)/2 possible pairs of points. For each possible pair, a count is made of all other points closer than a certain distance to a line formed between them. Pairs of points which could be considered to be part of a line of three or more (thus unlikely to be part of a base pair) score penalty points here. A further count is made of the number of points on either side of the above lines. Pairs of points having points on either side are penalised.

Thus, for each column-pair, the pairs of points whose combined scores are the lowest become candidates for the inter-column-pair comparisons. In these, all possible combinations of one pair from a given column-pair with another pair from a different column-pair are examined. If the four points are approximately aligned, each point involved scores bonus points. For each column-pair, the two highest scoring points are then identified. These, when combined, define the base line of the overall pattern.

Correcting for Distortion and Final Pattern Decoding

Several approaches have been developed for distortion correction and final decoding. One approach is to process one hole at a time in each column using a rectangular search area defined by the last verified hole. If the hole is found in this search area it is used to perform a shearing and scaling correction before the next search is mounted.

Another approach is based on adjusting each new search area in the light of what has been discovered, but without otherwise correcting for shear. This approach is more flexible in that it makes fewer prior assumptions. A refinement on these approaches would be to better exploit the ability to operate separately on the column pairs whereby localised fits for warping may be done. The resultant three fits must be mutually consistent, but not so strongly as a single fit to the whole pattern would demand.

14.6 Conclusion and Future Work

The development of technologies for the inspection, grading and identification of sheep pelts has presented many challenges. The work described in this chapter illustrates some of the added difficulties typically associated with dealing with natural products compared with man-made objects. No two products are the same, there is a wide variation in the defects encountered, the product is difficult to handle and there is a fair degree of subjectivity in the human grading process, which makes it difficult to even establish benchmarks for inspection.

A multi-disciplinary approach has been needed to develop a range of integrated technologies for marking, spreading, conveying and imaging the pelts and then interpreting the pelt images for both quality and identification purposes. A novel imaging approach has been taken to highlight the features of interest and thus simplify the inspection task. We have successfully integrated advanced electronic, electro-optical and computing technologies into a system which provides a unique capability. The pelt inspection system has the potential to increase the value of a significant NZ export product and to enhance New Zealand's reputation for producing quality products. However, a fully automated grading system is still a long way off. The research has spawned other related programmes of work in pattern recognition and in-depth studies of the fluorescent properties of other natural products.

In terms of return on investment, this project had a long-term focus on increasing the returns from pelts for both farmers and processors by improving the consistency of pelt grading, while also providing an incentive for farmers and processors to improve the quality of the product. As an example of the potential gains, if just 10% of all pelts were misclassified as 3rd rather than 1st grade, approximately NZ$30 million of value would be lost to the industry per annum.

The various technologies described have generated considerable industry interest and, while some aspects of this work are still at the research stage, in 2001 the pelt identification and area measurement technologies are being progressed to a commercial implementation.

14.7 Acknowledgements

The New Zealand Meat Research and Development Council (MRDC) funded development of the laser scanner and pelt branding technology. Other aspects of the work including the pattern recognition software and investigations into fluorescent properties of pelts have been funded by the New Zealand Foundation for Research Science and Technology (FRST). The authors gratefully acknowledge these sources of funding. The contributions made by various Industrial Research Limited staff members to this project, in particular Alister Gardiner, Michael Simpson and Richard Gabric, are also gratefully acknowledged.

14.8 References

[1] Hilton P.J., Gabric R.P. (1994) Multiple image acquisition for inspection of natural products, Optics in Agriculture, Forestry, and Biological Processing, George E. Meyer, James A. DeShazer, Editors, Proc. SPIE 2345, pp. 10–19.

[2] Bowman C.C., Hilton P.J., Power P.W. (1995) Towards automated sheep pelt grading using machine vision, Proc. Image and Vision Computing New Zealand, IVCNZ95, Lincoln, Canterbury, August, pp. 209–214.

[3] Bowman C.C., Hilton P.J., Power P.W., Hayes M.P., Gabric R.P. (1996) Sheep pelt grading using laser scanning and pattern recognition, Machine Vision Applications, Architectures and Systems Integration V, Susan S. Solomon; Bruce G. Batchelor; Frederick M. Waltz; Editors, Proc. SPIE 2908 pp. 33–42.

[4] Poelzleitner W., Neil A. (1994) Automatic inspection of leather surfaces, Machine Vision Applications, Architectures, and Systems Integration III, Bruce G. Batchelor; Susan S. Solomon; Frederick M. Waltz; Eds Proc. SPIE Vol. 2347, pp. 50–58.

[5] Branca *et al.* (1996) Automated system for detection and classification of leather defects, *Optical Engineering* 35(12), 3485–3494.

[6] Wang, Q *et al.* (1992) A new method for leather texture image classification Proceedings of the IEEE International Symposium of Industrial Electronics, pp. 304–307.

[7] Hoang K., Nachimuthu A. (1996) Image processing techniques for leather hide ranking in the footwear industry, *Machine Vision and Applications* 9: 119–129.
[8] Wen W., Huang K. (1995) Leather surface inspection using clustering criteria, DICTA 95 Conference Proceedings, pp. 479–484, 1995.
[9] Serafin. (1992) Segmentation of natural images on multi-resolution pyramids, Linking the Parameters of an Autoregressive Rotation Invariant Model. Application to Leather Defects Detection. 0-8186-2920-7/92 IEEE.
[10] Spiegelhalter J., Taylor C.C. (1994) Machine Learning, Neural and Statistical Classification, Ellis Horwood.
[11] Chen C.H., Pau L.F., Wang P.S.P. (1993) Handbook of Pattern Recognition and Computer Vision, World Scientific.
[12] Michie D., Spiegelhalter D.J., Taylor C.C. (1994) Machine Learning, Neural and Statistical Classification, Ellis Horwood.
[13] Power P.W., Clist R.S. (1996) Comparison of supervised learning techniques applied to colour segmentation of fruit images, Intelligent Robots and Computer Vision XV: Algorithms, Techniques, Active Vision, and Materials Handling, David P. Casasent; Ed. Proc. SPIE Vol. 2904, pp. 370–381.
[14] Ross Quinlan J. (1993) C4.5 Programs for Machine Learning, Morgan Kaufmann Publishers.
[15] Hsiao J., Sawchuk A. (1989) Supervised texture image segmentation using feature smoothing and probabilistic relaxation techniques, *IEEE Transactions on Pattern Analysis and Machine Intelligence*, Vol. II No.12.
[16] Tuceran M., Jain A., Lee Y. (1988) Texture segmentation using Voronoi polygons, IEEE vol. 2605.
[17] Connors R.W. *et al.* (1983) Identifying and locating surface defects in wood: part of an automated lumber processing system, *Transactions on Pattern Analysis and Machine Intelligence*, Vol. PAMI, No 6 .
[18] Bovik M., Clark W. S., Geisler. (1990) Multichannel texture analysis using localised spatial filters, *IEEE Trans Pattern Analysis. Machine Intelligence*, 12, 55–73.
[19] Power P.W. (1988) Progress in the automated recognition of sheep pelt branding, Proc. Image and Vision Computing New Zealand, IVCNZ98, Auckland, pp. 228–233.

Chapter 15

Editorial Introduction

Designing vision systems for use in the food industry requires a clear understanding of the application requirements and highlights the distinction between Machine Vision and Computer Vision. Inspecting products such as Jaffacakes and cream biscuits requires the use of robust image processing procedures that must be fast enough to keep pace with modern manufacturing techniques. To achieve high processing speed, it is sometimes necessary to devise new computational procedures. These may be based on well-established image processing operators, such as the Hough transform, convolution or morphology. Alternatively, they may be entirely new and are designed to take into account the special characteristics of the application, or be derivatives of non-imaging procedures, such as graph matching. By carefully adjusting a convolution mask to suit the problem requirements, it is often possible to achieve superior performance on such tasks as detecting insects, rodent droppings and other contaminants in grain. Algorithm design should be based, as far as possible, on sound scientific reasoning. However, algorithms cannot be designed without regard to other issues, such as the precision of placement of the object to be examined in front of the camera. Moreover, the transport mechanism must not jam when defective items are encountered. As always, lighting plays a crucial role in obtaining good-quality images. Inspecting cold (i.e., "solid") rather than liquid chocolate, which suffers from glinting, is an excellent illustration of the general principle that considering the application in its entirety is essential. There is a hint here of a much more difficult type of problem: designing inspection systems that are able to examine products, such as the texture of the chocolate on Jaffacakes, for their aesthetic appearance. This will surely become more important in the future.

Chapter 15

Design of Object Location Algorithms and Their Use for Food and Cereals Inspection

E.R. Davies

15.1 Introduction

It is many years since computers were first used for analysing digital images, and thousands of industrial, medical, space, forensic, surveillance, scientific and other applications have now been recorded. This huge subject area is called machine vision and is characterised by the use of artificial vision to achieve actions in the real world. Thus it is a rather separate subject from computer vision, which endeavours to understand visual processes by experiments which test algorithms to see how well they permit vision to be achieved, and to determine what principles are involved in the process. In a sense, computer vision is the science and machine vision is its engineering colleague. However, this does not mean that machine vision has to be unscientific. Far from it: machine vision necessarily takes on board all the methodology and conclusions of computer vision, but also aims to arrive at cost-effective solutions in real applications. As a result, there is some emphasis on producing more efficient solutions, which are not only effective but to some extent minimal, with viable but rigorous short cuts being taken to save computation and hardware cost, but without significant invasion of robustness, reliability, accuracy, and other relevant industrial parameters. The way in which this end can be achieved is by noting that many applications are constrained in particular ways, so the full panoply and power of the most comprehensive algorithms will not always be needed. Naturally, this implies that some loss of adaptability in the chosen algorithms may have to be accepted in any one case. However, this is by no means unusual in other types of application, and in everyday life we are used to selecting appropriate tools to handle any task we tackle. Thus machine vision is intimately concerned with the selection of software tools and algorithms, and furthermore it is concerned with the specification and production of such tools.

Unfortunately, although the subject has developed vastly and even impressively over the past decades, it cannot yet be regarded as a fully mature subject. Software specification and design are not yet possible to the extent that they are in other areas of engineering. The blame for this lies in the complexity and variability of the image data that machine vision systems are expected to handle. Not only are the image signals highly variable, but also the noise within the images is highly changeable. For instance, noise may be white or "coloured", Gaussian or impulse, position or time dependent. In addition, it may arise in the form of clutter, i.e. it may take the form of other objects, which partly occlude those being observed, or other objects forming a variegated or even a time-varying background. In fact, the variability of real imagery is more often due to the effects of irrelevant objects than it is to noise, though noise itself does play a significant part in complicating the overall situation. Lastly, the objects being observed can vary dramatically, not only in their intensity, colour or texture, but also in their shape (especially when the 3D viewpoint is variable) and in changes that are the result of defects or damage.

To develop the subject further, it is necessary to take account of all these factors, and to try to understand the specification–application problem in more detail. However, theory on its own is unlikely to permit this to be achieved. The way forward seems to be to take a careful record of what happens in particular applications and to draw lessons from this. As more such records become available, sound all-embracing theories and guidelines will gradually emerge. However, at the very least, such case studies will permit design by analogy and by example, so very little should be lost. The case study approach is a powerful one, which is particularly relevant at this point in the development of the subject. Accordingly, this chapter describes a number of case studies from which lessons can be drawn.

In fact, the subject of machine vision is very wide indeed, and there is a limit to what a single chapter can achieve. Hence it will be useful to consider a restricted range of tasks: by focusing attention in this way, it should be possible to make more rapid progress. On the other hand, many tasks in different areas of vision bear significant similarity to each other – notice the ubiquity of techniques such as edge detection, Hough transforms, neural networks, and so on – so the possibility of transfer of knowledge from one part of the subject to another need not be substantially reduced. For these reasons, this chapter is targeted at automated visual inspection applications, and in particular some tasks which relate to food products. Foodstuffs have the reputation of being highly variable, and so the tasks described here cannot be considered trivial – especially, as we shall see, because quite high rates of inspection are required.

In the following section we first describe the inspection milieu in rather more detail, and then we proceed to a number of case studies relating to food and cereals. Selected theoretical results are considered at relevant junctures, though one or two of the more important ones are left until after the case studies, as they provide useful lessons of their own. Finally, the concluding section of the chapter underlines the lessons learnt from both the case studies and the theory, and shows that both are relevant to the development of the subject.

15.2 The Inspection Milieu

Computer vision is widely considered to have three component levels: *low-level vision*, *intermediate-level vision*, and *high-level vision* [1,2]. Low-level vision consists of primitive processes such as removal of noise and location of image features such as edges, spots, corners and line segments. Intermediate-level vision involves the grouping of local features into boundaries or regions, and also the identification of geometric shapes such as circles. High-level vision analyses the purely descriptive groupings resulting from low and intermediate-level vision and makes careful comparisons with objects in databases: relational analysis, inference, reasoning and statistical or structural pattern recognition techniques are used to help achieve viable matchings, and the image features are finally identified as actual objects – together with various attributes such as motion parameters and relative placements.

As mentioned in Section 15.1, machine vision aims to go further and identify actions, which should be initiated as a result of visual interpretation. However, it is also able to limit the practical extent of the algorithms at a level below that represented by the computer vision ideal, so that efficiency and speed can be improved. This means that instead of focusing on low, intermediate and high-level vision, we should concentrate on achieving the task in hand. In the case of industrial inspection, this means attending first to object location and then to object scrutiny and measurement. However, the very first task is that of image acquisition. This is especially important for inspection, as: a. careful illumination can ensure that the image analysis task is much simplified, so fewer advanced algorithms are needed, thereby saving computation and speeding up implementation; and b. careful illumination and acquisition can ensure that sufficient information is available in the input images to perform the task in hand [3]. Here it has to be borne in mind that object recognition is less problematic when only one type of object can appear on a specific product line: the natural restrictions of machine vision permit the whole process to be made more efficient.

Overall, we now have the following four-stage inspection process:

a. image acquisition;
b. object location;
c. object scrutiny;
d. rejection, logging or other relevant actions.

Here the 'object' may be the entire product, or perhaps some feature of it, or even a possible contaminant: note that in the first of these cases the object may have to be measured carefully for defects, while in the last case, the contaminant may also have to be identified. The situation is explained more fully in the case studies.

Set out in this way, we can analyse the inspection process rather conveniently. In particular, it should be noticed that object location often involves unconstrained search over one or more images (especially tedious if object orientation is unknown or if object variability is unusually great), and can thus involve considerable computational load. In contrast, object scrutiny can be quite trivial – especially if a pixel by pixel differencing procedure can be adopted for locating defects. These considerations vary dramatically from one application to another, as the case studies will show. However, the speed problem often represents a significant fraction of the system design task, not least because products are likely to be manufactured at rates well in excess of 10 per second, and in some cases such as cereal grain inspection, inspection rates of up to 300 items per second are involved [4].

15.3 Object Location

In the previous section the problem of limiting the computational load of any vision algorithm was mentioned, with the indication that the demands for meeting real-time processing rates could be quite serious in food processing applications. In fact, this difficulty is inherent in the object location phase of any inspection process. The reason is that unconstrained search through one or more images containing some millions of pixels may be required before any object is encountered. Furthermore, at each potential location, a substantial template may have to be applied with a large number of possible orientations. If objects of 10,000 pixels are being sought, and 360 orientations can be defined, some 10^{12} pixel operations may have to be performed – a process which will take a substantial time on any computer. This clearly rules out brute force template matching, but turns out not to eliminate template matching *per se*, as it can instead be applied to small object features. Often the template can be as small as 3 by 3 pixels in size, and there will be a correspondingly small number of definable orientations, so the computational load can immediately be reduced to around 10^7 pixel operations for a 256 by 256 pixel image – a much more realistic proposition.

While searching for objects via their features represents a vital principle for reducing the computational load to viable levels, it does not automatically solve all the problems because special procedures are then required to infer the presence of the objects from their features. Inference differs from deduction in that it is not a purely logical process: whereas a deduction follows necessarily from the input assertions, inference is only consistent with these assertions, and some type of probabilistic reasoning is required for determining the most likely solution. Nevertheless, in many cases inference can be efficiently implemented – as we shall see in Sections 15.3.2 and 15.3.3.

15.3.1 Feature Detection

Before proceeding to discuss inference procedures in detail, we briefly examine how feature detection can be achieved. The most commonly used approach involves the use of convolution masks. If a feature of a certain shape is to be detected, this is achieved by applying a mask of the same intensity profile – insofar as this is realisable within a digital (pixel) representation. The method is amply illustrated by the case of corner detection. If corners are to be detected, convolution masks such as the following may be used[1]:

$$\begin{bmatrix} -4 & -4 & -4 \\ -4 & 5 & 5 \\ -4 & 5 & 5 \end{bmatrix}$$

Clearly, such a mask will only be able to detect corners of this particular orientation, but application of eight similar masks whose coefficients have been 'rotated' around the central pixel essentially solves the problem. Notice how such masks are given a zero sum so that they can achieve a modicum of invariance against variations in image illumination [5]. Interestingly, while corners and other general features require eight such masks within a 3 by 3 window – and rather more in larger windows – edge detection only requires four masks of the following general types:

$$\begin{bmatrix} -1 & 0 & 1 \\ -1 & 0 & 1 \\ -1 & 0 & 1 \end{bmatrix}$$

This is because four of the eight masks that might have been used differ from the others only in sign. However, if we permit ourselves to change the strategy for edge detection from pure template matching by noting that edges are vectors with magnitude and direction, it becomes possible to employ just two masks: these then detect the horizontal and vertical components of intensity gradient. The pattern of ±1 coefficients employed in the above edge detection mask is not the best possible, since it leads to inaccurate estimates of edge orientation – a parameter that is of particular importance for Hough transform and related methods of object location. Within a 3 by 3 window the most appropriate edge detection masks are thus those of the well-known Sobel operator:

$$S_x = \begin{bmatrix} -1 & 0 & 1 \\ -2 & 0 & 2 \\ -1 & 0 & 1 \end{bmatrix} \quad S_y = \begin{bmatrix} 1 & 2 & 1 \\ 0 & 0 & 0 \\ -1 & -2 & -1 \end{bmatrix} \tag{15.1}$$

for which magnitude and direction can be estimated from the gradient components g_x, g_y using the equations:

$$g = (g_x^{2} + g_y^{2})^{-} \tag{15.2}$$

[1] To be rigorous, the convolution mask *enhances* the feature, and this then has to be *detected* by thresholding or related operations [1]. In this section we follow common practice by referring to feature detection without further qualification of this sort.

$$\Theta = \arctan(g_y/g_x) \tag{15.3}$$

Next, we come to line segment detection. Line segments are not vectors as symmetry ensures that any vector that is associated with them will have zero magnitude. This suggests that we should employ an eight-mask set for line segment detection, though symmetry considerations reduce this set to just four masks. However, recent research has shown that if the following pair of masks is used [6]:

$$L_0 = \begin{bmatrix} 0 & -1 & 0 \\ 1 & 0 & 1 \\ 0 & -1 & 0 \end{bmatrix} \quad L_{45} = \begin{bmatrix} -1 & 0 & 1 \\ 0 & 0 & 0 \\ 1 & 0 & -1 \end{bmatrix} \tag{15.4}$$

line features can be detected by combining the two outputs in a vectorial manner. The situation is curious, but this operator acts within an abstract space where the normal 360° rotation is transformed to 180°, with the result that the operator predicts a line segment orientation Θ q, which is twice as large as expected. However, dividing Θ by 2 leads to sensible results as symmetry only permits line orientations to be defined in the range 0° to 180°.

Finally, note that hole and spot detection masks only require a single convolution mask to be applied:

$$\begin{bmatrix} -1 & -1 & -1 \\ -1 & 8 & -1 \\ -1 & -1 & -1 \end{bmatrix}$$

15.3.2 The Hough Transform

The Hough transform is a model-based inference technique which may be used for locating objects of known shape. It has been used in many applications for locating lines, circles, ellipses, and shapes composed of these basic structures. It is also known how to apply the Hough transform to locate arbitrary shapes [1]: however, these shapes must first be defined unambiguously, and this may be achieved using analytical formulae or with the aid of look-up tables. In this section we outline how the Hough transform may be used to detect circles.

The Hough transform is at its simplest when detecting circles of known radius R. The basic idea is to accumulate votes for candidate centre locations in a separate image space known as the parameter space, and then to search the parameter space for the highest peaks which could represent circle centres.

To find the candidate centre locations, it is only necessary to start at each edge point and move along the local edge normal direction a distance R (Figure 15.1) – a procedure that is straightforwardly achieved using the Sobel edge detector. Clearly, this intermediate-level vision procedure only *infers* that a circle will exist at the given location, though further evidence can if necessary be obtained by examining the intensity values inside and outside the cited region. In practice, the latter type of procedure may be unnecessary if the total number of votes cast at any centre location tallies closely with the number of edge pixels expected on a circle of radius *R*.

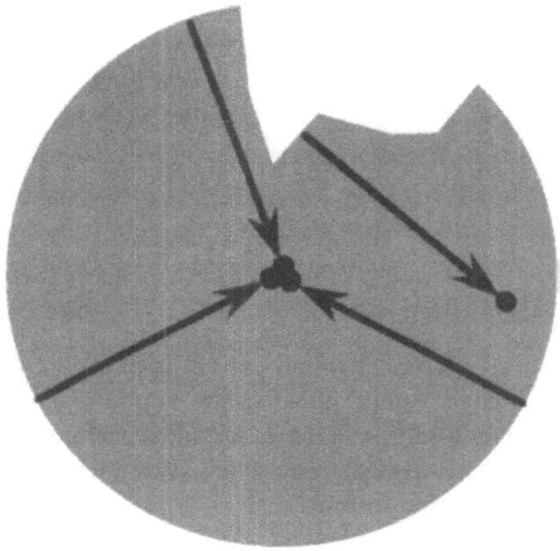

Figure 15.1. Circle detection using the Hough transform. This diagram shows how votes for the circle centre location are accumulated. Note how distortions lead to votes being placed at random positions, though sufficient votes appear at the true centre to permit the circle to be located robustly.

The particular advantage of the Hough transform lies in its impressive robustness against noise, occlusion and object distortion – or even breakage. It is achieved because edge pixels arising from noise, occluding objects or distorted parts of the object boundary give rise to candidate centre locations that are spread out in parameter space: these are unlikely to focus on particular locations and to give rise to misleading solutions. In any case, quick tests will usually eliminate false positives, leaving only true circles. The same applies for the shapes detected by alternative types of Hough transform and by more recent projection-based transforms [7].

15.3.3 The Maximal Clique Graph Matching Technique

Graph matching provides another approach to model-based inference, which may be used for object location. However, the graph matching approach is aimed at locating objects from point features such as corners and small holes rather than from edge or line segments which are only localised in a single dimension.

The concept of graph matching is to imagine the point features in the image as forming a graph, and those in the model as forming a separate graph. The matching problem then amounts to finding the isomorphism with the largest common subset in the two graphs, as this represents the greatest degree of agreement between the model and the image data. The solution is represented by a subset of the model graph because part of any object in the image may be occluded or broken away; likewise, it is represented by a subset of the image data as there will usually be features from other objects or from noise which have to be ignored (Figure 15.2).

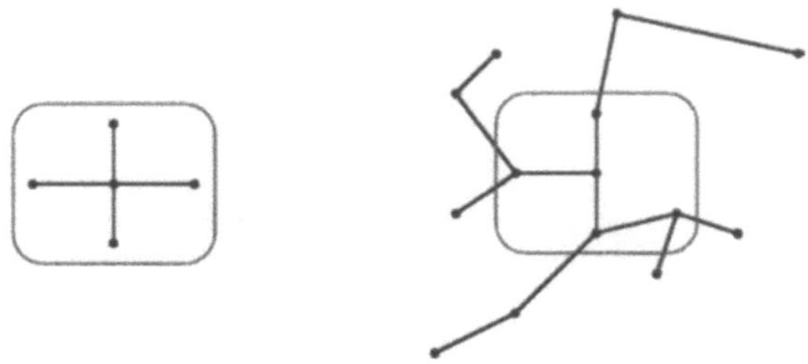

(a) (b)

Figure 15.2. The basis of graph matching: (a) the graph representing the object; (b) the graph representing the image. Subgraph-subgraph isomorphism is needed to cope with missing parts of the object and additional noise or clutter features appearing in the image. In this figure unnecessary complexity has been avoided by including only those graph edges (lines) which connect neighbouring features.

One way of achieving graph matching is by the maximal clique technique. This involves setting up a match graph whose vertices are the possible interpretations of image features. For example, it there are N image features F_i, $i = 1, 2, \ldots, N$, and M model features A_j, $j = 1, 2, \ldots, M$, then there will be NM possible interpretations of the image features: these may be labelled F_iA_j. Thus the match graph will have NM vertices. To determine which set of vertices gives the most likely match within the image, we first draw lines in the match graph between mutually consistent vertices: one common way of testing for consistency is to determine the distances between pairs of features in the image and in the model, and to flag consistency when these are equal. Note that any case of distance consistency leads to *two* lines being drawn in the match graph, as it will not be known *a priori* which way around the two features are in the image. Once all possible lines have been added to the match graph, maximum cliques[2] are sought as these represent the most likely object matches. The reason for this is that if full agreement exists within a subgraph, then there is no conflict in the overall interpretation, and also it explains the largest amount of image data. Clearly, the largest maximal clique will lead to the most likely object identification. Removing the image features for this object and repeating the whole process will lead to identification of the next most likely object, and so on – though at some stage the support for any further object identifications will be too slight for reliance to be placed on it without further tests being made on the original image. Note that because the technique searches for valid subsets of the input data, it is, like the Hough transform, robust against noise, occlusion and object distortion [1].

---

[2] A clique is a subgraph of maximal connectivity.

15.4 Case Study: Jaffa Cake Inspection

Jaffa cakes are examples of food products with high added value, as a significant number of processes are required during their production [8]. In addition, they are enrobed with a substantial layer of chocolate – an expensive commodity. These factors mean that, ideally, each stage of manufacture should be monitored by an inspection system to ensure that further value is not wasted on any defective products. However, inspection systems themselves cost money, both in the necessary hardware and software, and also in the need for careful setting up, calibration, and continual re-validation. In such cases it is usual to employ a single inspection system: there is some advantage from placing this at the end of the line so that the product is given a full check before it finally goes off to the consumer. This was the situation for the Jaffa cake inspection system described here. Note, however, that in other applications a different judgment may be made – as in the case of cereal grain inspection (see Sections 15.6–15.8), where it is crucial to inspect the raw commodity.

Jaffa cakes are round shallow cakes resembling biscuits which are topped by a small quantity of jam before they are enrobed with chocolate (Figure 15.3). Cooling is necessary before packing to ensure that the chocolate has solidified: this also helps the final inspection process as liquid chocolate gives rise to glints which could confuse the inspection software.

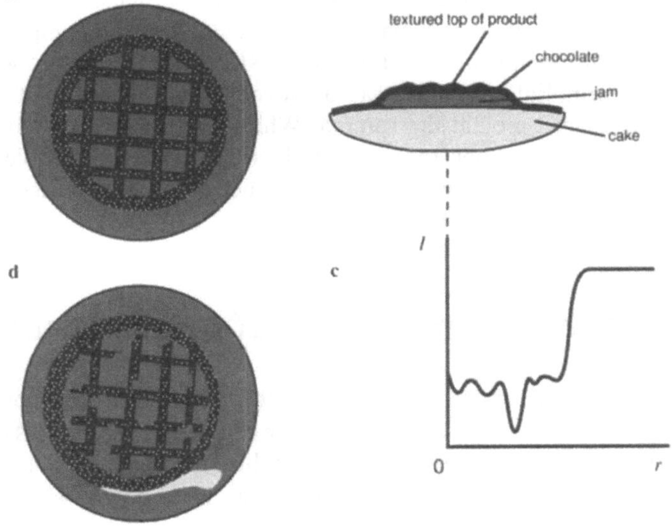

Figure 15.3. Jaffa cake characteristics: (a) an idealised Jaffa cake on which the boundary of the jam appears as a slanted, darker region; (b) its cross section; (c) its radial intensity histogram to the same scale as in b; (d) a less ideal Jaffa cake in which some 'show-through' of the cake appears, and the surface texture is more fuzzy.

Many problems arise with such products. First, the spread of cake mixture depends on temperature and water content, and if this is not monitored carefully the products may turn out too large or too small. While 10% variation in diameter may be acceptable, larger variations could lead to packing machines becoming jammed. Second, the jam or the chocolate could cause products to stick together, again causing problems at the packing stage. These factors mean that a variety of parameters have to be monitored – roundness, product diameter, the diameter of the spot of jam, the presence of excess chocolate, and any 'show-through' (gaps in the chocolate). In addition, the product has to be examined for general appearance, and in particular the textural pattern on the chocolate has to be attractive.

To achieve these aims, the products first have to be located and then suitable appearance matching algorithms have to be applied. In fact, the Hough transform provides an excellent means of locating these round products, and is sufficiently robust not to become confused if several products are stuck together, as all can be positively identified and located. To perform the matching several processes are applied: assessment of product area, assessment of degree of chocolate cover (area of dark chocolate region), computation of radial intensity histogram[3] to assess the jam diameter and offset (determined from the region bounded by the very dark ring on the chocolate) and the overall appearance of the product (measured from the radial intensity histogram correlation function) (Figure 15.3).

Considerable speed problems were encountered during this work, and special hardware accelerators had to be devised so that inspection could take place to perform 100% inspection at around 20 products per second. Hardware design was eased by making extensive use of summation processes, such as area measurements, computation of radial intensity histograms, and computation of the radial intensity histogram correlation function [8]. Finally, some care was required to obtain uniform lighting: this was achieved by using a set of four lights symmetrically placed around the camera. While lighting did not have to be set critically, because of the robustness of the algorithms used, reasonable uniformity in the lighting was useful to minimise the number of separate measurements that had to be made on each product.

15.5 Case Study: Inspection of Cream Biscuits

This case study describes the inspection of cream biscuits, which consist of two wafers of biscuit with cream sandwiched between them (Figure 15.4). There are two particular reasons for inspecting these biscuits: one is to check on the alignment of the two biscuit wafers; the other is to check whether there is substantial leakage of cream from between the wafers. Clearly, both of these aspects have a distinct influence on the appearance of the product and its acceptability to the consumer. They also reflect the possibility of jamming the packing machine.

---

[3] This sums and averages the intensities of pixels within concentric rings around the centre of the product [1].

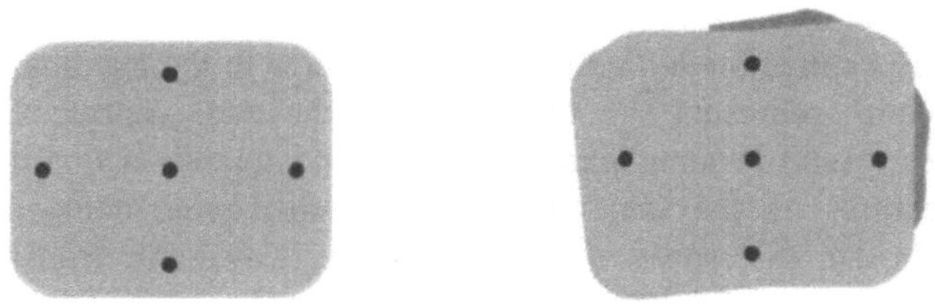

(a) (b)

Figure 15.4. Problems encountered in inspecting cream biscuits: (a) plan view of an ideal cream biscuit; (b) a less ideal product with misaligned wafers and leakage of cream. The 'docker' holes by which the upper wafer can be located are clearly indicated.

Conventional shape analysis techniques can be used to check the overall shape of the final product and thus detect these defects. However, the overall shape can be quite complex even if small amounts of misalignment or cream are evident (Figure 15.4). Hence considerable simplification of the overall algorithm results from adopting an alternative strategy – to locate the biscuits by the small 'docker' holes on the upper wafer, and then to check the overall shape against a template [1]. In fact, a further simplification is to determine how much of the product lies outside the region of the upper wafer as computed from the positions of the docker holes (Figure 15.4).

To achieve this, it seemed appropriate to employ the maximal clique technique, as it is suited to location of objects from point features. While this worked well with biscuits containing up to six docker holes, it was found to be far too slow for biscuits with many holes (in some cases biscuits are decorated with patterns of well over 100 holes) [4]. The reason for this is that the maximal clique computation is 'NP-complete', and this results in its having a computational load which varies approximately exponentially with the number of features present. Because of this, other procedures were tried, and it was found that the Hough transform could be used for this purpose. To achieve this, a reference point G was selected on the biscuit pattern (the centre of the biscuit was a suitable point), and all pairs of features located in the image were used to estimate the position of G, the votes being placed in a separate parameter space. This permitted the positions of all biscuits to be found with much the same degree of effectiveness as for circle location. The only problem was that it was not known *a priori* which was which of the holes in each pair, so two votes had to be cast in parameter space in each case [9]. While this gave a considerable number of false votes, no difficulty was experienced in determining the centre of each biscuit by this method.

15.5.1 Problems with Maximal Cliques

The computational problems with the maximal clique approach have been known for some time, though as noted above they were not a particular problem for biscuits with up to six features. However, in this application another problem with the technique was identified. This arose because in the end the technique only makes use of features that are part of the largest available maximal clique: features which are not compatible with all the other features in the maximal clique that is finally taken to represent the object are in the end ignored. Unfortunately, distortions can result in some features not being compatible with a limited number of other features on the same object, while still being compatible with the remainder. In that case some information which is available from these features and which would be useful for interpretation is ignored.

As the method is potentially highly robust, it may still result in the object being recognised. However, when only a small number of features are available on any object, there is little latitude for ignoring any of them and some loss of robustness then results. A further factor is the loss of location accuracy that results from the decrease in position data. In general, then, the maximal clique method does seem to lose out significantly relative to the Hough transform approach when locating distorted objects. Food products are quite variable and are often slightly distorted during manufacture, so they are especially affected by these problems. Specifically, changes in the fluidity of the biscuit mixture may lead to size and shape changes which can push a number of inter-feature distances outside accepted tolerance limits.

Naturally, many distortion problems can be overcome by permitting a certain tolerance in the inter-feature spacings. However, if too much tolerance is permitted, this ultimately results in a large increase in the number of false alarms (whether these are recorded as votes in parameter space or lines in the match graph). Hence in practical situations, with given tolerance levels, the maximal clique approach results in reduced levels of robustness and location accuracy.

15.6 Case Study: Location of Insects in Consignments of Grain

Automated visual inspection has now been applied to a good many food products, and as indicated in Section 15.4, it is often appropriate to apply it at the end of the production process as a final check on quality. Nevertheless, there is a distinct need to attend to the quality of the raw products, such as the grain from which flour, breakfast cereals and many other derived products are made. In the case of grain, attention has so far been concentrated on the determination of grain quality [10], the determination of varietal purity [11,12] and the location of damaged grains [13]. Relatively little attention has been paid to the detection of contaminants. In fact, there is a need for a cheap commercial system that can detect insect infestations and other important contaminants in grain.

This case study describes research that has been orientated towards the design of such a system. Of especial interest is the location of insects amongst the grains, as these have the potential for ruining a large store of grain in relatively few weeks, because of the rather short breeding cycle. This means that a highly reliable method is needed for detecting the insects, involving a far larger sample than human inspectors can manage: in fact, a sample of some 3 kg (~60,000 grains) for each 30 tonne lorry load of grain appears to be necessary. As lorries arrive at intervals ranging from 3 to 20 minutes, depending on the type of grain terminal, this represents a significant challenge from both the effectiveness and the computational speed points of view [14].

Preliminary analysis of the situation suggested that insects would be detected highly efficiently by thresholding techniques, as they appear dark when compared with light brown wheat grains. However, careful tests showed that shadows between the grains and discolorations on them can produce confusing signals, and that further problems can arise with dark rapeseeds, fragments of chaff, and other artefacts (Figure 15.5). This meant that the thresholding approach was non-viable, though further analysis showed that line segment detector algorithms (Section 15.3.1) were well adapted to this application because the insects approximated to short rectangular bars. It turned out that high orientation accuracy is not required in this instance, though it is attractive to use just two vectorial masks to save computation.

The next significant problem was to design larger versions of these masks so as to obtain good signal-to-noise ratio at higher resolution with insects about 3 pixels wide. This was achieved with 7 by 7 masks which approximated to rings of pixels, for which about half of each mask would be in the background and half within the insect region. Geometric arguments showed that this would be achieved when the mask radius was about $1/\sqrt{2}$ of the width of an insect (Figure 15.6).

Tests showed that the strategy was successful on images containing real grains and insects such as *Oryzaephilus surinamensis* (saw-toothed grain beetle). To perform the final detection, thresholding of the line segment operator signal was required; in addition, a test had to be applied to check that the signal resulted from a dark rather than a light patch in the image, as the line segment detector is sensitive to both. This procedure resulted in a false negative rate of around 1% against a total of 300 insects: minor detection problems arose from insects being viewed end-on or being partly obscured by grains. Another problem was a small false positive rate arising from the dark boundaries on some of the grains and particularly from 'black-end-of-grain'. Other problems occasionally arose from pieces of chaff resembling insects (Figure 15.7): one solution to this problem would be to increase resolution to improve recognition, though this would give rise to a drop in the processing rate and a faster, more expensive processor would be needed. Such issues are beyond the scope of this case study. However, the line segment detector approach seems the right one to solve the problems of insect detection; and the vectorial strategy seems capable of limiting computation with the larger masks required in this application [14].

It would be interesting at this point to compare the work described above with that of Zayas and Flinn [16] who adopted a discriminant analysis approach to the insect detection problem. However, their work is targeted primarily at a different type of insect, *Rhyzopertha dominica* (the lesser grain borer beetle) and it would appear to be difficult to make a realistic comparison at present. More generally, the reader is referred to an interesting appraisal of different ways of finding pests in food [17].

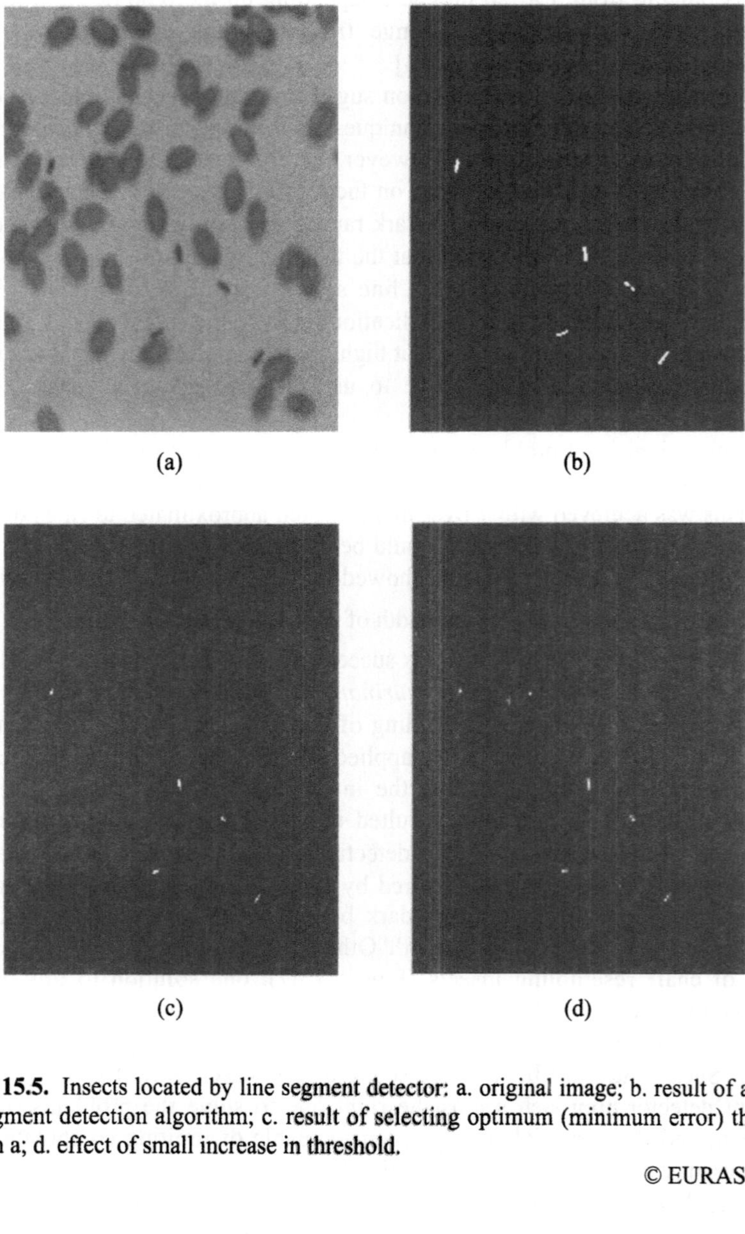

Figure 15.5. Insects located by line segment detector: a. original image; b. result of applying line segment detection algorithm; c. result of selecting optimum (minimum error) threshold level on a; d. effect of small increase in threshold.

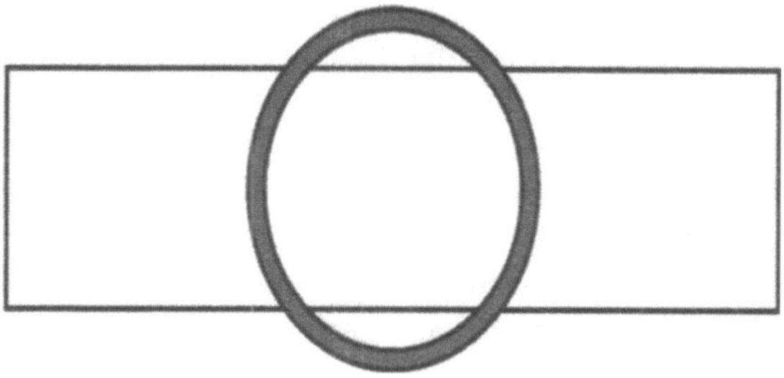

Figure 15.6. Geometry for design of a line segment detector using ring masks. For optimum results the ring must be symmetrically placed and of such a size that it lies half within and half outside the rectangular object to be detected.

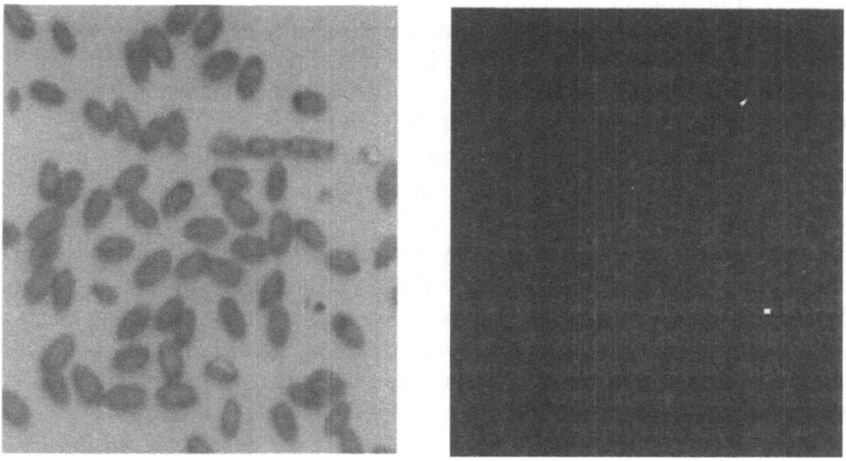

(a) (b)

Figure 15.7. Incidence of false alarms: (a) original image; (b) thresholded output from line segment detector. The false alarms are due to chaff and the darkened boundaries of grains: all are easily detected and eliminated via their shapes. For the specific type of line segment detector used here, see [15].

15.7 Case Study: Location of Non-insect Contaminants in Consignments of Grain

As noted above, there is a need for a commercial system for the inspection of consignments of grain. In fact, grain is subject to many contaminants other than insects. In particular, rat and mouse droppings have been reported, and moulds such as ergot are especially worrying as they are toxic to humans. There is also a need to keep a check on foreign seeds such as rape, though their presence is not especially worrying at low concentrations. It turns out that rodent droppings and ergot are dark and have characteristic elongated shapes, while rape seeds are dark and round: all these contaminants are readily detected by human operators. This again suggests that such contaminants could be detected automatically by thresholding. However, rodent droppings are often speckled (Figure 15.8), while ergot is shiny and tends to exhibit highlights, so again it has light patches – albeit for different reasons. There is also the possibility of confusion between the speckled regions and the grainy regions, which are light, but with a variety of dark patches caused by shadows and some dark patches on the grains themselves [18].

These considerations suggest that morphological techniques [19] should be used to help with detection of the dark contaminants mentioned above. More particularly, as human operators are able to recognise these contaminants from their characteristic shapes and spatial intensity distributions, a suitable combination of morphological operators should be sufficient to achieve the same end.

An obvious way forward is to apply erosion operations to eliminate the shadows around, and dark patches on the grains. These erosion operations should be followed by dilation operations to restore the shapes of the contaminants. While these actions are reasonably effective in eliminating the dark patches around and on the grains, they leave holes in some of the contaminants and thus make it difficult to identify them unambiguously (Figure 15.8).

An alternative approach is to consolidate the contaminants by performing dilation operations followed by erosion operations, but this has the predictable effect of giving significant numbers of false positives in the grain background. Eventually a totally different strategy was adopted – of consolidating the whole image using very large median filters. This procedure is able to virtually eliminate the speckle in the contaminants, and is also successful in not generating many false alarms in the grain regions. However, a certain number of false positives do remain, and it has been found that these can be eliminated by applying additional erosion operations (Figure 15.8).

Once good segmentation of the contaminants has been achieved in this way, with good preservation of their original shapes, recognition becomes relatively straightforward. Perhaps the main disadvantage of the method is the need to apply a large median filter. However, many methods are known for speeding up median operations, and this is not especially problematic. In addition, it could be argued that the methods described here constitute a form of texture analysis, and that there are many alternative ways of tackling this type of problem. However, texture analysis also tends to be computation intensive, and a considerably larger number of operations would be needed to segment the contaminants in this way. Overall, the approach adopted here seems to have the advantages of simplicity, effectiveness and also to be sufficiently rapid for the task. Perhaps the important lesson to be learnt from this case study is that the usual types of morphological operator are not as powerful as is commonly thought, so they have to be augmented by other techniques if optimal solutions are to be found [18].

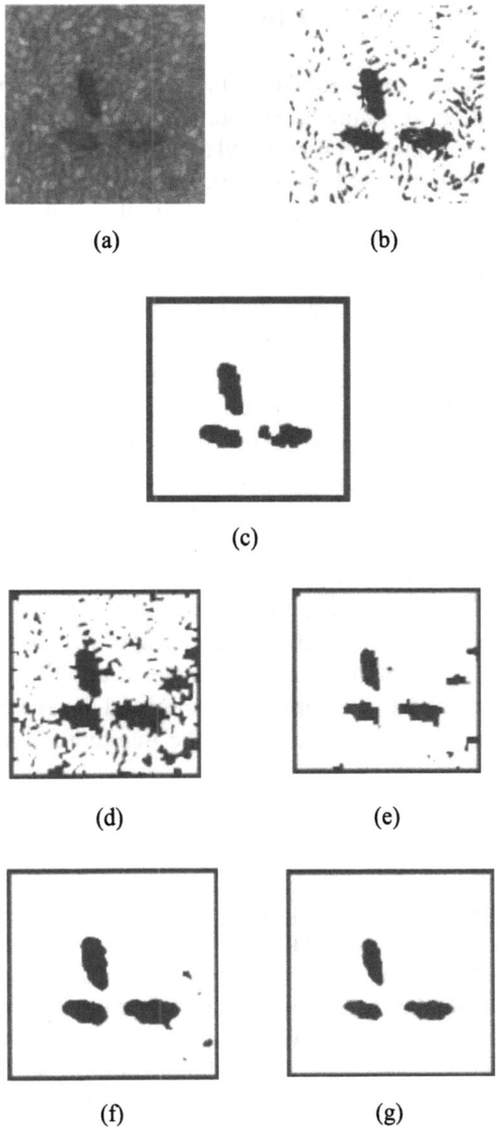

Figure 15.8. Effects of various operations and filters on a grain image: (a) Grain image containing several contaminants (rodent droppings); (b) Thresholded version of a; (c) Result of erosion and dilation on b; (d) Result of dilation and erosion on b; (e) Result of erosion on d; (f) Result of applying 11 by 11 median filter to b; (g) Result of erosion on f. In all cases, 'erosion' means three applications of the basic 3 by 3 erosion operator, and similarly for 'dilation'.

15.7.1 Problems with Closing

The above case study was found to require the use of a median filter coupled with a final erosion operation to eliminate artefacts in the background. An earlier test similarly involved a closing operation (a dilation followed by an erosion) followed by a final erosion. In other applications, grains or other small particles are often grouped by applying a closing operation to locate the regions where the particles are situated. It is interesting to speculate whether, in the latter type of approach, closing should sometimes be followed by erosion, and also whether the final erosions used in our tests on grain were no more than *ad hoc* procedures or whether they were vital to achieve the defined goal.

To analyse the situation, we consider two regions containing small particles with occurrence densities ρ_1, ρ_2, where $\rho_1 > \rho_2$ [20]. Clearly, the mean distances d_1, d_2 between particles will depend on the occurrence densities. Dilation will tend to group the particles, but the real aim is to dilate sufficiently to group the particles in region 1 but not those in region 2 (Figure 15.9). Assuming that this is possible, a subsequent erosion should give a good approximation to region 1. However, if any region 2 particles are near the boundary of region 1, they will be enveloped by region 1, which will then become too large, and a subsequent erosion will be needed to restore region 1 to its proper size and shape. The shift δ can be calculated on the basis that it is essentially a one-dimensional effect. However, corrections need to be applied as the particle density in the second dimension in region 2 is not unity and produces a reduction relative to the one-dimensional shift. The final result is:

$$\delta_{2\mathrm{D}} = 2ab\rho_2(a+b) \tag{15.5}$$

where a is the radius of the dilation kernel and b is the width of the particles in region 2. It is important to notice that if $b = 0$, no shift will occur, but for particles of measurable size this is not so. Clearly, if b is comparable to a or if a is much greater than 1, a substantial final erosion may be required. On the other hand, if b is small, it is possible that the two-dimensional shift will be less than 1 pixel. In that case it will not be correctable by a subsequent erosion, though it should be borne in mind that a shift has occurred, and any corrections relating to the size of region 1 can be made during subsequent analysis. While in our work the background artefacts were induced mainly by shadows around and between grains, in other cases impulse noise or light background texture could give similar effects, and care must be taken to select a morphological process which limits any overall shifts in region boundaries. In addition, it should be noticed that the whole process can be modelled and the extent of any necessary final erosion estimated accurately. More particularly, the final erosion is a necessary measure, and is not merely an *ad hoc* procedure [20].

15.8 Case Study: High-speed Grain Location

It has already been noted that object location can be significantly more computation intensive than the scrutiny phase of inspection, as it involves unconstrained search over the entire image area, and sometimes over several images. This means that careful attention to high-speed object location procedures can be extremely useful. Of course, hardware accelerators can be designed to bring about any required improvement in speed, but if inspection systems are to be cost-effective, rapidly operating software is required.

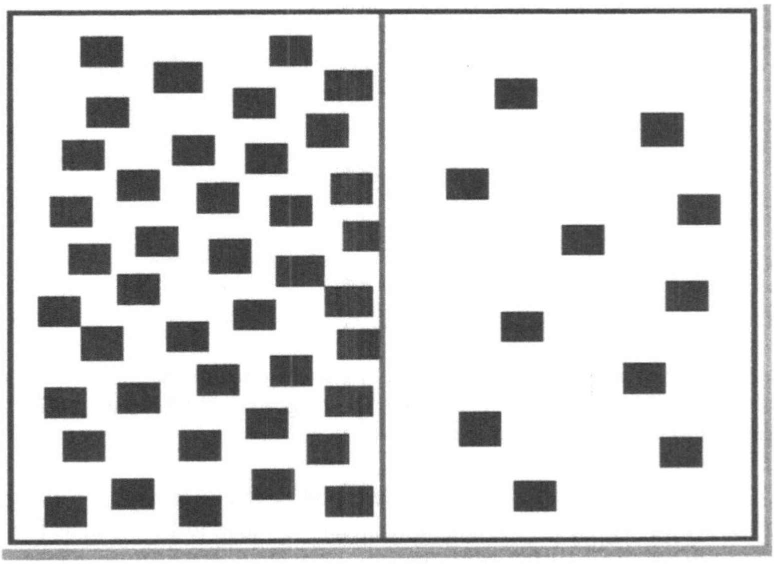

Figure 15.9. Model of the incidence of particles in two regions. Region 2 has sufficiently low density that the dilated particles will not touch or overlap.

In his earlier work on biscuit inspection (see the Jaffa cake case study in Section 15.4 for the general milieu), the author solved this problem by scanning the image along a restricted number of horizontal lines in the image, and taking the mid-points of chords across objects [21]. Suitable averaging in the x and y-directions then permitted the centres of circular objects to be located highly efficiently, with speedup factors as high as 25. While some loss in robustness and accuracy occurred with this approach, performance degraded gracefully and predictably.

In the cereal grain inspection problem, even higher speed-up factors were required, and an even faster type of algorithm was required. The only way of obtaining further substantial improvements in speed appeared to be to concentrate on the object regions rather than on their boundaries. Accordingly, a sampling technique was developed which sampled the whole image in a regular square lattice of spacing s pixels. This provided an intrinsic speedup of s^2 coupled with a loss in accuracy by the same factor [22].

Not surprisingly, the technique has significant limitations. In particular, the ideal saving in computation only occurs for square objects with their sides orientated parallel to the image axes. However, if a square object lies at 45° to the image axes it can miss all the sampling points, with the result that the latter must be placed two times closer together than the ideal to ensure intersecting the objects (Figure 15.10). Similarly, circular objects can miss the sampling points unless their spacing is twice the circle radius.

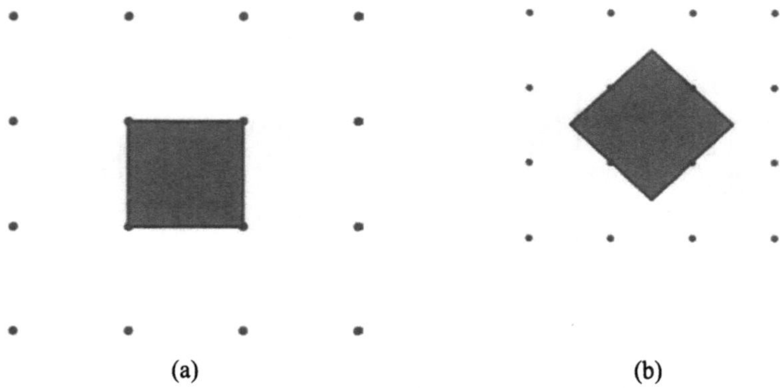

Figure 15.10. Sampling points for location of square objects: (a) Spacing of sampling points for locating square objects aligned parallel to the image axes; (b) Reduced spacing required for square objects aligned at 45° to the image axes.

To understand the sampling process properly it is necessary to imagine the image space being tiled with shapes identical to those to be located. For simple shapes such as squares, the tiling can be perfect, but for other shapes such as circles, the image space will not be entirely covered without expanding the tiles so that they overlap – thus making the sampling procedure less efficient than the supposed ideal. In fact the situation is even worse when the varying orientations of real objects are taken into account: i.e. a greater number of tiles has to be used to cover the whole image space, and thus an increased number of sampling points is needed; a full explanation appears in [4, 23]. However, although the ideal is not achievable, the results can still be extremely useful, at least for convex objects, and impressive speed-up factors can be obtained.

Another limitation of the technique is that accuracy of location is necessarily limited. However, this problem can be overcome once an object's approximate position has been found by sampling. At worst, a region of interest can be set up: within this region a conventional object location method can be applied, and the object can be located with the usual accuracy. An overall gain in speed results because large tracts of image space are quickly by-passed by the sampling procedure. Interestingly, there is the possibility of finding an accurate location technique whose speed matches that of the sampling procedure. For circles and ellipses the *triple bisection* algorithm operates extremely rapidly, and has some resemblance to the earlier chord bisection strategy. It involves taking two parallel chords near to the sampling point, which originally located the object, bisecting them, forming the chord joining the two bisecting points, and then bisecting this chord (Figure 15.11). The final bisector is theoretically at the centre of the ellipse, though minor errors (~1 pixel) may exist which can only be eliminated by further averaging. For a proof of the validity of this procedure, see [4, 23].

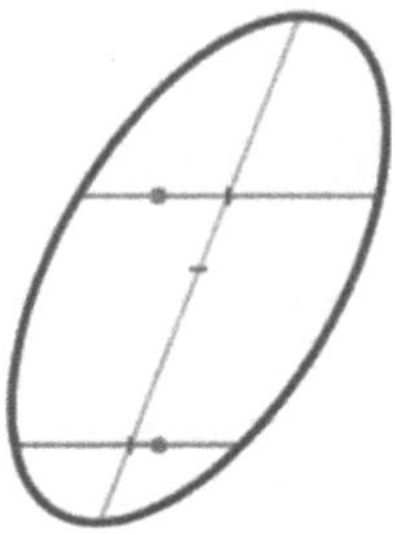

Figure 15.11. Illustration of triple bisection algorithm. The round spots are the sampling points, and the short bars are the midpoints of the three chords, the short horizontal bar being at the centre of the ellipse.

The technique has been applied to the location of peppercorns, which are approximately circular, and wheat grains, which are approximately elliptical with an aspect ratio of about 2:1. The method is highly efficient when used to detect circles, as is evidenced by the fact that ideally only one or two sampling points intersect with each object. For elliptical wheat grains, the method is rather less efficient and in the ideal case, up to three sampling points may appear in each object, though the average is close to two (Figure 15.12).

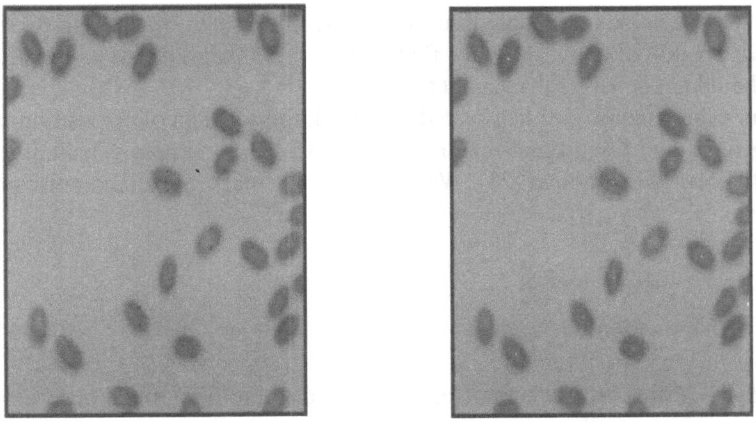

(a) (b)

Figure 15.12. Image showing grain location using the sampling technique: (a) Sampling points; (b) Final centre locations.

Overall, the technique works well, and saves considerable computation, but only if the objects are well separated. However, alternative procedures can be devised for accurately locating the centres of objects when these are in contact: the basic sampling procedure does not fail in such cases: it is merely more difficult to find accompanying procedures for accurate location whose speed matches that of the basic sampling technique.

15.9 Design of Template Masks

In the earlier sections of this paper we have seen that much of the weight of object location devolves onto the use of template masks for feature detection. While design of such masks can be dealt with intuitively and checked by experiment, design principles are hard to come by. The *zero mean rule* [5] is a generally useful one, but important questions remain – such as how large the masks should be and how the pixels should be divided between the foreground and background regions.

In fact, it is generally the case that greater sensitivity of detection results from the additional averaging achieved by larger masks. However, there is a limit to this which is determined by the probability of disruption introduced by impulse noise [24]. Specifically, if one pixel within a mask succumbs to impulse noise, this may well falsify the response from that mask, and (for example) cause a vote to be placed at a totally erroneous position in the Hough transform parameter space. Clearly, the probability P of this happening increases rapidly with the number of pixels within the mask. If the probability of a pixel intensity being corrupted in this way is p, and the area of the mask is A, we find [24]:

$$P = (1 - p)^A \tag{15.6}$$

If we set P at some reasonable limit of acceptability, such as 0.75, and if p is taken to have a value of about 0.01, we find that $A \approx 28.6$, and in that case the window size should not be larger than about 5 by 5.

Once the window size has been determined, the problem of disposition between background and foreground remains (Figure 15.13). A simple calculation can be used to solve this problem [25]. We first write the mask signal and noise power in the form:

$$S = w_f A_f S_f + w_b A_b S_b \tag{15.7}$$

$$N^2 = w_f^2 A_f N_f^2 + w_b^2 A_b N_b^2 \tag{15.8}$$

where A_f, A_b are the numbers of pixels allotted to the foreground and background regions, w_f, w_b are the respective mask weights, S_f, S_b are the respective signal levels, and N_f, N_b are the respective noise levels. Optimising the signal-to-noise ratio for a zero mean mask of fixed total area $A = A_f + A_b$ now leads to the equal area rule:

$$A_f = A_b = A/2 \tag{15.9}$$

though complications arise when $N_f \neq N_b$, a situation that becomes important when the foreground and background textures are different. Finally, when the *equal area rule* does apply, the mask weights must follow the rule:

$$w_f - w_b \tag{15.10}$$

Thus theory can help substantially with the design of masks, once the design constraints are known reliably.

Figure 15.13. General geometry of a mask M containing two regions of respective areas A_1 and A_2. 1 is the background region and 2 is the object region.

15.10 Concluding Remarks

This paper has described a number of case studies relating to the location of objects in images containing food and cereals. These have ranged from location of Jaffa cakes, cream biscuits and cereal grains to location of insects, rodent droppings, moulds and foreign seeds amongst the cereal grains. In all cases the aim has been inspection, though in the first three of these cases the inspected objects had to be scrutinised for defects, whereas in the other cases the main commodity, cereal grains, had to be scrutinised for contaminants.

It is worth remarking that the reason for the emphasis on object location is that this is often, as in all these cases, the most computation-intensive part of the inspection task. Scrutiny itself takes a fraction of the computational effort taken by object location – even when this is carried out by feature detection followed by inference reference to the need to ensure that packing machines are not jammed of the presence of the whole object. In fact, the problem of efficient object location for real-time implementation is so serious that in applications related to the Jaffa cake inspection task it was necessary to develop an especially fast line-sampling technique. In the later cereal grain inspection problem an even faster technique had to be developed, based this time on a fast point-sampling method. Though sampling was not performed in the other case studies, very careful attention nevertheless had to be paid to achieving high speeds.

In the cases studied, the reasons for inspection ranged from checking that the products were the right size and shape to ensuring that the products were attractive in appearance. While this criterion obviously benefits the consumer, it also reflects the quantity of product that can be sold and the profit that can be made. However, there are other reasons for checking that products have a normal attractive appearance: specifically, there is then significantly reduced likelihood of foreign objects having alighted on or having been mixed into the product. With raw materials such as cereal grains, the possibility of contaminants is significantly higher. Indeed, cereal grains start their lives on the land, and are thus liable to be contaminated by mud, stones, twigs, insects, droppings, chaff, dust, and foreign grains. In addition, they are susceptible to moulds, sprouting and physical damage. Thus there are distinct health and safety aspects to the inspection task – the more so as certain moulds such as ergot are poisonous to humans. These factors make it all the more essential to inspect consignments of grain for insect infestations and excessive amounts of non-insect contaminants. This would seem to constitute sufficient reason for including these particular case studies in this paper.

From the academic point of view, these case studies are mere vehicles for the exploration of the problems of inspection, and in the wider field, of machine vision. The similarities between the methods used to search for objects and their features vary very little from topic to topic in machine vision, there are a good number of distinct transferable techniques. This point is illustrated by the theoretical sections of the paper, which help to bridge the gaps between the case studies and to extend the area of application of the ideas they embody far more widely. In particular, it has been shown how speed can be optimised, how problems with matching algorithms (particularly the paradigm maximal clique technique) can arise and be by-passed, how an apparently *ad hoc* solution to a morphological technique (closing) can be rigorously solved, and how template matching masks can be designed to optimise signal-to-noise ratio and to prevent over-exposure to impulse noise or small amounts of occlusion.

In an ideal world it would be possible to design vision algorithms on paper, starting from careful specifications, but the present reality is that the situation is too complex and such specifications are seldom available, and in many application areas such as food the data images are too variegated and variable to permit us even to approach the ideal. In such circumstances case studies have to be the main route forward, though on occasion this route can be bolstered up and its elements bound more firmly together by theoretical analysis. It is hoped that this paper will help show the way forward, and firmly demonstrate, with the other papers in this volume, the enormous value of the case study approach. The reader will also find useful techniques and case studies in [1–3, 26–28].

15.11 Acknowledgements

The author is grateful to Dr J. Chambers and Dr C. Ridgway of Central Science Laboratory, York, UK for useful discussions on the needs of the grain industry, and for providing the original images used to test the high-speed object location algorithms. Figure 15.3 is reproduced from [8] with permission from the M^2VIP'95 Organising Committee. Figures 15.5, 15.11 and 15.12 are reproduced from [14, 23] with permission from EURASIP. Figures 15.7, 15.8, 15.9 and 15.13 are reproduced from [15, 18, 20, 25] with permission from the IEE.

15.12 References

[1] Davies E.R. (1997) Machine Vision: Theory, Algorithms, Practicalities, Academic Press, London (2nd edition).

[2] Ballard D.H., Brown C.M. (1982) Computer Vision, Prentice-Hall, Englewood Cliffs, NJ.

[3] Batchelor B.G., Hill D.A., Hodgson, D.C. (1985) Automated Visual Inspection, IFS (Publications) Ltd, Bedford, UK/North Holland, Amsterdam.

[4] Davies E.R. (2000) Image Processing for the Food Industry, World Scientific, Singapore.

[5] Davies E.R. (1992) Optimal template masks for detecting signals with varying background level, *Signal Process.* 29, no. 2, 183–189.

[6] Davies E.R. (1997) Vectorial strategy for designing line segment detectors with high orientation accuracy, *Electronics Lett.* **33**, 21, 1775–1777.

[7] Davies E.R. and Atiquzzaman, M. (eds) (1998) Special issue on projection-based transforms, *Image Vision Comput.* 16, nos. 9–10, pp. 593–725.

[8] Davies E.R. (1995) Machine vision in manufacturing – what are the real problems?, Invited paper in Proc. 2nd Int. Conf. on Mechatronics and Machine Vision in Practice, Hong Kong (12–14 Sept.), pp. 15–24.

[9] Davies E.R. (1992) Locating objects from their point features using an optimised Hough-like accumulation technique, *Pattern Recogn. Lett.* 13, no.2, 113–121.

[10] Ridgway C. and Chambers J. (1996) Detection of external and internal insect infestation in wheat by near-infrared reflectance spectroscopy, *J. Sci. Food Agric.* 71, 251–264.

[11] Zayas I.Y. and Steele J.L. (1990) Image analysis applications for grain science, *Optics in Agric.*, SPIE 1379, pp. 151–161.

[12] Keefe P.D. (1992) A dedicated wheat grain image analyser, *Plant Varieties and Seeds* 5, 27–33.

[13] Liao K., Paulsen M.R. and Reid J.F. (1994) Real-time detection of colour and surface defects of maize kernels using machine vision, *J. Agric. Eng. Res.* 59, 263–271.

[14] Davies E.R., Mason D.R., Bateman M., Chambers J., Ridgway, C. (1998) Linear feature detectors and their application to cereal inspection, Proc. EUSIPCO'98, Rhodes, Greece, 8–11 Sept., pp. 2561–2564.

[15] Davies E.R., Bateman M., Mason D.R., Chambers J., Ridgway, C. (1999) Detecting insects and other dark line features using isotropic masks, Proc. 7th IEE Int. Conf. on Image Processing and its Applications, Manchester (13–15 July), IEE Conf. Publication no. 465, pp. 225–229.

[16] Zayas I.Y. and Flinn P.W. (1998) Detection of insects in bulk wheat samples with machine vision, *Trans. Amer. Soc. Agric. Eng.* 41, no. 3, 883–888.

[17] Chambers J. (1997) Revealing invertebrate pests in food, Crop protection and food quality: meeting customer needs, Proc. BCPC/ANPP Conf., Canterbury, Kent (16–19 Sept.), pp. 363–370.

[18] Davies E.R., Bateman M., Chambers J., Ridgway C. (1998) Hybrid non-linear filters for locating speckled contaminants in grain, IEE Digest no. 1998/284,

Colloquium on Non-Linear Signal and Image Processing, IEE (22 May), pp. 12/1–5.

[19] Haralick R.M., Sternberg S.R., Zhuang, X. (1987) Image analysis using mathematical morphology, *IEEE Trans. Pattern Anal. Mach. Intell.* 9, no. 4, 532–550.

[20] Davies E.R. (2000) Resolution of problem with use of closing for texture segmentation, *Electronics Lett.* 36, no.20, 1694–1696.

[21] Davies E.R. (1987) A high speed algorithm for circular object location, Pattern Recogn. Lett. 6, no.5, 323–333.

[22] Davies E.R. (1997) Lower bound on the processing required to locate objects in digital images, *Electronics Lett.* 33, no.21, 1773–1774.

[23] Davies E.R. (1998) Rapid location of convex objects in digital images, Proc. EUSIPCO'98, Rhodes, Greece, 8–11 Sept., pp. 589–592.

[24] Davies E.R. (1999) Effect of foreground and background occlusion on feature matching for target location, *Electronics Lett.* 35, no. 11, 887–889.

[25] Davies E.R. (1999) Designing optimal image feature detection masks: the equal area rule, *Electronics Lett.* 35, no.6, 463–465.

[26] Chan J.P., Batchelor B.G. (1993) Machine vision for the food industry, Chapter 4 in Pinder A.C. and Godfrey G. (1993), pp. 58–101.

[27] Naim M.M., Chan J.P., Huneiti A.M. (1994) Quantifying the learning curve of a vision inspection system, Proc. IEE 4th Int. Conf. on Advanced Factory Automation, York (Nov.), IEE Conf. Publication No.398, pp. 509–516.

[28] Graves M., Smith A., Batchelor B.G. (1998) Approaches to foreign body detection in foods, *Trends in Food Science and Technology* 9, no. 1, 21–27.

Chapter 16

Editorial Introduction

A chicken-processing factory is not a pleasant place to work; it is cold and wet. Manual de-boning, which is prevalent at the moment, can lead to repetitive strain injury. Present-day detection rates for bone in supposedly "boneless" chicken meat, achieved by human inspectors, are far worse than retailers seem prepared to admit. For these reasons, it is often difficult to recruit and retain workers for chicken processing. Thus, there is a need for improved inspection and processing systems. The latter needs visual sensing, to cope with the inherent uncertainty. Inspection systems using x-rays for detecting bone and other foreign bodies in food products have been studied for almost two decades but earlier earned an unenviable reputation for being inaccurate and unreliable. By careful design and diligent attention to the needs of the poultry industry, it has been possible to make x-ray inspection systems that overcome many of the problems of earlier designs. Chickens are not all alike, even if they are carefully bred and are fed in an identical way. The age and diet of the bird is critical to the process of ossification: fully calcified bone, from older birds, can be detected more easily than from young chickens. In order to build a successful Machine Vision system for this, or any other application, it is necessary to pay attention to numerous details, each of which may seem to be of minor importance but which can individually undermine an otherwise excellent design. A system must be x-ray safe. It should be capable of providing management information on demand, without distracting the system from its primary function of inspection. It should be easy to strip and clean the system. The equipment must not be damaged by either water-jet or chemical cleaning. A system must be easy to use, or it won't be used at all! Switching on should be a simple process, but should automatically initiate a calibration procedure. The mechanical handling system must be safe. Among the other key features that have been incorporated into the system described in this chapter, we should mention the air conditioning for the internal components. This maintains the stability of the x-ray sensor. This is a dual-energy detector, which has been designed specifically to optimise for the task of detecting bone in chicken meat. A new power supply to drive the x-ray source can be controlled via a serial (RS-232) data link. A special *accept/reject* mechanism had to be designed for this application, since the chicken meat is sticky. Novel image processing methods were also necessary. Simple techniques, such as fixed-level thresholding do not work reliably enough. Automated thresholding, incorporating morphology is generally superior. In the past, attempts were made to make the detection of bone more reliable, by mechanically compressing the meat during inspection. Other designers chose to immerse the meat in water during inspection, so that the x-ray path length is nearly constant everywhere. However, by using modern decision-making methods, based on Neural Networks, and self-adaptive learning, it has been possible to avoid mechanical "tricks", which always introduce new difficulties, in this case reduced levels of hygiene. The important point to note is that diligent design has produced a system which offers real potential for making this type of produce safer.

Chapter 16

X-ray Bone Detection in Further Processed Poultry Production

M. Graves[1]

16.1 Introduction

The presence of bones left in poultry products by the de-boning process is one of the major concerns in the poultry industry today [1, 2]. This concern is growing as more use of poultry meat is made in further processed products. Whilst consumers may (for now at least) only complain mildly when finding a bone in a supposedly boneless chicken fillet, the same consumers take a very different attitude with products such as nuggets, sandwiches, ready meals containing chicken meat and chicken soup. For such products consumers demand and expect a bone-free product. Poultry processing companies routinely hand check every chicken fillet for bones (in some cases twice with different people) yet despite this the occurrence of bones is much more widespread than the typical levels which the food retailers admit [3].

X-rays have been used to inspect food for the presence of foreign bodies since the early 1970s [4] and specifically to inspect chicken meat for bones since 1976 [5]. During the 1980s and 1990s a number of commercial organisations attempted to develop x-ray technology for the inspection of bones in chicken based on their general x-ray packaged food inspection systems [1]. The vast majority of these systems are no longer in use today and were not successful because the technology required for poultry meat inspection is very different from that required to inspect a coffee jar or packet of biscuits.

As a result of these numerous and very expensive failures, the use of x-ray technology within the poultry processing industry has a very bad reputation for being unreliable and inaccurate. This is an unfair assessment. As we will show in this chapter, x-ray technology can be used very successfully as part of a bone detection program within a modern poultry processing plant. However, for this to be the case the x-ray technology must be designed for this application from the start and the designers must have a good understanding of the poultry processing industry. In this chapter we will highlight our experiences at successfully installing over twenty BoneScan™ x-ray machine vision systems in the poultry processing industry for the inspection of chicken breasts, chicken nuggets, cooked stripped chicken meat and chicken thigh meat.

16.2 The Extent of the Problem

16.2.1 Data for the Occurrence of Bones in Poultry Meat

There is a great deal of disagreement within the industry as to the extent of the problem of the occurrence of bones in deboned poultry meat. Most factories undertake a random hand inspection of their meat and from this, estimate the amount of bones remaining in the product. However, usually this random sampling is not a thorough destructive test and usually just involves a quick check over the fillet and therefore it is not surprising that such checks often give a false level of security. This is especially true with the impacted wish bones (also known as pully or clavicle bones). These bones are embedded deep within the meat and a quick manual check will usually miss these.

Recently researchers from Campden and Chorleywood Food Research Association undertook a large destructive test of poultry meat purchased from the shelves of five of the UK's leading supermarket retailers [2]. The results of their finding are shown in Table 16.1

Table 16.1 Occurrence of bones found in UK supermarket poultry.

Superma rket	No. of bones found	Weight of meat inspected
1	113	100.1 kg
2	87	100.1 kg
3	81	102.5 kg
4	92	109.5 kg
5	97	105.9 kg

These results show a staggering increase in bone contamination levels over the supermarkets own guidelines of 1–3 bones per 100kg.

A recent study in USA [3] found levels of bone contamination to be significantly lower than these. However, in this study the meat was only given the usual manual plant inspection and was not subject to the destructive test of the CCFRA study [2].

16.2.2 Future Trends Which Will Increase the Problem

At present in the developed world it is increasingly hard to find a labour force willing to work in the conditions of a modern poultry plant. The conditions are often cold and damp and repeated actions, especially cutting actions, can lead to conditions such as repetitive strain injury (RSI) [6]. As a result of this there is considerable interest in the use of automated equipment to reduce labour, especially on the manual de-bone line because of the risk of wrist injury caused by repeated cutting actions. A number of companies manufacture commercially available equipment to replace a large percentage of humans on a manual cone de-boning line. However the automated de-boning cutting machines do not have the same tactile feedback as a skilled human de-boner. Therefore the incidence of bones occurring on lines with automated de-bone equipment is much greater than manual de-bone equipment, especially the incidence of embedded pully/wish bone which is the most dangerous [1] and difficult for a human inspector to find.

In addition there is the potential threat of litigation especially in the United States. At present there are very few cases of poultry companies being sued for bones occurring in their finished bone-free product although recently there have been several multi-million dollar law suits for fast food products containing a chicken head, rat head and two instances of hyperdermic needles [7]. The poultry companies use the "due diligence" argument claiming that they have taken "all reasonable measures" to ensure that their food product is safe. Whilst there was no technology to automatically find bones in poultry this argument was valid but now that technology has been proved to reliably find bones automatically this argument will be shown to be invalid.

16.3 The Technical Challenge

X-ray imaging has been used for many years to find foreign bodies in general packaged food products and containers such as glass jars and metal cans [8,9,10]. The principal is quite straightforward. A source of x-rays passes through the product moving on a conveyor belt to a sensor underneath the conveyor belt which converts the x-ray signal into a digital signal. This digital signal corresponds to the x-ray absorption image of the product and it is this image which is processed to make the decision as to whether the product contains a foreign object or not. An outline schematic of the x-ray imaging system is shown in Figure 16.1.

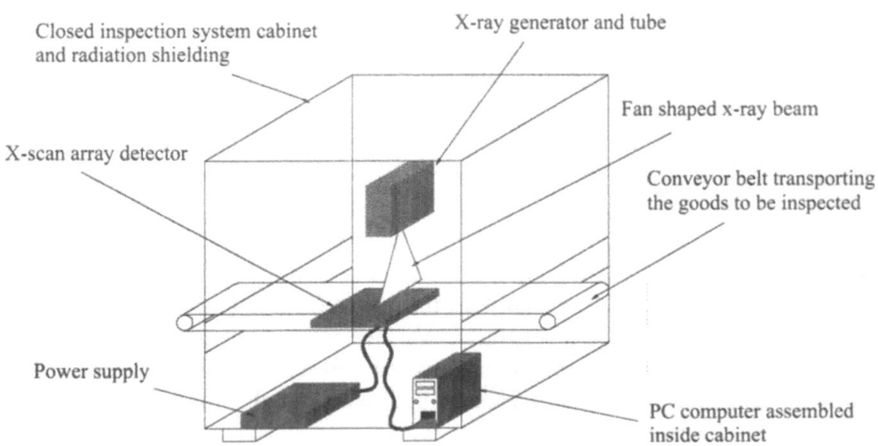

Figure 16.1. Schematic of an x-ray imaging system.

In many applications the foreign body to be detected will be denser than the food product in which it is embedded and therefore a simple thresholding applied across the whole image will enable the foreign object to be detected. This is the basis on which most commercial x-ray foreign body detection systems work today. Such an approach is fine for detecting dense contaminants when the food background is perfectly homogeneous and therefore such threshold-based systems work fine for detecting metal, stone or glass in products such as margarine. For less homogeneous products it is necessary to use more complicated techniques. One technique widely used in a number of commercially available systems is dual-level thresholding with morphology. In this technique two thresholds are applied to the image. The first is set at a value such that there will be little false segmentation, the resulting segmented regions are then used as seed points for a second threshold. Binary morphology is used on the second thresholded image so as to remove those segmented regions considered too small to be a defect. A more robust technique is to change the threshold level dynamically based on the entropy of the image histogram [11].

The problem with the inspection of poultry products is that they are not perfectly homogenous, there is a distinct thickness profile across a piece of chicken meat which requires the use of techniques beyond thresholding and binary morphology [12]. There have been a number of attempts to overcome this difficulty using mechanical techniques. We will briefly describe these before outlining our machine vision solution to the problem.

16.3.1 Attempts to Homogenise the Poultry Product

The requirement to make the x-ray inspection of poultry products easier by mechanically homogenising the product was first recognised in 1976 [5]. In this patent Ramsay *et al.* first tried to float the chicken pieces in a container of water so that when inspected by the x-ray imaging system the chicken/water combination represented a uniform combined thickness and any variation would be due to a bone. The physics of what they were trying to achieve made sense but unfortunately the practical problems associated with placing the chicken meat in water-filled containers meant the system was not successful in practice.

Since then there have been two attempts to mechanically homogenise poultry meat. The first was to squash the meat between two rollers prior to the x-ray system. The second attempt was to pump the meat through a tube with water. Both of these systems have been commercially implemented and are available today. The problem with both techniques is that they mechanically deform the product, which means that it can no longer be used for a high-value product. Additionally both systems are difficult to clean, thus there is the real possibility that the poultry producer reduces his bone levels but introduces serious microbiological problems because the equipment cannot be adequately cleaned. A third problem with the pipeline approach is that the rejected product is in the form of a slurry and therefore it is extremely difficult for any human checker to hand inspect the rejected meat to see if there were actually any bones rejected. There are numerous anecdotal stories in the industry of pipeline systems randomly rejecting a slurry of product every few minutes with nobody ever really knowing whether it actually was successful in finding any bones.

16.4 The BoneScan™ Solution

Between 1993 and 1997 the author and his colleagues at Spectral Fusion Technologies developed their first BoneScan™ machine specifically for bone detection in poultry products. Since that first prototype a number of technological advances have been made resulting in the twenty plus systems that are now out in the field successfully scanning chicken breast meat, stripped cooked chicken meat, chicken nuggets, chicken thigh and fish fillets. These machines most of which are still in operation today, have been upgraded during their life to incorporate developments in the technology and in the rest of this chapter we will describe some aspects of this unique technology.

16.4.1 Design Requirements

There were a number of key design requirements from the outset:

- accuracy of bone detection;
- low level of false rejection;
- robustness of detection over time;
- cleanability of the system;
- high volume throughput;
- ease of use of the system;
- robust rejection technology;
- ability to provide management information;
- future upgradability of the system.

16.4.2 Accuracy of Bone Detection

It might appear obvious that a system designed for automatic bone detection in poultry products should have as its primary design goal a high level of detection accuracy. However, many commercial x-ray systems designed for packaged food inspection have been used for bone detection and found to be capable of detecting bones to a minimum thickness of 3mm. However, a careful analysis of typical chicken bones will reveal that most lie in the 1.5mm to 2mm thickness range and so to be able to detect these reliably required a major development on from the standard commercially available x-ray packaged food imaging systems.

There were three major developments which we undertook in order to make our system more accurate especially in the detection of the small bones typically left in after the de-boning process.

Dedicated X-ray Sensor

A few early systems for jar inspection used x-ray image intensifiers as the chosen x-ray sensor [8,13]. However, the disadvantages of limited viewing geometry and the requirement to pulse the x-ray source [14] meant that most developers opted for solid-state linear photodiode arrays when they had overcome their initial technological difficulties [15].

A linear diode array consists of a series of buttable photodiodes with a scintillator and filter material positioned above the photodiodes. The filter material acts as a high-pass filter to remove those low-energy parts of the x-ray beam which do not include useful signal. The optimum filter material depends on the x-ray energy being used. The scintillator material converts the x-ray photons to light photons and is usually based on a rare earth oxide with a doping agent. The choice and thickness of scintillator material is a very important part of the overall systems design. The material should be chosen according to the x-ray energy being used in the application. The doping agent should be chosen to optimise the speed of response of the scintillator. Different doping agents will be used depending on the sampling speed of the sensor. The thickness of the scintillator is a critical parameter. A thicker scintillator will be more efficient in converting the x-ray photons to light photons but will give more scatter thus reducing the contrast of any defect. Similar trade-offs exist for the grain size of the scintillator material.

With off-the-shelf (OTS) sensors there is no scope for optimisation of any of these parameters for a specific application. Typically OTS sensors will have x-ray energy inputs from 30 to 140 keV. Whilst these sensors give a good general overall response they cannot by nature be optimal for any specific application. We designed our own LDA with the filter material, scintillator material, scintillator thickness and grain size and doping material all optimised for the x-ray energies and conveyor speeds typically required in poultry meat inspection. Since the spectral response of this sensor was optimised for the application and the x-ray tube output we have named this sensor SpectraLine™.

Another major disadvantage of commercially available LDAs is their drift with temperature. Typically an OTS sensor will require recalibration for every 0.5° C change in temperature [24]. Whilst it is possible to continue to operate without a recalibration, overall system performance is compromised. For the BoneScan™ system we designed a fully automatic computer-controlled air conditioning based temperature management system. This system ensures that the temperature within the cabinet remains at 25° (+/–°C) independently of external temperature fluctuations. However even this level of temperature stability is still not optimal if photodiode drift is to be minimised. To overcome this problem we designed into SpectraLine™ a closed-loop temperature feedback system. As a result of this the analogue gain and offset parameters are automatically adjusted according to the exact temperature profile change since the calibration at the beginning of the day.

A final improvement in image quality resulted from the use of a linear rather than switch mode power supply. Although the linear power supply was much bulkier than the commercially available switch mode types we found an improvement of 0.5 bit on our 12 bit output.

A schematic of an x-ray tube is shown in Figure 16.2.

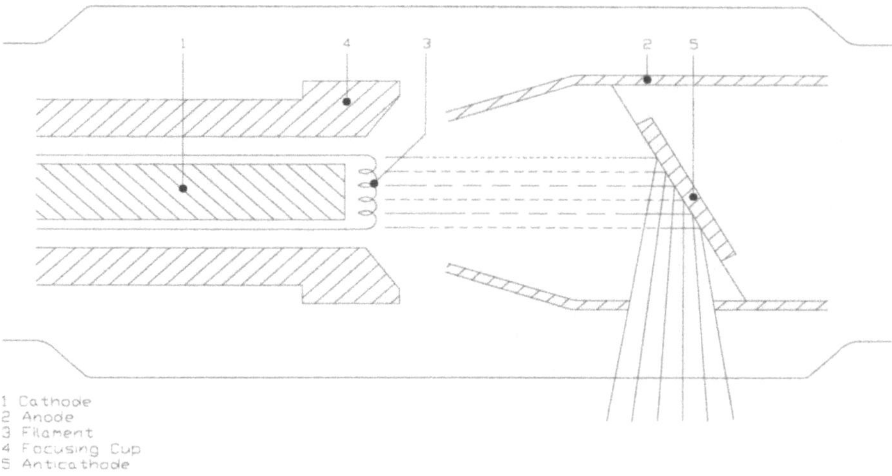

Figure 16.2. Schematic of an x-ray tube

The tube operates by having a high potential voltage applied between the anode and the cathode by an external x-ray power supply. In our system this is under computer control (RS232) so that it can be varied for each application. This voltage, defined as the KeV_{max}, determines the maximum x-ray photon energy. The tube will actually emit a range of x-ray energies up to the maximum known as *Bremstrahlung*. This spread of energies is exploited in our multispectral x-ray imaging system described in Section 16.5.2.

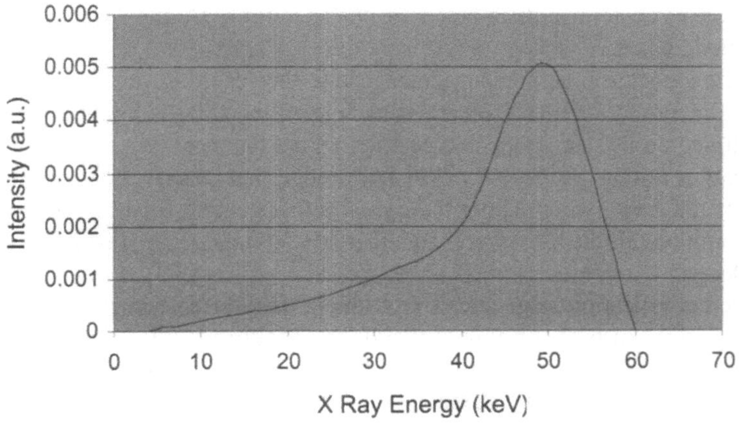

Figure 16.3. Bremstrahlung spread of x-ray energies.

A separate power source supplies a current to the filament. As the filament warms up by the flow of current there is an electron cloud which forms by thermionic emission around the filament. This electron cloud is attracted to the anticathode because of the electrical potential between the anode and cathode. As electrons from the electron cloud collide with the target material x-ray photons are emitted. The energy of the x-ray photons will be controlled by the electrical potential between the anode and cathode. The number of x-ray photons will be controlled by the filament current. The anticathode target material has several impacts on the usefulness of the resulting x-ray photons. Firstly the smaller the focal spot upon which the electrons are focused onto the anticathode the sharper the resulting image [22, 26]. However, the smaller the focal spot the greater the amount of heat energy given off as a result of the sudden loss of momentum experienced by the electron as it impinges into the anticathode target. As a result there is a trade-off between image sharpness and expense of the tube cooling requirements. Secondly, the choice of target material will have an impact on the resulting spread of x-ray energies. The most common target material is tungsten because of its high melting point but other more exotic materials offer the potential to match the spectral response of the tube with that of the sensor for a specific application.

Image Processing Algorithms

There are three aspects to the image processing algorithm; image enhancement, segmentation and classification.

The image enhancement algorithm consists of initial calibration of the 12 bit data from the LDA followed by noise reduction filters. The calibration consists of offset correction (for photodiode dark current) and gain correction (for non-uniform response across the photodiodes). Most commercial LDAs include these offset and gain correction factors as part of the sensor. We found that by reading the data from our LDA and performing these operations in software on the PC we were able to customise them better for the specific LDA and specific applications being undertaken.

Following the image enhancement stage the resulting image was segmented using a hierarchy of filters, where at each level the filter is looking for a specific bone based on its size, shape and position within the meat. We are fortunate in this application in that we have *a priori* knowledge that specific bones lie in specific parts of the meat (which is not the case in other general food inspection) and our segmentation algorithms were considerably enhanced by making use of this knowledge.

The classification stage consists of taking data for each segmented region and feeding this into a previously trained backpropagation neural network. Classification accuracy was found to be very dependent on the quality of data gathered during the training phase.

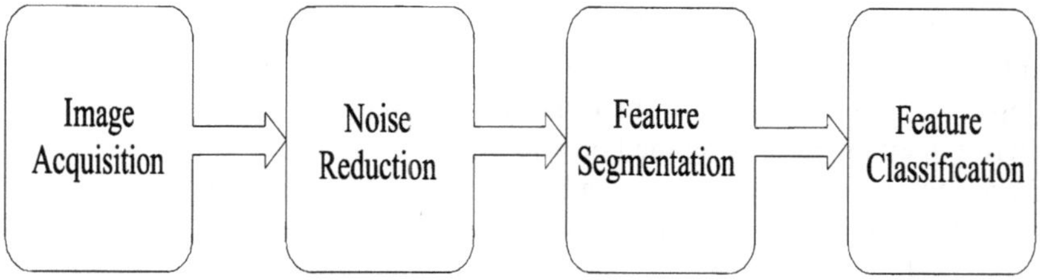

Figure 16.4. Overview of image processing algorithm.

16.4.3 Low Level of False Rejection

There is a trade-off between detection accuracy and false rejection level in any inspection process. This trade-off is especially important in bone detection in poultry meat. There is a huge amount of anecdotal evidence of people using a machine in limited laboratory conditions and being relatively satisfied with the detection accuracy only to install the machines on-line and find false rejection levels approaching 40% and making the machines completely useless. Equally, stories abound of people having machines set to high sensitivity for an important demonstration only to desensitise the system after the visit for normal factory working conditions.

Usually the false rejections occur when there is a locally dense piece of flesh. This is due to the fillets being trimmed roughly or to the fillets not being placed perfectly flat on the conveyor belt. Detection of perfectly trimmed flat fillets (especially those that have been flattened and then trimmed by a water jet cutter) is so much easier than fillets placed roughly on a conveyor belt. For the vast majority of processors it is not acceptable to mechanically flatten the fillets and therefore false rejection is a major problem with systems which cannot be trained to cope with the trimming and loading that will be experienced in practice.

We have found that the best approach is to ensure that the development of the image processing algorithms and neural network training takes place on images that reflect exactly the nature of the meat that is being inspected. Very often this means that fine tuning of the system will take place in the plant because the actual trim and loading quality cannot be simulated in the laboratory.

Even allowing for this it is inevitable that certain false rejections will occur. We have tried to devise ways of minimising the effect of these by novel mechanical handling solutions. These vary on an application basis and are described in Sections 16.5 and 16.6.

16.4.4 Robustness of System Performance over Time

It is important that the level of detection achieved in the initial setup of the system is achieved over the expected lifetime of the machine. There are several reasons why x-ray bone detection systems for poultry meat inspection have not been robust over time.

First, the nature of the meat can change. A variety of factors can cause the meat/bone calcification to change over time, these include diet of the bird, breed of the bird, age at slaughter, transportation to slaughter, stunning method, de-feathering method, de-boning method and, for cooked meat, cooking method. Each of these factors may make a very subtle change to the meat and/or bones which may not be visible to the human eye but can dramatically affect the performance of an x-ray system which does not have software algorithm models which can cope with these variations [1].

Secondly the presentation of the meat to the system may change over time. Different people feeding the meat on the infeed conveyor can make a dramatic difference to the performance of the system. Typically the level of bone detection will remain constant but if the meat presentation is not so uniform then the false rejection rate will increase

Finally the x-ray tube and sensor have a finite life. The tube will gradually deteriorate and its output for a given input energy will not be the same. Tube life can be extended by reliable water cooling. We designed a custom air conditioning unit to use recirculated cooling water to maintain tube cooling. This was found to greatly extend tube life rather than relying on cooling water supplied by the factory which was subject to a range of pressure and temperature fluctuations. Similarly the response of the photodiodes in the LDA will gradually drop off as they suffer from radiation hardening effects. We are still gathering data to see if these effects can be accurately modelled. At present we rely on periodic testing with a standard test block either at a service visit or via the modem connection in order to give us the data necessary to make the necessary correction factors to allow for deterioration of the tube and sensor.

16.4.5 Cleanability of the System

Any equipment designed for the meat processing industry will have to be designed to endure severe cleaning with high-pressure water jet hoses and caustic chemicals. It is vital that such considerations are allowed for from the beginning of the design process and not just added on as an afterthought. We had the advantage that several members of the design team had experience of meat factory conditions.

The mechanical design was done in accordance with best practice for meat processing equipment design. The design was such that the conveyor system could be pulled out of the frame of the machine and completely broken down in a matter of a few seconds. The floor of the framework was pented so that all excess water flowed off. Welding was such that there were no seams which could cause microbiological contamination.

Figure 16.5. BoneScan™ Machine

16.4.6 High-volume Throughput

An important requirement is that the system could cope with the typical throughput rates of a busy meat processing plant The viewing width of the conveyor belt is an important design parameter. As the viewing width increases the x-ray tube needs to go higher up to get the coverage of the x-ray beam across the belt. As the tube to sensor height increases so does the power requirement that is supplied to the tube. Therefore additional viewing width can have a dramatic cost on the overall costs of components inside the system.

As the speed of the conveyor system increases it is necessary to increase the sampling frequency of the LDA if the resolution of the system in the direction of conveyor travel is to remain the same as the resolution across the conveyor viewing width. There are numerous anecdotal stories of customers being demonstrated a certain level of detection performance at one belt speed but then never having these detection levels achieved in the plant when a higher belt speed was required. The reason for this is that to maintain the same resolution, the sampling frequency has to increase and in order to receive the same signal on the LDA the x-ray current must increase.

Therefore the viewing width and sampling frequency are important design parameters which will dictate the x-ray power supply requirements. For our system we decided on give a viewing width of 50cm, which equated to two lanes of chicken butterflies or three lanes of chicken thighs/fillets, with a typical conveyor belt speed of 30m/minute. Under these conditions we could inspect 12,000 chicken butterflies per hour, 20,000 chicken fillets per hour, 125,000 chicken nuggets per hour.

16.4.7 Ease of Use of the System

One major difference in the inspection of natural materials compared with manufactured goods is the skill level of staff in the respective factories. The skill level of the technical staff in an electronics/turbine blade/automotive factory is far higher than the skill level in a typical meat processing factory. As a result is it often the case that the machine vision system has to be far more automated with less reliance on human interaction for many of the systems described in this book compared with industrial machine vision systems. In many of the factories in which we have installed equipment production line staff were immigrant labour and were unable to read and write in the language of the country in which the equipment was installed.

We designed the BoneScan™ system to be fully automatic. A keyboard is supplied mainly for diagnostics and engineering use but for everyday operation the system calibrates itself on startup and will run automatically in inspection mode without any human involvement. A modem connection with our facility is available enabling help, data gathering, diagnostics and software downloads. Although the system runs under the Windows NT operating system, we modified the registry settings to prevent users gaining access to the operating system. Simple diagnostics that the user can attempt are under password protection. However, we learnt early on that passwords become public over time and therefore any advanced diagnostic operations required by service engineers can only be undertaken by inserting appropriate floppy disks inside the PC.

16.4.8 Robust Rejection Technology

As with any inspection technology it is not sufficient just to find the defect, one also has to take some action to remove it from the process line. With packaged food goods of known shape, size and which are dry, rejection is often undertaken with a pneumatic push arm, which simply knocks the product off the line. With a wet, sticky flat product such as chicken butterflies rejection is not so easy. There is no point developing complex image processing algorithms and x-ray sensing technology which find bones to a 99% accuracy if the rejection technology only works at 90%. We chose to use a high-speed retracting belt conveyor as shown in Figure 16.6.

(a)

(b)

Figure 16.6. High speed retracting belt conveyor for poultry meat rejection: a. plan view; b. underside view.

This conveyor mechanism is placed directly after the inspection conveyor mechanism but is programmed to run faster so as to create a gap between the fillets.

The belt was split into two/three lanes (depending on application) so that each lane of the belt could retract and take out a single piece of meat from each of the two/three lanes running through the BoneScan^{TM} machine.

Each time an image was acquired the rejection timing was offset based on the distance of the leading edge of the fillet within the image frame. This was done independently for each lane of the system. The reject signal was delayed (according to the respective belt speeds) and sent just ahead of the arrival of the leading edge of the fillet at the point of rejection. The rejection mechanism retracts for a fillet containing a bone allowing the fillet to fall onto a rejection conveyor underneath. The mechanism then extends the next time a bone-free fillet arrives.

16.4.9 Ability to Provide Management Information

When a defect occurs in a man-made product it is usually as a result of a specific fault at some point within the manufacturing process. This action is usually corrected immediately and production resumes without the defect occurring. In the inspection of a man-made product such as poultry meat the situation is far more complex. Bones can occur in the final product as a result of a problem with any of the following processes:

- manual inspection after deboning;
- deboning;
- carcass damage during eviseration;
- defeathering;
- stunning;
- transport to point of slaughter;
- collection prior to transportation.

Indeed the poultry industry has funded many studies to investigate the effects of specific parts of the poultry production process on the occurrence of bones [16,17,18].

These studies are, however, of limited use without accurate data on the actual occurrences of bones being left in the poultry meat. As part of our BoneScan™ system we developed a Management Statistical Information (MSI) software package. This package records a complete history of when the bones were found, their type, their size, their location within the meat, etc. By plotting this data over time it is possible to spot trends and thereby take corrective action at other parts of the process.

16.4.10 Future Upgradability of the System

The x-ray system is capable of detecting ossified bone of thickness greater than 1.5mm. Bones thinner than this, or not fully ossified bone will not be detectable with current x-ray sensing technology. In addition other defects such as cartilaginous material, keel, blood spots and bruising will not be detectable with x-ray technology but are on the surface of the meat and are therefore potentially detectable with camera-based technology. It was therefore a design requirement to allow the system to be upgraded to enable a future camera system to act as an additional input in order to find defects not detectable with the x-ray system. This is described in more detail in Section 16.7.

16.5 Applications Overview

There are four main applications of the technology: the inspection of chicken breasts (fillets and butterflies), chicken nuggets (and patties), chicken thigh meat and cooked stripped chicken meat. The inspection of cooked stripped chicken meat requires some additional technology and is described in detail in Section 16.6.

16.5.1 The Inspection of Chicken Breast Butterflies

Figure 16.7 shows an x-ray image of a chicken breast butterfly containing six bones.

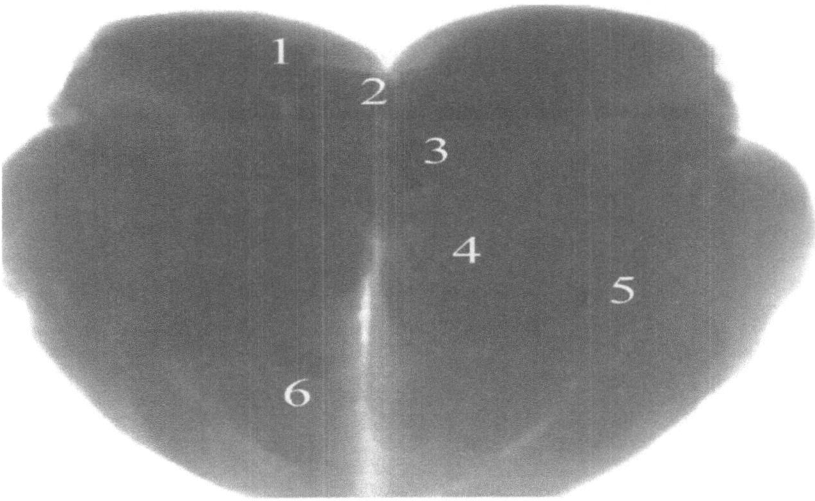

Figure 16.7. X-ray image of chicken breast butterfly containing six bones.

After application of custom noise filters, initial segmentation of the bones based on a shape, size and contrast criteria reveals segmented image shown in Figure 16.8.

From Figure 16.8 it can be seen that the small bones have been segmented but that additional non bone data points have been segmented. A neural network trained over hundreds of images of chicken meat containing bone and non bone samples is used to classify the segmented blobs into bone and non bone categories. The result of the trained neural network is shown in Figure 16.9.

It can be seen from Figure 16.9 that the small bones have seen correctly identified but that the large bones have been missed.

These larger bones are segmented using a different filter the output of which is shown in Figure 16.10 and correctly after classification with a suitably trained neural network in Figure 16.11.

Figure 16.8. First segmented result from the image in Figure 16.7.

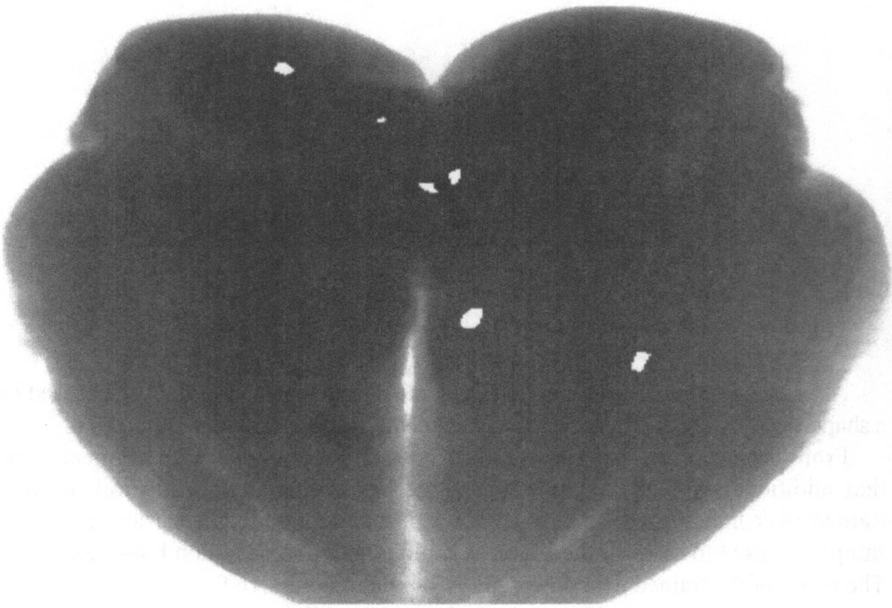

Figure 16.9. Chicken butterfly after first segmentation and neural network classification.

Figure 16.10. Result of applying second segmentation on Figure 16.8.

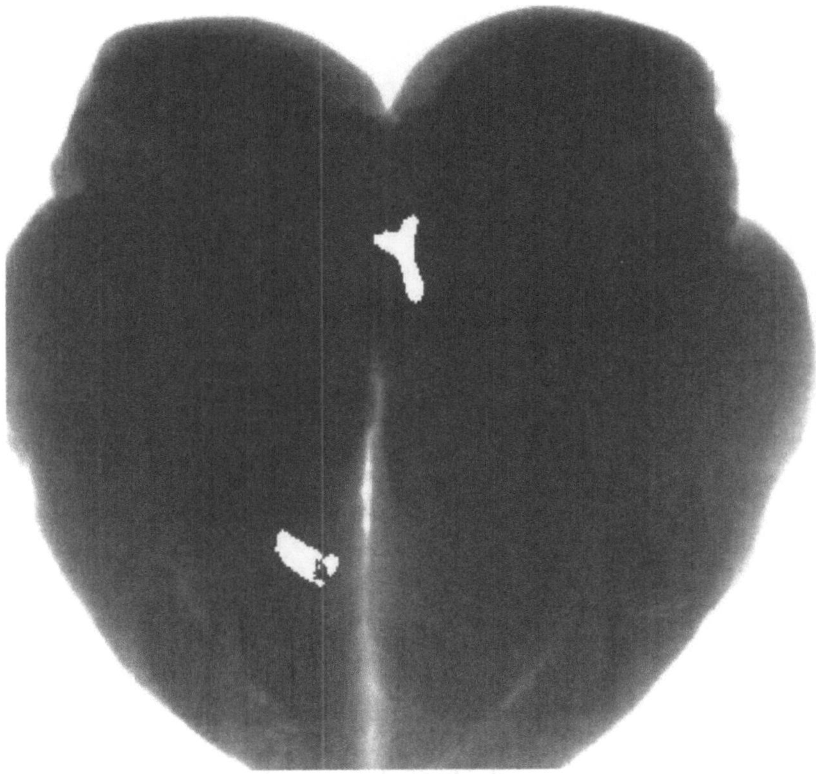

Figure 16.11. Chicken butterfly after second segmentation and classification with second neural network.

Very often the chicken butterflies are not trimmed as nicely as shown in Figure 16.7. Figure 16.12 shows a more typical chicken butterfly image. From this image it can be understood why even with a well trained neural network it is possible that false rejections will occur.

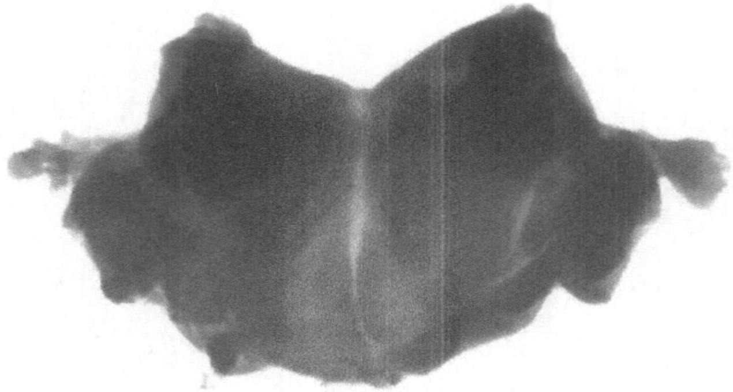

Figure 16.12. Typical poorly trimmed chicken butterfly containing a bone (image supplied by Harbir Chahal)

Figure 16.13. Chicken butterfly after initial segmentation

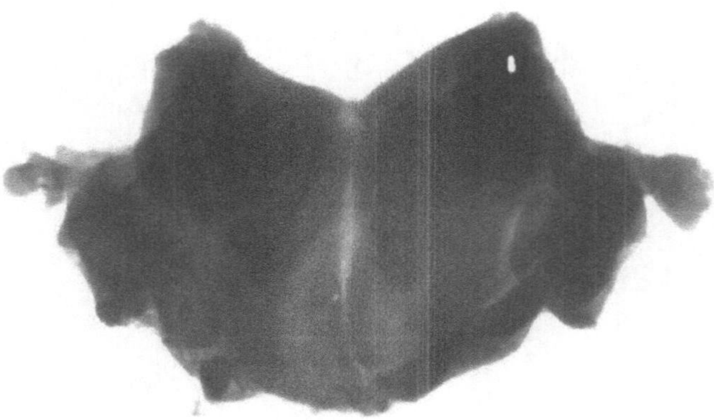

Figure 16.14. Bone correctly identified after neural network classification

16.5.2 Chicken Thigh Meat Inspection

Chicken thigh meat inspection is considerably more complex than chicken breast meat inspection because the muscle structure of the meat is far less homogenous. Such an image of a piece of chicken thigh meat containing a bone is shown in Figure 16.15.

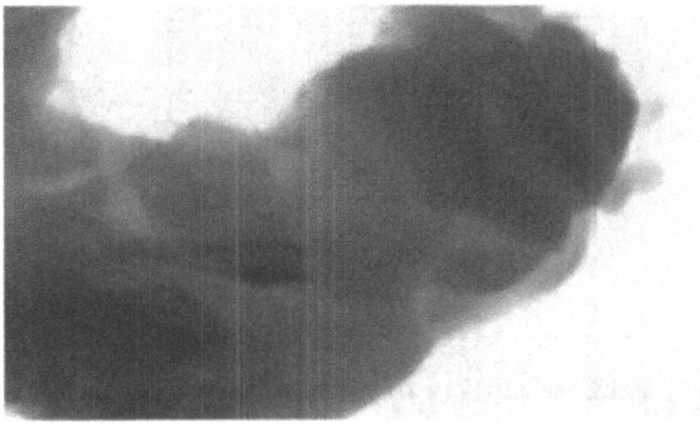

Figure 16.15. X-ray image of chicken thigh meat containing a bone.

Applying the standard segmentation techniques which we use on breast meat would not be successful for chicken thigh meat as shown in Figure 16.16.

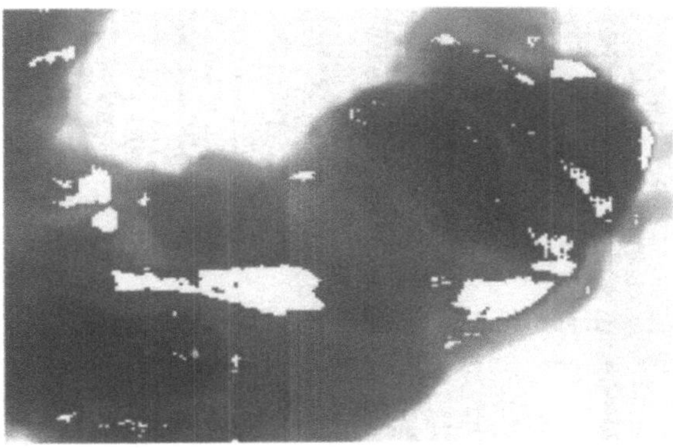

Figure 16.16. Segmentation of chicken thigh image with standard breast algorithm.

In order to solve this problem we developed a novel x-ray sensor. This x-ray sensor, which we have named SpectraLine™, enables the simultaneous acquisition of two x-ray images. These images correspond to the high- and low-energy parts of the multispectral x-ray beam. By combining these two images it is possible to normalise out the background muscle variations leaving only the bones identified.

Figure 16.17 shows an image of chicken thigh meat from the low-energy part of the sensor. Notice how the bone is visible but also there are confusing muscle structures which would make segmentation of the bone using just this image extremely difficult. Figure 16.18 shows the same piece of chicken thigh meat but this time taken from the high-energy side of the sensor. In this image the fine detail present in Figure 16.17 is lost and the image represents the general profile of the chicken thigh. Figure 16.19 shows the output of a non-linear combination of the two images, where the background image of Figure 16.18 has been used to "normalise" the fine detail image of Figure 16.17 to leave just the segmented bone.

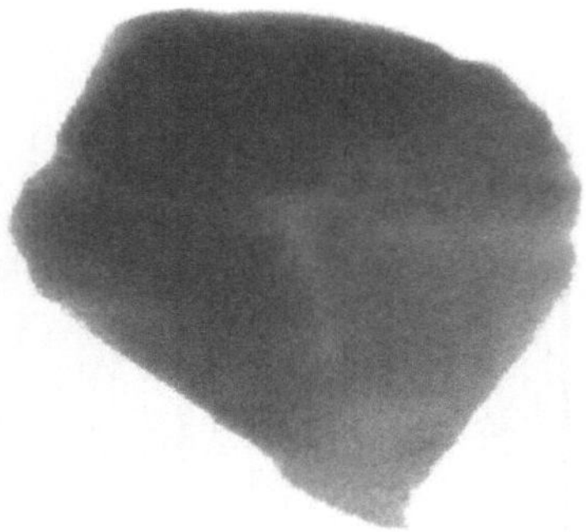

Figure 16.17. X-ray image of chicken thigh meat taken from the "low"-energy side of the SpectraLine™ x-ray sensor.

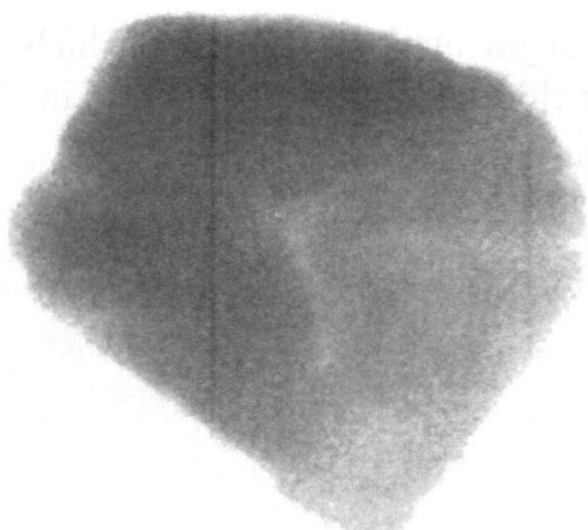

Figure 16.18. X-ray image of chicken thigh meat taken from the "high"-energy side of the SpectraLine™ x-ray sensor.

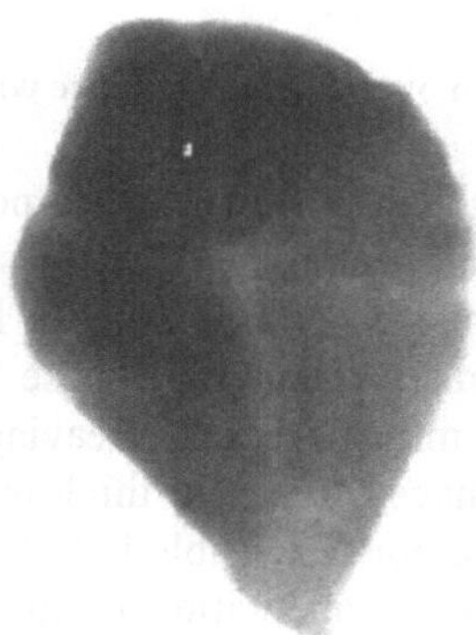

Figure 16.19. Output of a non-linear normalisation of Figure 16.17 by Figure 16.18.

16.6 Stripped Cooked Chicken Meat Inspection

In this application the chicken is cooked as a whole bird and then human workers pull the chicken meat from the warm carcass. This meat is separated into different categories (breast, thigh, wing, etc.) and then sent for further processing, often for soups, pies, etc. The difficulty in inspecting such meat is that the meat is made up of a large number of pieces all of which overlap with each other. Figure 16.20 shows a typical image of the cooked stripped chicken meat containing two bones.

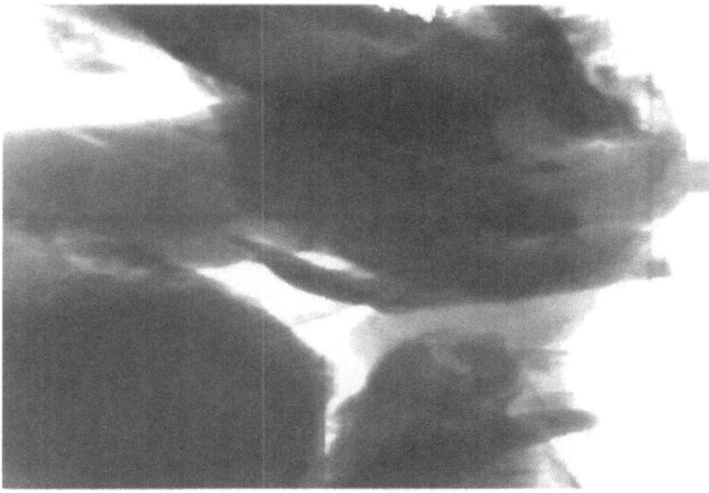

Figure 16.20. X-ray image of cooked stripped chicken containing two bones.

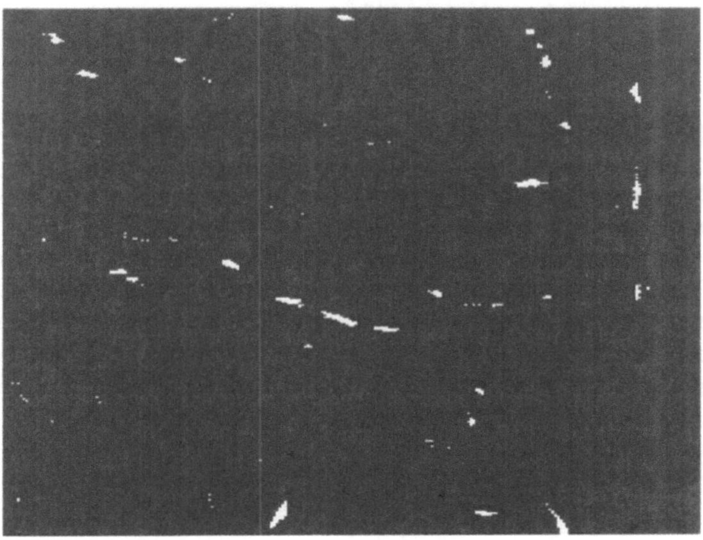

Figure 16.21. Segmentation of Figure 16.19.

Figure 16.21 shows the output of the first stage of segmentation and Figure 16.22 shows the subsequent correct classification by the trained neural network.

The difficulty with the inspection of cooked stripped chicken meat was not so much the inspection as the rejection. Since the meat is not in discrete pieces it is not possible to remove the individual piece containing the bone. In our first application we had a punch arm which removed a slug of meat from the conveyor every time a bone was found. Given that cooked stripped chicken meat contains many more bones than raw meat (the warm bones are more pliable and tend to break off in the hand as the meat is pulled from the bird) the system resulted very quickly in large piles of meat somewhere within which were bones. In order to overcome this difficulty we developed a tray-based inspection system as shown in Figure 16.23 in which each tray was uniquely tracked around the system.

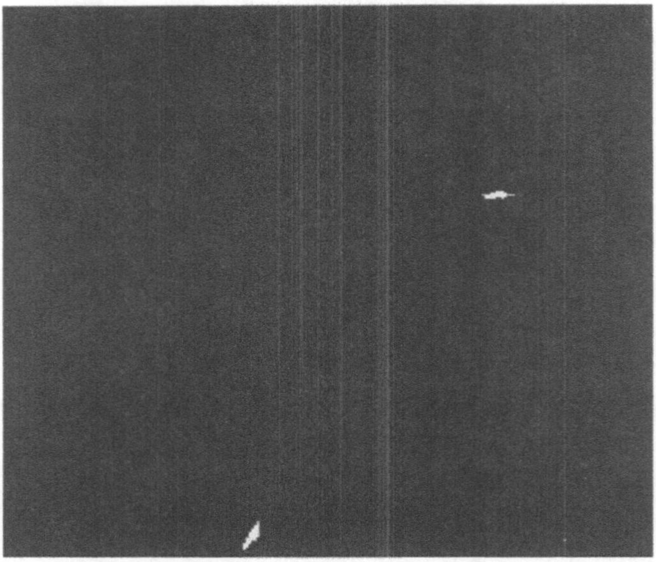

Figure 16.22. Output of segmented image following neural network classification.

In this system the meat to be inspected is placed in a plastic tray by a human operator as shown in Figure 16.23.

The plastic tray has a pocket to wish a plastic card with a printed bar code is added. The bar code contains information such as the type of bird (hen or broiler), the type of meat (leg, breast, wing, mixed) and the person who carried out the meat de-boning and inspection process. Applications so far have concentrated on cooked chicken meat. Additional information could also be added for other applications such as beef, pork or turkey meat inspection.

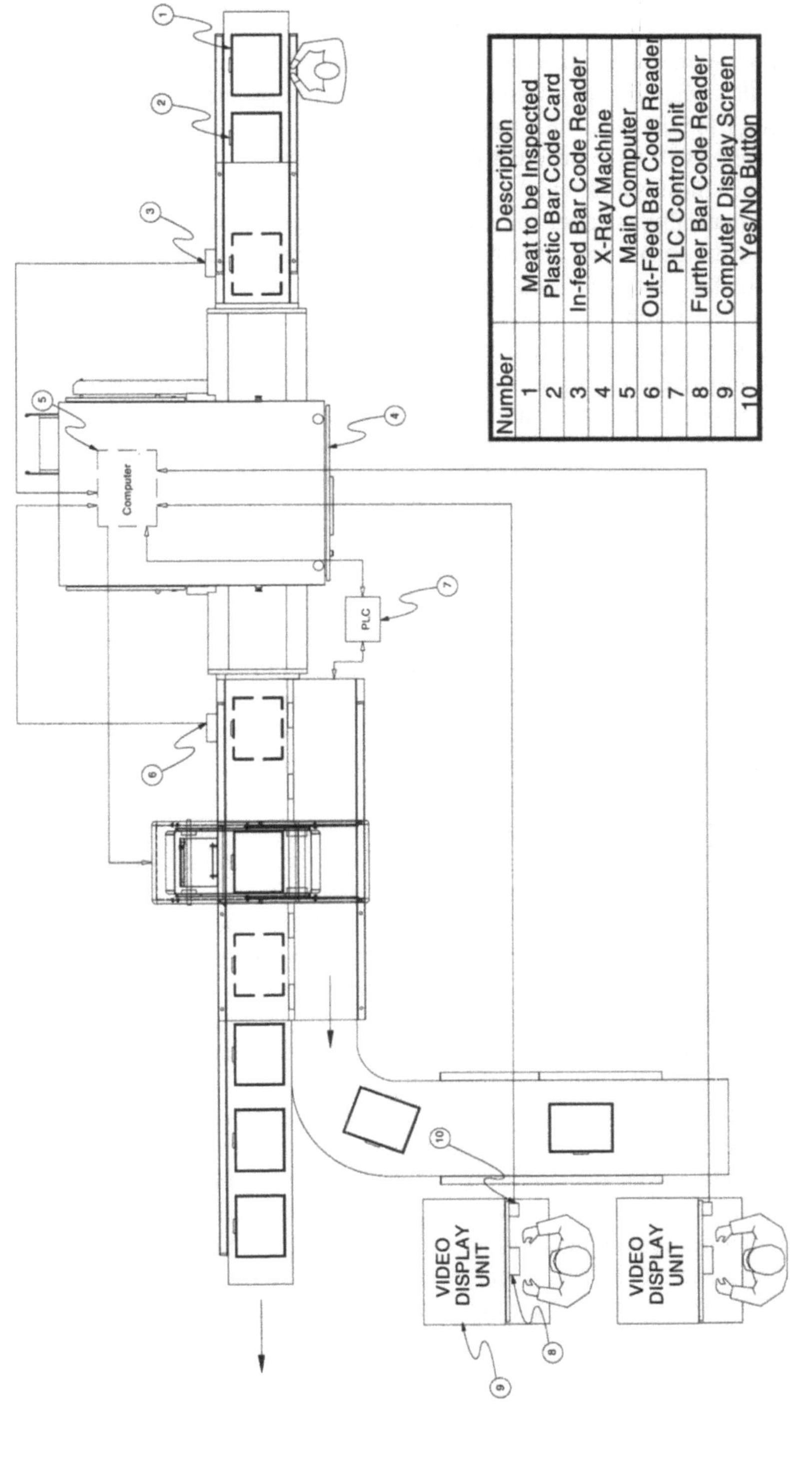

Number	Description
1	Meat to be Inspected
2	Plastic Bar Code Card
3	In-feed Bar Code Reader
4	X-Ray Machine
5	Main Computer
6	Out-Feed Bar Code Reader
7	PLC Control Unit
8	Further Bar Code Reader
9	Computer Display Screen
10	Yes/No Button

Figure 16.22. Tray based x-ray inspection system

The tray is then placed on a conveyor belt along which it passes until it reaches a bar code reader at the entrance to the x-ray machine. As the tray passes by the bar code reader the bar code is read in and the bar code is passed to the main computer inside the x-ray machine. The computer then adjusts the settings of the x-ray power supply so that when the tray passes under the x-ray beam the x-ray power settings are optimum for the particular tray under inspection. The reason for this is that different meats will have different thicknesses and the optimum x-ray setting for breast chicken meat is different for wing chicken meat, etc.

By automatically adjusting the x-ray power settings on the basis of the bar code information the inspection system ensures that each image acquisition event is optimum for the particular food product to be inspected. In addition to changing the x-ray power setting the computer also changes to different image processing and neural network programs on the basis of the bar code read in for the specific item to be inspected. In this way different computer programs can be run for the different types of meat passing through in the trays.

After the tray has passed under the x-ray inspection region the image will be read into the computer memory and the specific program for that type of meat will be run. The output result for that particular image will then be stored and associated with that unique bar code. In addition to giving information on the type of meat each printed bar code has a unique number so that each tray of meat can be uniquely identified.

The tray of meat will then pass onto an outfeed conveyor and will pass by another bar code reader. When this bar code is read in the result for that particular tray will cause the tray to be deflected if the tray of meat contains a defect (bone) or to continue undeflected if the tray of meat does not contain a defect (bone). Timing and control for the pneumatic deflection unit is carried out via a PLC control unit.

Trays which do not contain a defect (bone) will pass onto the next stage of production. Trays which contain a defect (bone) are deflected onto a second conveyor which then transport them to a workstation unit.

Initially when the inspection system is installed the system is pre-programmed with neural network data derived from historical data sets. Although this gives good starting results the systems cannot be fine tuned with data from the specific inspection application. Usually data is obtained by spending considerable time in the installation of the system at a particular customer installation. This data is then added to the pre-programmed neural network data. This has the disadvantage that it requires considerable time to be spent during the installation stage and also it does not respond to changes over time caused by product or system changes.

This problem was overcome by using feedback from the humans at the inspection workstations. At the workstation unit the human inspector places the tray into a holder containing a further bar code reader. As the tray is placed in the holder the image stored on the computers memory is displayed on a computer display screen and the area where the algorithm identified a bone is highlighted. At this point the human inspector examines the tray and presses a yes/no button depending on whether the tray actually contained a bone in the highlighted position. The computer program then records this feedback information to produce statistical information on the amount of bones in the different types of meat and to enable identification of those de-boners and checkers that left the bones in the meat. The feedback information from the human checker at the workstation is used to update the neural networks training samples used in the image processing feature classification algorithm.

Such an adaptive training algorithm means that the neural network algorithm will be able to adapt to changes in the meat over a long period of time – such as the meat being thicker, thinner, different texture, different method of presentation on the tray, different breed of bird, different diet of bird prior to slaughter, different method of processing bird at the slaughterhouse, radiation damage of the x-ray sensor, etc. Such an adaptive learning algorithm also means that the neural network can be applied to applications with little or no previous training since the training now occurs on-line rather than off-line.

16.7 Future Work

The x-ray system described in this section has been proved to reliably detect ossified bones in chicken fillets, thighs and nuggets. We are now currently working in collaboration with researchers at Georgia Tech, USA, to incorporate camera-based technology which can find soft bones and cartilage. This work will incorporate both stand-alone vision modules for specific soft tissue inspection and also combined systems where the x-ray and camera images will be fused together to produce a more intelligent combined result. Results obtained in the laboratory are promising and we are shortly to commence commercial trials of the technology in partnership with Georgia Tech.

16.8 Conclusions

The automatic inspection of poultry meat using x-rays is a challenging application for machine vision. The product is highly variable and moving at high speed and the bones are of low contrast. In order to solve this application it was necessary to develop novel software techniques, custom x-ray linear photodiode array sensing technology, closed-loop air conditioning technology and to integrate these components to engineer a system that was robust, safe and easily cleanable. The development of this technology has taken over eight years and work is still ongoing to further refine the system and to integrate camera technology to produce a combined camera and x-ray vision system. Throughout the course of this development it has been necessary for the engineers involved to spend large amounts of time in poultry processing plants both to gather data and to sufficiently understand the application requirements so as to be able to develop practical and robust solutions.

16.9 Acknowledgements

The author would like to thank the following colleagues for general discussions and providing some of the images used in this chapter; Andrew Marshall, Harbir Singh-Chahal, Anwar Hussain, Gordon Hart and Dr Craig McIntyre.

16.10 References

[1] Graves M. (2000) X-ray Machine Vision for on-line Quality Control in Food Processing, PhD Thesis, University of Cardiff, Wales.

[2] Smith, D.P. (2001) Defect of pre and post deboned broiler breast, *J. Appl. Poultry Res.* 10:33–40.

[3] George M., Sant C., and Wood F. (2001) UK Supermarket Poultry Fillet Contamination and Safety Assessment Study, Report No. PPD/REP/63574/1 pp. 1–24.

[4] Palmer J., Kitchenman A.W., Milner J.B., Moore A.B., Owen G.M. (1973) Development of a field separator of potatoes from stones and clods by means of X-radiation, *J. Agric. Eng. Res.* 18, 293–300.

[5] Ramsay J.D., and Del Rossi, G. (1976) Method and apparatus for the detection of foreign material in food substances, United States Patent 3,995,164

[6] Colar A., and Wyvill J.C. (2000) HIMP project shows value of collaboration in food safety inspection, *Poultry Tech.* Vol. 12, no. 3, 1–4.

[7] www.cbsnewyk.com/eatrisk/StoryFolder/story_215402679_html/index_html

[8] Penmann D., Olsson O., and Beach D. (1992) Automatic X-ray inspection of canned products for foreign material, *Machine Vision Applications, Architectures and Systems Integration*, SPIE, Vol 1823, pp. 342–347.

[9] Graves M., Batchelor B.G., and Palmer, S. (1994) 3D X-ray inspection of food products, Proc. SPIE Conf. on Applications of Digital Image Processing 17, Vol 2298, pp. 248–259.

[10] Patel D., Hannah I., and Davies E.R. (1994) Texture analysis for foreign object detection using a single layer neural network, Proc. IEEE Int. Conf. On Neural Networks, Florida, vol. vii, pp. 4265–4268.

[11] Hannah I., Patel D., and Davies E.R. (1995) The use of variance and entropic thresholding methods for image segmentation, *Pattern Recognition* 28, no. 8, pp. 135–1143.

[12] Graves M., Marshall A., and Ahmed S. (1996) X-ray quality control in the chicken processing industries, Proc. Campden and Chorleywood Conf. On Applications of Image Analysis in the Food Industry.

[13] Dykes G.W (1985) Automated inspection of food jars for glass fragments, Proc. Society of Manufacturing Engineers, Conf. on Vision.

[14] Graves M., Smith A., Batchelor B.G., and Palmer S. (1994) Design and analysis of x-ray vision systems for high-speed detection of foreign body contamination in food, Proc. SPIE Conf. on Machine Vision Applications, Architectures and Systems Integration III, vol 2347.

[15] Munier B. and Hause W.R. (1990) X-ray detection of foreign bodies, *Food Analysis and Control.*

[16] Knowles T.G. and Wilkins L.J. (1990) Broken bones in chickens: Effect of stunning stunning and processing in broilers, *Br. Poultry Science*, 31:53–58.

[17] Bilgill S.F. (1995) Wish bone breakage in processed broilers: Evaluation of strain-cross, sex and stunning conditions, *Poultry Sci.* 74 (Suppl. 1): 179.

[18] Mohan Raj A.B., Greogry N.G. and Austin S.D. (1990) Prevalence of broken bones in broilers killed by different stunning methods, *Vet. Rec.* 127: 285–298.

[19] Graves M., Marshall A., De Lange S., Chahal H., (2000) Dynamic bone contamination detection through X-ray machine vision, Proc. Int. Conf. On Meat Automation, Malaga, Spain, pp. 87–91.

Chapter 17

Final Remarks

B.G. Batchelor and M. Graves

17.1 General Comments

Machine Vision is already established as a valuable tool for inspecting engineering artefacts. While there is still some scope for innovation, many industrial systems perform routine inspection tasks. Integrating a vision system into a well-controlled manufacturing environment may require little more than bolting the camera and lights in place and writing a small amount of application software. However, the inspection of natural materials using Machine Vision has not yet reached the same level of user acceptance and much research is still needed. As a result, the invention of novel solutions is still progressing apace. Each of the applications studies described in this book required a high level of skill and ingenuity to obtain satisfactory results. This is evidence of the fact that there are very few standard procedures that are able to accommodate more than a narrow range of natural products. Nevertheless, we can discern some common themes emerging from the development of our subject.

Designing the mechanical handling sub-system for highly variable products often presents a considerable challenge. Large variations in size and shape must be tolerated. Moreover, organic materials are often flexible. Organic materials are often slippery, or sticky. Mineral materials, and products derived from them, are often dusty. We have seen situations in which an all-round view of an object is needed. Rotating an object that is easily damaged by mis-handling and which has an unpredictable shape and size is never easy. Animals are uncooperative and may actually be hostile, perhaps through fear of the strange environment created by the inspection equipment. At the very least, novel mechanical arrangements, firmly based on animal psychology, are required, to guide/coax fish and land animals into a position where they can be viewed properly. While vision engineers are used to making their equipment robust enough to continue working in a hostile environment, few locations for a vision system can match the hazards found in a milking parlour! Systems for food inspection require regular and vigorous cleaning, often involving high-pressure water jets and corrosive chemicals. Although Machine Vision is an inherently hygienic technology, the need to maintain high levels of cleanliness within the inspection system imposes severe constraints on its mechanical construction. In other words, hygiene is as big a hazard as dirt! To summarise, designing the mechanical handling system is one of the most difficult tasks encountered during the construction of an automated visual inspection system for natural products.

Vision systems for inspecting natural materials are often expected to make subtle judgements about size, shape, colour, texture, or the vaguely defined concepts of aesthetic appearance. (Does the product look "right"?) Natural objects, such as chickens, pigs, fish, fruit, vegetables and wood, do not have a "standard" measure or form. Hence, decision-making techniques based on statistics, fuzzy logic, expert systems and neural networks are often employed to make judgements about the subtle variations that occur in natural materials/objects. Such subtleties are often overlooked. To avoid our taking a superficial view of an application, we must consult an expert who has a deep understanding of the subtle variations of product appearance that can occur in practice. Many of the contributors to this book have spent large periods of time in the factories, farms, or other places where vision systems were to be installed. To be most effective, vision engineers often need to become quite knowledgeable about the industries they are serving. This inevitably leads to specialisation; a vision engineer used to dealing with engineering components cannot, for example, switch immediately to food inspection, without learning about the environment, requirements and practices in the new area of application.

As in all areas of application of Machine Vision, it is very important that we take advantage of any physical phenomena that can help make defects, or other features of interest, more obvious. It is always worthwhile devoting a considerable amount of effort to designing the lighting/illumination system, as this will very often make the task of the image processing and analysis sub-system much easier. Many researchers with a background in Computer Vision tend to over-emphasise the importance of algorithm design, without paying due regard to the illumination. Here, as in industrial Machine Vision, we must always pay particular attention to the image-acquisition sub-system. In these pages, we have seen several situations where clever design of the illumination–viewing system has enabled the solution of applications that would otherwise have been impossible. Relying on sophisticated image analysis algorithms to compensate for poor lighting is *never* justified.

Ultimately an image is digitised and is usually stored, in computer memory, prior to processing. In certain applications, however, higher processing speed is required and dedicated electronic hardware is used instead of a conventional computer. In either event, an algorithm to analyse the digital image must be chosen/designed. The range of available image processing operators is perplexingly wide. We have seen that we can re-use many of the algorithmic techniques used in engineering applications, as well as employing some additional ones. Certain algorithms, for example correlation and template matching, are not suited to inspecting highly variable objects. We can confidently predict that as the subject develops, we shall see the wider use of techniques that are able to cope with subtle variations of natural materials in terms of shape, texture and colour. Most relevant of all for inspecting natural products are self-adaptive learning, fuzzy logic and rule-based reasoning. However, there are fundamental problems inherent in generating properly representative data sets. The first is the ubiquitous problem of labelling product samples reliably and consistently. The second is that it may not be possible to collect a truly representative set of samples from the "faulty" class.

At this stage in its development, natural product inspection is still an adolescent subject, worthy of further academic study: it has not yet reached the level of maturity that engineering product inspection has achieved. However, we anticipate that it will shortly progress through a "growth spurt", as customers come to realise the benefits of the technology and understand the hazards to be avoided. As always in situations like this, it is impossible to predict either the pace of development, or even the precise direction that it will take in the long term. Many lessons have been copied directly from our subject's well-established "older relative", while others have been learned the hard way: by trial and error. It is still too soon to predict whether future systems will have to be carefully designed "by hand", with the vision engineer required to exercise great ingenuity, or whether intelligent design tools can be developed that will make the task of finding solutions more methodical. For the foreseeable future, solutions to vision applications involving natural products will continue to rely heavily on the skill and experience of the vision engineer. The applications case studies presented in these pages should therefore be seen as providing some important general lessons that can guide the design of future systems.

We finish with a compilation of observations, comments and suggestions, which encapsulate the most important points that we have learned about applying Machine Vision to natural products. The following list complements the Machine Vision "Proverbs", originally devised as a guide to inspecting engineering parts. [*http://www.eeng.dcu.ie/~whelanp/proverbs/proverbs.html*]

17.2 Proverbs

- *Careful sample preparation can greatly assist the inspection process, by reducing the product variability.*
  The more "pre-processing" that can be done mechanically, the simpler will be the inspection algorithm.
- *Remove leaves and roots before inspecting fruit and vegetables.*
  Removing excess foliage and roots from plants, fruit and vegetables makes it easier to obtain an unobscured view of the surface.
- *Keep it clean.*
  Washing removes soil and dirt, which might otherwise be mistaken for surface blemishes.
- *If inspection is difficult, modify the product.*
  For example, adding traces of fluorescent dyes to food products can assist inspection. (Cake decoration patterns can be made more obvious in this way.)
- *Don't try to inspect a damp product.*
  Drying the sample before inspection may make it easier to view.
- *Just add water for easier inspection.*

Spraying the sample with water, or oil, before inspection may make it easier to view.

- *Serve hot for best results.*
  Briefly heating the sample surface, using microwaves, or hot air, just prior to inspection, may make it easier to obtain a high-contrast image, using a thermal imaging camera. (IR imagers are sensitive to moisture content.)
- *There are many ways to see the world.*
  Fluorescent or other staining techniques may be used to improve contrast. Staining is widely used in microscopy. It is also used to detect cracks in ferrous forgings and other engineering e
- *Movement is critical*
  The mechanical handling sub-system is critical for presenting the sample being inspected to the camera in the appropriate manner. It is likely to be rather more sophisticated than the transport mechansim used for engineering artefacts.
- *Faulty product come in all shapes and sizes.*
  The transport mechanism must be able to cope with objects of variable shape and size. The product may be flexible, slippery or sticky.
- *Defects must not jam the transport mechanism.*
  The sample being examined must travel through the inspection machine smoothly, without jamming, even when it is over-sized and severely malformed.
- *Smooth movmement gives best results.*
  The transport mechanism (e.g., conveyor-belt) should run smoothly at constant speed, without significant juddering or jerking, in any direction.
- *Know the speed of movment.*
  The speed of a conveyor system should be stable and measured to a high level of precision so that the effective size of the pixels can be calculated and is constant. (If we are using a linescan camera, this point is particularly important.)
- *A clean machine is a happy machine.*
  Regular cleaning is essential for correct operation. The machine should be designed with this in mind. A cleaning protocol should be established as part of the design process. The workforce must be educated to accept this.
- *The inspection machine must be clean but the product is always dirty.*
  The transport mechanism must be able to withstand contamination from soil, dust, sap, oils and other chemicals exuded by the product being inspected. It may be necessary to build in automatic cleaning facilities.
- *Hygiene is a major hazard.*
  The transport mechanism and environmental enclosure for the lights, optics and camera must be able to withstand water-jet cleaning and a variety of harsh sterilising/cleansing chemicals.
- *Design the machine for easy cleaning.*

The transport mechanism for organic materials and food products must be capable of being stripped down quickly and easily, for cleaning

- *Gently does it.*
  The transport mechanism must handle the product gently, to avoid damaging it.
- *Two views are better than one.*
  A low-resolution vision system might be used initially, to determine the size, position and orientation of the product to be inspected. This information is then used to guide an intelligent handling robot, which places it in standard position and orientation, prior to more detailed examination.
- *Sophisticated movement.*
  The transport mechanism may be quite sophisticated mechanically. It may be required to align the product automatically, rotate it in front of the camera, and trim it automatically after inspection, to remove small blemishes.
- *Vision engineers like a good yarn.*
  Stretc.hing yarn, fabrics, leather and other flexible, fibrous/sheet materials, so that they lie flat and wrinkle-free, makes illumination and viewing much easier
- *Optically active materials.*
  Many biological materials are optically active in solution (i.e., they rotate the plane of polarisation). This fact can be used to measure optical activity objectively.
- *Physical parameterscan be measured visually.*
  Viscosity can be measured optically; the speed of ascent and size of small air bubbles in a column of stationary liquid can be measured by a vision system. A simple formula is then used to compute the viscosity of the liquid. The surface tension of a liquid can be measured by observing drop size, either as it falls from a nozzle, or when it falls onto a flat plate. The droplet/particle density within an aerosol, or smoke plume, can be measured optically. Air/fluid flow can be visualised by injecting smoke, or dye, into the stream. Many other physical phenomena can be observed by observing their optical effects.
- *Don't use electronics to do what optics can do.*
  Everything can be viewed in many different ways. We can use electro-magnetic waves of any convenient wavelength to form an image: microwaves IR, VIS (visible wavelengths), UV, x-rays, or gamma rays. We can also make use of a variety of physical effects: fluorescence (x-ray-to-VIS, x-ray-to-UV, VIS-to-VIS, UV-to-VIS, UV-to-IR, VIS-to-IR), phosphorescence, diffraction, polarisation, bi-refringence, refraction, scattering (in a suspension or emulsion), specular reflection, diffuse illumination, diffusion through a transluscent material and ducting through a transparent medium by total internal reflection. (The most familiar examples of ducting occur in cut and moulded glassware and cut gemstones.) Optical filters are useful for detecting/blocking

certain specific optical wavelengths. In any given application, any of these phenomena might simplify the image processing.

- *Keep it cool.*
  The applied light must be "cold", containing low levels of IR, to avoid damaging the product.
- *Seemingly innocent features can indicate serious problems.*
  Subtle changes of colour or texture may be very significant in organic materials. For example, bruising in apples.
- *If it's obvious, it ain't natural.*
  While industrial artefacts often exhibit discrete colours, many natural products show marked colour blending. For example, a "red" apple will almost certainly show red, green and intermediate colours too.
- *Keep a sense of proportion.*
  We can use structured lighting for 3D shape analysis. This may rely on a single or multiple stripes, arrays of spots, grids, concentric circles, or other projected patterns chosen to detect certain specific contours and make the image processing easier. However, natural products often have steep-sided fissures and other cliff-like features which make it difficult to obtain a continuous map of the surface geometry. Some intelligent interpretation is needed to cope with "unknown" values in a depth map.
- *Things look different under water.*
  We can view transparent materials immersed in water, or oil, to compensate for the high refractive index. For example, it is easier to view the bubbles in an ice-cube if it is floating in water. This causes the silhouette of the ice-cube to disappear, since the refractive indices of ice and water are nearly equal. Even materials with a very high refractive index, such as uncut gemstones, can benefit from this approach, making it possible to view internal flaws more easily. (The water/oil bath must, of course, be kept clean and free from algae and other micro-organisms.)
- *Glowing in the dark is not healthy.*
  Certain micro-organisms fluoresce when irradiated with UV. This fact can be useful for an inspection system, since it provides the basis for discriminating between healthy and infected tissue. (This particular example emphasises the importance of application knowledge.)
- *Even the vision of natural objects presents to us insurmountable difficulties.*
  Both the vision engineer and the machine he is designing need to be intelligent. A Machine Vision system used to inspect highly variable products often requires a greater level of intelligence than industrial systems do.
- *If it works use it.*
  Inspection procedures are likely to be heuristic, rather than algorithmic, in nature. As a result, we may have to be content with satisfactory, rather than optimal, solutions.

- *Divide and conquer.*
  An inspection procedure may consist of a series of partial solutions that each covers some but not all of the situations that can occur in practice. An "intelligent switch" is then used to select the most appropriate one. This is an ideal situation for using a rule-based system. Other computational methods and techniques borrowed from Artificial Intelligence and Pattern Recognition (e.g., Neural Networks, Fuzzy Logic, Genetic Algorithms, Simulated Annealing) are likely to be used to inspect natural products
- *Hygiene is of paramount importance.*
  A high level of hygiene is, of course, essential in preparing food and pharmaeutical products. For this reason, the mechanical handling sub-system must be designed so that is easy to keep clean.
- *The inspection system must be compatible with existing working practices.*
  Automated visual inspection provides a completely new range of quality control tools and may provide information about the product that was not available hitherto. However, the vision system must be compatible with previous inspection and manufacturing techniques. Failure to follow this maxim will result in a system that is not accepted by the workforce and will therefore be rejected in favour of older, tried and tested methods.

Index

MIX
Papier aus verantwortungsvollen Quellen
Paper from responsible sources
FSC® C105338

If you have any concerns about our products,
you can contact us on
ProductSafety@springernature.com

In case Publisher is established outside the EU,
the EU authorized representative is:
Springer Nature Customer Service Center GmbH
Europaplatz 3, 69115 Heidelberg, Germany

Printed by Libri Plureos GmbH
in Hamburg, Germany